Delaware 1850-1860 Agricultural Census

Volume 1

Linda L. Green

WILLOW BEND BOOKS
2006

WILLOW BEND BOOKS

AN IMPRINT OF HERITAGE BOOKS, INC.

Books, CDs, and more—Worldwide

For our listing of thousands of titles see our website
at
www.HeritageBooks.com

Published 2006 by
HERITAGE BOOKS, INC.
Publishing Division
65 East Main Street
Westminster, Maryland 21157-5026

International Standard Book Number: **0-7884-3826-3**

Introduction

This census names only the head of the household. Often times when an individual was missed on the regular U. S. Census, they would appear on this agricultural census. So you might try checking this census for your missing relatives. Unfortunately, many of the Agricultural Census records have not survived. But, they do yield unique information about how people lived. There are 48 columns of information. I chose to transcribe only six of the columns. The six are: Name of the Owner, Improved Acreage, Unimproved Acreage, Cash Value of the Farm, Value of Farm Implements and Machinery, and Value of Livestock. Below is a list of other types of information available on this census.

Linda L. Green
13950 Ruler Court
Woodbridge, VA 22193

Other Data Columns

Column/Title

6. Horses
7. Asses and Mules
8. Milch Cows
9. Working Oxen
10. Other Cattle
11. Sheep
12. Swine
14. Wheat, bushels of
15. Rye, bushels of
16. Indian Corn, bushels of
17. Oats, bushels of
18. Rice, lbs of
19. Tobacco, lbs of
20. Ginned cotton, bales of 400 lbs each
21. Wood, lbs of
22. Peas and beans, bushels of
23. Irish potatoes, bushels of
24. Sweet potatoes, bushels of
25. Barley, bushels of
26. Buckwheat, bushels of
27. Value of Orchard products in dollars
28. Wine, gallons of
29. Value of Products of Market Gardens
30. Butter, lbs of
31. Cheese, lbs of
32. Hay, tons of
33. Clover seed, bushels of
34. Other grass seeds, bushels of
35. Hops, lbs of
36. Dew Rotten Hemp, tons of
37. Water Rotted Hemp, tons of
38. Other Prepared Hemp
39. Flax, lbs of
40. Flaxseed, bushels of
41. Silk cocoons, lbs of
42. Maple sugar, lbs of
43. Cane Sugar, hunds of 1,000 lbs
44. Molasses, gallons of
45. Beeswax, lbs of
46. Honey, lbs of
47. Value of Home Made Manufactures
48. Value of Animals Slaughtered

Delaware Table of Contents

Kent County Delaware
1850 Agricultural Census

This agricultural census was filmed from original records in the Delaware State Archives in Wilmington Delaware by the Delaware State Archives Microfilm office.

There are some forty-eight columns of information on each individual. Only the head of household is addressed. I have chosen to use only six columns of the information because I feel that this information best illustrates the wealth of the individuals. These are shown below:

1. Name of Owner
2. Acres of Improved Land
3. Acres of Unimproved Land
4. Cash Value of the Farm
5. Value of Farm Implements and Machinery
13. Value of Livestock

Thus, the numbers following the names represent columns 2, 3, 4, 5, 13.

The following symbol is used to maintain spacing where information in a column is left blank (-). This symbol is used where letters, names or numbers are not legible (_).

Pages 65 and 67 of Murderkill Hundred used O (owner) and T (tenant) following the names of the individuals). However, there is no explanation for the use or what the actual O and T mean.

The last page of this county (page 85) was totally blackened. I don't know whether it was poor microfilming or whether the page was actually that bad. This film was negative rather than positive and that may have had an impact.

Manlove Cally, 300, 100, 6000, 75, 998
James Wallace, 206, 50, 3000, 50, 703
Taylor Meridy, 117, -, 500, 108, 301
Wm. Atkins, 100, 10, 600, 120, 415
Wm. Anderson, 39, 7, 400, 4, 130
John Hobson, 150, 25, 5000, 100, 385
Charles Short, 125, 21, 7000, 150, 618
James L. Havsin, 350, 50, 10000, 300, 1658
Saml. D. McGonigal, 310, 40, 7000, 75, 785
Wm. McGonigal, 150, 20, 700, 75, 462
John Callen (Cablen,Cabler), 260, 50, 1000, 150, 900
John Bell, 240, 60, 8000, 200, 965
Outten Heverin, 260, 40, 10000, 200, 944
James Ratledge, 60, -, 2000, 40, 225

Timothy Slaughter, 110, 25, 4000, 50, 273
Mathew Artis, 70, 10, 1700, 60, 155
James G. Waples, 175, 40, 6000, 150, 684
Nathaniel D. Wilds, 200, 80, 10000, 120, 790
Eliza Woodall, 100, 300, 6000, 120, 658
Edward West, 207, 45, 10000, 75, 500
Jacob H. Hill, 125, 25, 5000, 75, 430
Daniel C. Hoffaker, 160, -, 8000, 300, 866
David G. Lewis, 70, 30, 2000, 60, 205
Meriah Molaster, 60, 40, 1000, 60, 30
Isaac W. Terer (Tenes), 100, 60, 2500, 100, 80
Thomas Pierce, 110, 20, 2500, 25, 250
Wm. Webb, 400, 200, 10000, 60, 225
Charles Aigne, 220, 80, 4000, 50, 375
Alexander Collins, 100, 20, 3000, 40, 200
Robert Powell, 210, 74, 3408, 12, 100
Henry Ridgley, 250, 150, 20000, 400, 866
Moses America, -, -, 400, 60, 475
Peter Heins, 200, 100, 4250, 100, 200
Timothy C. Kellen, 180, 20, 5000, 150, 572
Clement Reed, 200, 100, 3100, 50, 356
Isaac Tester, 125, 100, 3000, 50, 356
Stephen Howard, 120, 100, 6600, 75, 492
Benjamin White, 50, 40, 2000, 75, 200
Samuel Emery, 33, 7, 700, 40, 230
John Denney, 75, 75, 5000, 75, 330
Benjamin Waitman, 120, 90, 3500, 50, 400
John P. M. Denny, 105, 50, 2500, 100, 370
Wm. Morgan, 60, 50, 2000, 25, 100
Charles Washington, 160, 40, 350, 50, 210
John Raughly, 200, 100, 5000, 50, 270
Samuel Kinchey, 100, 20, 5000, 200, 275
Ebin Long, 100, 30, 3000, 50, 275
Manlove Frasier, 74, 75, 3000, 50, 170
George M. Manlove, 85, 26, 400, 50, 270
Curtis Anderson, -, -, -, 40, 225
Jame Eunice, 100, 75, 2000, 50, 200
Robert Poor, 47, 153, 1000, 50, 60
John Mosely, 4, -, 150, 40, 100
Ezekiel Clark, 200, 200, 8000, 55, 275
Richard Jacobs, -, -, -, 45, 185
John Dunning, 315, 25, -, 50, 340
Joseph Warasin, 125, 375, 5000, 150, 365
Zachariah Johns, 70, 44, 700, 50 185
George Hill, 30, 40, 500, 40, 100
George Hall, 150, 50, 2000, 50, 220
John Doherity, 114, 86, 1800, 50, 295
Wathe Farrow, 80, 100, 1000, 100, 175
John Morgan, 100, 33, 700, 60, 190
Wm. Burts (Rurts), 70, 10, 600, 30, 114
Noble Fireacres, 90, 15, 1000, 40, 110
Charles Emery, 50, 17, 500, 60, 100
Thomas Fireacres (Foreacres), 40, 20, 400, 20, 96
Isaac Jacobs, 75, 55, 800, 15, 90
Roger Molaster, 25, 15, 300, 15, 75
Capril Carlisle, 75, 65, 700, 25, 175
John Hargadine, 150, 65, 2500, 100, 525

Charles Miller, 57, 40, 1000, 35, 213
John Wiley, 60, 76, 600, 30, 100
John Moore, 32, 31, 1200, 75, 267
Heary Dasham, 24, 26, 500, 30, 40
Phillip Hoffacker, 70, 50, 2500, 50, 190
Jonathan Shaham, 60, -, 1000, 40, 85
John Ratledge(Rutledge), 35, 15, 500, 20, 100
Wm. Casson, 90, 110, 1500, 50, 290
John Sanders, -, -, -, 35, 75
Nathaniel Ratledge, 40, 70, 600, 30, 137
Wm. Tomlinson, 300, 150, 9000, 100, 1023
Thomas Hobson, 560, 100, 10000, 100, 823
Samuel Layton, 120, 50, 250, 130, 1100
Peter Wilson, 300, 50, 6000, 55, 564
Isaac Herrington, 500, -, 8000, 300, 1078
Stephen Taylor, 200, 38, 2500, 25, 218
Peter Meredith, 104, 21, 1522, 100, 595
Robert Reynolds, 175, 75, 4000, 200, 750
Thomas Cooper, 40, 60, 600, 35, 127
Benjamin Reed, 100, 100, 1500, 35, 236
Benjamin Cohee, 190, 50, 9500, 75, 387
Robert Mitchell, 33, -, 450, 10, 226
Steven Levick, 200, 150, 5000, 75, 505
Joseph Millway, 80, 35, 1800, 75, 258
Jackson Lafferty, 45, 14, 900, 50, 100
Steven Cates, 150, 50, 3000, 40, 230
Francis Register, 150, 100, 3750, 100, 814
Balitha Wharton, 160, 81, 3500, 40, 250
Samuel Wharton, 400, 129, 7000, 50, 600
Benjamin Hinsor (Hinson), 150, 40, 3800, 10, 150
George Kelly, 125, 5, 3000, 90, 235
Charles Wharton, 175, 25, 8000, 150, 701
James Kimmey, 180, 44, 6000, 100, 600
Benjamin Hinson, 110, -, 2200, 60, 320
James Raymond, 275, 40, 5000, 80, 480
Caleb Sipple, 240, -, 7200, 150, 810
Robert Planewell, 80, -, 2400, 100, 400
Wm. J. Clark, 25, -, 1000, -, 100
Thomas Postles (Potter), 200, 60, 7000, 150, 1000
Wm. Walker, 75, -, 2500, 125, 392
Robert H. Moore, 10, -, 1500, -, 35
James Rabin, 30, -, 1500, 20, 182
Daniel Cowgill, 10, 5, 1000, 30, 170
James M. Rabin, 33, -, 1500, 30, 325
Martin W. Bates, 30, -, 2000, 60, 210
Isaac Jump, 320, 36, 6000, 125, 475
Joseph McDaniel, 178, 200, 6740, 110, 725
Zadock Potter (Postles), 160, 10, 4000, 70, 675
James Slaughter, 200, 200, 10000, 120, 465
Elizabeth Palmer, 126, 19 4500, 35, 332
Wm. _. Hutchinson, 90, 15, 3500, 100, 215
Darling Rash, 50, 47, 500, 25, 131
Daniel Mason, 100, 40, 1000, 50, 240
Hugh M. Martin (Mastin), 75, 75, 1500, 125, 360
Wm. Raughley, 50, 80, 1000, 40, 204
John Taylor, 12, 18, 400, 30, 50
Isaac Burlington, 200, 160, 3600, 120, 550

Wm. Taylor, 25, 71, 700, 30, 110
James Simmons, 40, 10, 300, 40, 115
John Taylor 150, 150, 2000, 100, 275
Henry Moore, 45, 60, 600, 50, 285
Samuel Biles (Biler), 136, 154, 1600, 50, 290
James Barens, 75, 80, 1000, 45, 165
Isaac Johnson, 70, 60, 1000, 45, 185
Henry Beaman (Bearman), 100, 45, 1500, 100, 291
Charles Griffin, 75, 127, 7000, 40, 80
Elisha Counseler, 240, 200, 3000, 30, 207
Simon Lootman (Loadman), 30, 15, 200, 25, 100
Peter Carlisle, 70, 130, 1500, 15, 140
George Mitchell, 40, 47, 1000, 40, 175
Henry Goodwright, 40, 50, 900, 30, 65
Samuel Gibson, 120, 40, 1000, 40, 100
Robert Rawley, 100, 50, 2000, 20, 135
James Dennis, 200, 50, 4000, 75, 473
Brainard Wheatley, 250, 100, 6000, 75, 310
Rachel Craig, 260, 40, 4000, 30, 260
Isaac Pratt, 90, 50, 1000, 50, 192
Jonathan Stiles, 190, 60, 7000, 100, 385
James Hill, 75, 100, 1500, 30, 110
Edward Conseler (Consider), 75, 75, 2000, 35, 100
Nathan Slaughter, 230, 390, 4000, 50, 425
John Brown, 40, -, 400, 25, 50
Washington Selvy, 16, 10, 200, 20, 75
Eugene Ridgely, 200, 130, 5500, 200, 300
Henry Todd, 45, 3, 2000, 386, 137
Myars Casson, 100, 80, 1500, 100, 260
Ruth VanBurklow, 100, 80, 1500, 75, 185
Jackson Course, 70, 52, 600, 50, 95
Augustus Shorts, 100, 46, 600, 30, 165
Levick Hegrous, 100, 118, 1500, 60, 350
Mark Rash, 60, 100, 3000, 100, 270
Vincent Cohee, 60, 40, 800, 35, 140
Wm. Thomas, 50, 40, 1000, 100, 85
James Collins, 30, 74, 1000, 30, 185
Richard Fargason, 9, 24, 300, 20,83
Nathan Gibbs, 40, 560, 4000, 5, 110
Moses Rash, 35, 20, 2500, 150, 275
Wm. Pearson, 150, 40, 6000, 100, 550
James Rash, 60, 77, 1500, 75, 230
Isaac Pearson, 50, 50, 1000, 75, 160
John Wilson, 100, 60, 1000, 75, 160
John Ware, 100, 60, 1000, 10, 165
Nathan H. Green, 75, 58, 800, 50, 225
Martin Rash, 70, 76, 1800, 30 180
David Gron, 50, 37, 1000, 30, 225
Wm. Darling, 30, 10, 500, 35, 100
Thomas Sevey (Levey), 100, 70, 2000, 200, 285
James Powell, 100, 60, 1500, 75, 300
Wm. Voshell, 75, 30, 2000, 75, 275
Phillip Marsel, 100, 300, 4000, 100, 400
Wm. Ware, 120, 80, 1500, 40, 295
Wm. Porter, 110, 150, 1500, 20, 80
George Johnson, 40, 83, 600, 42, 146
Wm. Biddle, 40, 26, 1000, 20, 260
Wm. Bedwell, 30, 30, 700, 70, 95
George Wallace, 175, 83, 3000, 100, 527
Thomas Slay (Hay), 33, 19, 1000, 35, 150
Thomas Wallace, 90, 35, 1200, 40, 165
John S. Kersey 39, -, 1500, 100, 236

John Slay (Hay), 176, 57, 3500, 100, 241
John Voshel, 150, -, 3000, 150, 520
Wm. Aron, 158, 50, 5000, 100, 590
John Reed, 75, 25, 700, 35, 60
John Seward, 100, 50, 1000, 50, 200
John Caulk, 200, 120, 4000, 50, 300
Zadock Lofland, 100, 84, 2000, 60, 350
Jonathan Green, 137, 49, 1800, 130, 331
James Powel, 100, 56, 3000, 130, 390
Nathan Jones, 100, 200, 1000, 60, 250
Timothy Irons, 50, 61, 2000, 15, 120
Joseph Fireacres, 80, 82, 3000, 23, 310
John Hutcherson, 120, 100, 1000, 50, 265
Wm. Thompson, 60, 53, 1200, 70, 355
Thomas Huchens, 65, 30, 2000, 100, 363
James Clark, 50, 17, 600, 45, 230
Daniel Cox, 60, 40, 1000, 90, 330
Abner Roe, 40, 237, 2000, 50, 205
Thomas S. Moore, 60, 22, 600, 75, 150
Moses Ford, 100, 80, 2000, 100, 285
Joshua Lodine, 50, 10, 600, 25, 100
G. Joseph Whiteley, 25, 25, 500, 5, 55
Daniel Moore, 40, -, 400, 20, 85
Daniel George, 150, 50, 2500, 150, 460
Littleton Reed, 90, 10, 1000, 50, 320
James Streets, 90, 60, 2000, 100, 247
Amos Kinesly, 58, 50, 1600, 125, 170
Newton Hubbert, 125, 50, 2000, 35, 375
Obadiah Vossel, 40, 16, 500, 35, 180
John Hutchens, 60, 40, 1000, 60, 265
Ruben Marbell, 100, 155, 1500, 50, 260
Martha Day, 200, 100, 4000, 100, 620
Thomas Clemens (Cleuress, Clenress), 180, 52, 2500, 100, 400
Noah Moore, 15, 15, 450, 15, 95
Wm. Marsh, 80, 28, 600, 20, 100
John B. Hersey, 100, 130, 1000, 100, 236
Martin Ford, 70, 70, 1200, 20, 211
James Francis, 12, -, 700, 6, 95
James Duhedaway, 17, 5, 400, 40, 145
Wm. Wilson, 125, 75, 2500, 125, 455
Hadden Smith, 90, 125, 1500, 20, 225
Samuel Powel, 100, 200, 1500, 45, 150
John Couley (Conley), 20, 10, 300, 40, 100
Martha Ford, 75, 100, 2000, 75, 415
Joseph Smith, 100, 50, 1500, 40, 255
WM. Cursey, 150, 123, 1200, 150, 467
James Stevens, 35, 25, 500, 30, 145
John Harrington, 75, 25, 700, 35, 130
James Hinsley, 100, 100, 1800, 35, 275
Merrit Scotten, 100, 175, 2200, 100, 450
Philaman Scotten, 80, 40, 1000, 50, 240
Elias Perry, 60, 40, 700, 40, 300
Wm. Jones, 50, 30, 700, 40, 145
John Thornton, 72, -, 800, 20, 120
Wm Scotten, 30, 3, 600, 40, 130
James Walls, 30, 45, 700, 15, 100
Joseph Head, -, -, -, 25, 142
Robert Wilson, 200, 300, 6000, 200, 860
James Williams, 110, 60, 1000, 15, 70
John W. Davis, 75, 31, 2000, 50, 218
Trustin L. Davis, 75, 31, 2000, 50, 218

John L. Davis, 27, -, 2000, 25, 125
Christopher Ford, 15, 12, 600, 25, 125
James M. Rose, 73, 154, 2000, 100, 283
Samuel Derry, 20, 20, 500, 30, 230
Richard Gibbs, 32, 6, 300, 10, 50
Henry Lockwood, 60, 40, 600, 10, 115
Wm. Dean, 35, 10, 500, 40, 175
John Collins, 21, -, 300, 30, 160
Perry Notts, 70, 34, 1000, 40, 210
George Jones, 100, 100, 2000, 140, 305
Robert Rite, 90, 100, 3000, 100, 340
Edward Barber (Barbee), 105, 25, 1000, 20, 110
Samuel Herds (Heeds), 250, 100, 1000, 60, 195
Peter Hudson, 120, 60, 1000, 40, 150
Peter Miller, 190, 50, 2000, 100, 380
Nathaniel White, 100, 50, 300, 25, 50
Samuel Logan (Legar), 125, 100, 2000, 100, 435
John W. Clark, 60, 60, 1000, 40, 270
Powel Aron, 70, 60, 1500, 50, 335
Jonathan Bedwell, 30, 3, 300, 25, 50
David Smith, 75, 125, 2000, 100, 305
Joseph J. Smith, 60, 42, 1000, 50, 145
Jesse Montague, 175, 103, 2000, 125, 370
Samuel Vossel, 16, 10, 225, 5, 87
Peter Vossel, 16, 10, 225, 5, 87
Furier Vossel, 30, 10, 300, -, 40
John Taylor, 90, 80, 600, 25, 228
John Durham (Duhamel), 200, 123, 2000, 40, 287
Wm. Smith, 200, 110, 3600, 150, 467
George Cabbage, 100, -, 1600, 60, 179
James Amos, 15, 165, 150, 10, 69
Isaac Thomas, 115, 34, 2000, 35, 475
Samuel Carter, 60, 50, 4200, 100, 415
John Cossell, 70, 10, 1200, 50, 310
Daniel Duhamel, 20, -, 300, 20, 110
Thomas Hayes, 60, 65, 700, 50, 335
John Clark, 80, 60, 2000, 400, 420
Joseph Darling, 20, -, 200, 30, 90
David Marvel, 335, 50, 2500, 250, 605
John Williams, 330, 20, 2500, 100, 360
Wm. Manlove, 120, 30, 5000, 120, 335
Wm. Cole, 280, 30, 7000, 150, 670
Asa Lofland, 190, 46, 4500, 200, 517
Richard M. Jones, 100, 30, 2600, 60, 188
Alexander F. Barr, 220, 100, 4500, 105, 500
James Kimmey, 140, 30, 2000, 150, 525
Joseph Barrett, 40, 10, 600, 80, 255
James Johns, 120, 80, 1600, 60, 340
Joshua Nickerson, 100, 100, 2000, 35, 270
Samuel Cahall, 40, 160, 1000, 20, 75
Wm. Nickerson, 100, 400, 5000, 110, 450
Wm. Jackson, 20, 20, 400, 40, 50
George Notes, 100, 80, 1800, 50, 235
Samuel Powell, 30, 130, 2500, 150, 475
James Numern (Numen), 40, -, 400, 40, 100
Benjamin Greenage, 75, 37, 700, 30, 130
John Greenage, 34, 28, 700, 95, 150
Jonathan Powell, 75, 27, 1000, 100, 346
Joshua Brown, 80, 120, 700, 4, 80
Wm. Hawkins, 115, 115, 2300, 60, 362
Zedoc Franklin, 50, 50, 1000, 15, 70

James Foreacres, 25, 20, 400, 20, 125
John D. Conner, 25, 18, 500, 40, 90
Thomas Perry, 60, 46, 1100, 60, 100
Eli Kenton, 18, 12, 300, 20, 105
Thomas Chambers, 100, 40, 2000, 35, 270
John Fisher, 40, 60, 600, 15, 75
Wm. Slaughter, 100, 110, 3000, 250, 605
Jonathan Stewart, 50, 40, 700, 50, 230
Jesse M. Jones, 35, 20, 800, 60, 230
Thomas Herrington, 23, 27, 500, 40, 90
Thomas Marvel, 100, 80, 2000, 75, 285
John Anderson, 75, 75, 1500, 50, 190
Samuel Milburn, 170, 110, 2000, 100, 285
Wm. Slay, 100, 449, 4000, 150, 490
Jonathan S. Green, 150, 61, 2500, 100, 375
Philip Roberts, 50, 58, 1000, 75, 245
Benjamin Walker, 35, 13, 1000, 50, 170
Stephen Gibbs, 75, 75, 1200, 40, 260
James Greenwood, 80, -, 800, 100, 220
Nathaniel Morris, 100, 30, 1400, 40, 185
Wm. Verdon, 360, 100, 1240, 160, 1112
Thomas Scase, 30, -, 300, 40, 145
Wm. Carpenter, 20, -, 200, 5, 110
Isaac Foreacres, 50, 86, 800, 40, 100
Joseph Gibbs, 70, 30, 800, 40, 110
James Patton, -, -, -, 30, 105
James Jones, 145, 10, 4000, 150, 508
Alexander Hillen, 94, 12, 3000, 85,236
Henry Scout, 200, 46, 3500, 95, 405
Loadman Jones, 10, 40, 1000, 80, 340
Henry Hall, 90, 12, 2000, -, -
Enoch David, 155, 45, 3000, 60, 370
Alexander Jones, 160, 70, 10000, 800, 590
John Hickey, 163, 30, 6000, 150, 400
James Shahan, 80, 7, 1700, 50, 160
John D. Mattiford, 90, 90, 1500, 50, 540
James Chiltens, 150, 60, 2000, 40, 300
James Hall, 150, 33, 4000, 150, 346
James Boils, 127, 40, 4000, 100, 377
John Cleaver, 115, 55, 3500, 75, 250
Charles Numbers, 80, 16, 1500, 20, 210
Jeremiah Bior, 75, 10, 1000, 30, 235
George Hazel, 200, 50, 7000, 100, 475
Nathan J. Sevil, -, -, 107000, 300, 1125
John Mayberry, 75, 12, 2000, 50, 100
Thomas Taylor, 75, 100, 1700, 150, 375
James Boggs, 115, 70, 2000, 50, 230
William Ennis, 53, 20, 1200, 40, 325
Andrew N. Harper, 350, 59, 4000, 80, 350
Philip Cummins, 10, 12, 250, 30, 150
John Barcus, 100, 25, 1500, 25, 260
John Foreacres, 80, 100, 4000, 30, 250
Edward R. Helmbold, 270, 130, 10000, 200, 490
Sherry Greenage, 12, 9, 600, 40, 145
Levin Gibson, 100, 20, 1200, 50, 330
William Smith, 66, 100, 200, 75, 275
George Wilson (W. Ilson), 20, 7, 500, 60, 175
James Cammins(Cummins), 54, 25, 1000, 50, 165
Benjamin Husbands, 90, 93, 3000,75, 515
Thomas Pearson, 50, 36, 1000, 35, 175

John Parker, 160, 120, 4000, 100, 512
Henry Taylor, 140, 110, 3000, 20, 400
Isaac Denney, 75, 5, 1500, 50, 250
Willebers Daws, 160, 140, 4000, 60, 390
Samuel Craig, 160, 140, 4000, 60, 350
Nehemiah Moor, 100, 20, 2000, 35, 275
Truitt Melvin, 100, 75, 2000, 50, 160
Abraham Truax, 140, 50, 4000, 75, 360
John Maclary, 200, 150, 10000, 150, 695
James T. Slaughter, 30, 12, 520, 50, 100
Robert J. Fowler, 40, 10, 150, 125, 325
Caleb Clow, 30, 2, 1500, -, -
John Brown, 200, 76, 3200, 50, 470
Parker Hardy, 90, 66, 1500, 50, 175
William Fox, 130, 150, 4000, 100, 560
Henry K. Hazel, 80, 20, 3000, 50, 388
Francis B. Harper, 200, 100, 10000, 200, 625
Daniel Ford, 100, 20, 2400, 80, 125
Walker Mifflin, 200, 20, 8000, 150, 665
William Cowgill, 120, 30, 4000, 150, 465
Benjamin F. Haines, 120, 20, 5000, 125, 389
Charles Marin, 200, 28, 10000, 100, 350
William Duhamel, 300, 60, 10000, 400, 1145
William S. Emerson, 100, 30, 3000, 85, 465
Daniel Cowgill, 130, 30, 6000, 200, 720
Sarah Cowgill, 200, 50, 8000, 15, 270
Philip Cummins, 130, 8, 4000, 30, 160
John Slaughter, 360, 100, 12000, 170, 1175
William P. Smithers, 190, 40, 8500, 185, 560
John Goodnight, 130, 20, 3000, 35, 227
Thomas C. Green, 300, 60, 12000, 75, 580
William Slaughter, 100, -, 2000, 75, 230
John Woodall, 160, 40, 7000, 225, 8805
James Robinson, 50, -, 2000, 25, 247
Edward M. Wilson, 120, 25, 6000, 250, 490
Thomas N. Wilson 180, 56, 10000, 250, 896
Robert Collings, 33, -, 2000, 75, 452
Thomas S. Buckmaster, 40, 39, 1200, 100, 240
Thomas York, 206, 60, 3000, 75, 320
Thomas Holston, 189, 49, 4700, 50, -
Manlove Hays, 234, 20, 9000, 300, 835
Robert Lewis, 260, 84, 4000, 100, 696
Abraham Moore, 182, 34, 5000, 128, 842
James Mitchel, 140, 10, 1600, 189, 180
Timothy Brown (Bunn), 50, 10, 1200, 40, 325
Thomas Gibson, 150, 50, 8000, 100, 410
James Coker (Caker), 75, 5, 1000, 50, 207
Thomas L. Calhoun, 150, 20, 3500, 75, 400
James P. Snow, 150, 35, 4000, 100, 410
Joseph Snow, 130, 100, 3000, 200, 640

John Carrow, 135, 30, 6000, 150, 500
Washington Green, 224, 40, 12500, 200, 950
Charles Harper, 230, 60, 12500, 300, 1000
Jacob Stout, 200, 10, 7000, 30, 200
Henry H. Hoffecker, 153, 50, 7000, 250, 560
Joseph H. Anderson, 130, 40, 6000, 150, 425
Thomas Jones, 107, 20, 2020, 100, 500
Presley Hoffecker, 140, 40, 6500, 125, 445
James Caleb, 140, 40, 650, 175, 575
Joseph George, 125, 75, 4000, 100, 500
Henry F. Hill, 188, 45, 7000, 200, 720
John R. Griffin, 175, 60, 10000, 150, 300
David Pleasanton, 140, 66, 6000, 200, 720
Hugitt Layton, 183, 5, 3500, 50, 300
John Williams, 12, -, 250, 5, 100
John C. Wilson, 180, 50, 5020, 180, 535
Samuel Patterson, 140, 100, 2500, 50, 185
John Cummins, 50, 30, 1000, 25, 150
James Hoffecker, 100, 30, 3000, 100, 250
William Carrow, 75, 25, 800, 50, 310
James Raughley, 240, 50, 8000, 125, 745
Timothy Carrow, 220, 50, 6000, 150, 566
Jonathan Allee, 110, 20, 3000, 50, 285
Lambert Jackson, 160, 50, 5000, 100, 460
John Truax, 120, 90, 3500, 60, 2300
Jacob Williams, 166, 64, 8000, 200, 600
Robert D. Crockett, 150, 60, 4000, 60, 330
Jonathan Purnell, 38, 8, 4500, 60, 80
William W. Rash, 20, 10, 600, 20, 50
William C. Smith, 120, 80, 5000, 100, 500
James Truax, 200, 100, 5000, 100, 1100
Presly Ford, 160, 150, 6000, 250, 600
Robert George, 55, 15, 3000, 50, 316
Gamalin Garrison, 100, 25, 3000, 80, 330
William F. Thompson, 130, 17, 3000, 75, 390
David J. W. Thompson, 260, 40, 5000, 200, 510
Joseph Reynolds, 130, 8, 4000, 75, 350
Robert J. Moore, 30, 3, 1500, -, 175
John Cloak, 100, 44, 8000, 150, 520
Elijah Green, 240, 60, 10000, 10, 680
Joseph Huffecker, 300, 200, 15000, 350, 750
Abraham Allee, 118, 50, 10000, 100, 560
James S. Moore, 120, 80, 5000, 50, 4990
Peter S. Collins, 166, 134, 10000, 2, 820
John Thomson, 51, 50, 2000, 40, 240
Philip Denney, 120, 20, 2000, 50, 200
Peter Raughley, 120, 40, 3000, 100, 320
William S. Vane, 150, 60, 3500, 50, 1100
Mitchell Perkins, 90, 10, 1000, 40, 200
John Johns, 100, 50, 1000, 60, 300
George Slaughter, 100, 50, 1000, 50, 268

Benjamin Sammons, 150, 20, 3000, 25, 260
Perry Jones, 10, 20, 800, 20, 150
Thomas Palmatry, 20, 10, 1000, 65, 100
Isaac Williams, 27, 10, 600, 10, 100
Benjamin Collins, 10, 5, 200, 20, 100
Joseph Macy, 25, 5, 500, 20, 50
Isaac Short, 75, 70, 9000, 50, 140
Minas Short, 150, 70, 2000, 25, 220
John Sammons, 75, 25, 4000, 25, 80
James Cole, 50, 400, 1500, 50, 250
George Thornton, 80, 165, 6000, 25, 225
John Severton, 40, 40, 1000, 50, 200
John Goldsborough, 60, 100, 1000, 50, 270
Pery Hanson, 120, 30, 1500, 50, 350
Robert Thompson, 140, 100, 3000, 50, 225
William Windell, 180, 85, 8000, 200, 660
Benjamin Farmer, 120, 100,7000, 100, 400
John Cameron, 90, 100, 3000, 75, 570
Austin Tomlinson, 140, 50, 3000, 50, 520
Dennis Hudson, 100, 500, 2000, 25, 265
James Robinson, 117, 28, 2500, 50, 380
Thomas Comerton (Cowerton), 80, 28, 1800, 50, 120
Samuel Sipple, 17, 2, 400, 10, 100
John H. Rash, 200, 30, 8000, 125, 490
Joseph Richard, 120, 30, 2500, 100, 250
Jeremiah Bradley, 12, 12, 500, 25, 120
Robert Davis, 109, 9, 2500, 40, 196
Robert Wilson, 17, 27, 250, 15, 50
John VanGesel, 150, 50, 6000, 150, 300
William Manwaring, 150, 50, 6000, 150, 220
Edward Attix, 150, 55, 2200,70, 350
Stephen Attix, 70, 30, 1000, 40, 225
Benjamin Donoho, 70, 20 2000, 75, 260
John Reynolds, 200, 50, 5000,75, 300
Thomas S. Hager, 56, -, 2500, 1000, 200
John R. Dixon, 99, 26, 2500, 1000, 180
Reese Lewis, 75, 10, 1500, 40, 135
Samuel _. Griffin, 150, 22, 6800, 300, 475
Thomas Pierse, 100, 6, 1500, 50, 200
John C. Jones, 110, 25, 1500, 75, 385
Thomas Ratliff, 230, 90, 10000, 150, 740
Daniel Mager, 115, 21, 3000, -, -
Nathaniel T. Underwood, 160, 60, 5000, 200, 400
John Macy, 100, 60, 4000, 25, 135
Luke Coverdale, 150, 23, 3500, 100, 365
William Stephens, 100, 40, 3500, 100, 365
Richard Casperton, 100, 30, 2000, 40, 125
Caleb Jones, 49, 23, 1200, 40, 120
William Farrell, 140, 70, 4000, 50, 370
David Turner, 100, 20, 1200, -, 165
Isaac Brister, 55, 10, 1000, 20, 195
William Thompson, 215, 40, 10000, 250, 655
James Voshell, 90, 57, 2500, 100, 4707
Robert McMullen, 100, 12, 3000, 100, 230
Thomas Keith, 125, 8, 4000, 110, 416
Isaac Register, 200, 175, 8000, 125, 500

Henry Wrench, 100, 30, 2600, 40, 150
Christopher Hillyard, 50, -, 750, 25, 90
John R. Reese, 300, 77, 10000, 400, 666
William J. Blackiston, 190, 100, 6000, 100, 365
Benedict B. Harris, 140, 50, 4000, 100, 420
John Wright, 120, 35, 300, 75, 250
Samuel Coverdale, 140, 16, 12000, 375
Elizabeth Reese, 160, 40, 8000, 150, 400
James West, 110, 46, 4000, 150, 460
Robert Brocksup, 50, 25, 2000, 150, 250
William C. Mitchell, 200, 42, 8000, 350, 840
Ephraim Jefferson, 84, 16, 2000, 50, 395
Samuel Marshall, 50, 20, 1500, 25, 190
John P. Reese, 70, 110, 5000, 150, 353
Isaac Green, 146, 36, 2000, 40, 250
Aaron Morris, 103, 10, 1500, 40, 112
Charles Cowgill, 240, 110, 14000, 30, 955
William H. Holding, 220, 29, 8000, 200, 640
David J. Murphy, 120, 21, 6000, 225, 399
William Howell, 158, 67, 5500, 115, 450
Joseph Sutton, 150, 80, 3000, 300, 342
William Burgess, 184, 43, 5000, 100, 670
Jacob Hurlock, 220, 189, 8000, 150, 350
Benjamin F. Hurlock, 193, 45, 8000, 100, 250
Hemsley Holland, 60, 40, 1500, 50, 225
Isaac Hudson, 45, 24, 1000, 100, 265
William Jacobs, 150, 140, 2500, 50, 325
John S. Hurlock, 400, 500, 18000, 300, 1000
Jonathan Jones, 100, 50, 3000, 60, 433
James Hall, 100, 75, 3500, 40, 184
Daniel Ford, 200, 80, 6000, 150, 700
Nathaniel Hutchinson, 130, 196, 3000, 75, 500
Richard Graves, 90, 75, 1500, 50, 250
George Ward, 75, 25, 1000, 50, 225
William A. Hazel, 360, 240, 9000, 200, 230
Thomas Mayberry, 50, 35, 8500, 50, 220
Thomas Tucker, 60, 40, 1150, 50, 150
Jacob Crossby, 35,65, 800, 25, 100
Henry Burrows, 60, 50, 800, 50, 100
David B. Hall, 20, 40, 600, 50, 145
Solomon Hottel, 15, 11, 300, 40, 125
William Robinson, 30, 20, 500, 30, 110
James Conner, 85, 160, 1500, 100, 370
Slaughter Harris, 200, 227, 6000, 150, 346
David Carter, 300, 50, 3000, 75, 270
James Roe, 100, 45, 3600, 195, 275
J. & S. William Cloud, 100, 35, 2500, 100, 400
Israel H. Hall, 100, 60, 2000, 50, 244
John D. Griffin, 100, 50, 1500, 40, 225
Thomas Lamb, 100, 200, 6000, 200, 1400
John Eatonx, 150, 85, 6000, 130, 595
William Cummins, 27, 1, 2800, 100, 240
George W. Cummins, 168, 60, 15000, 325, 1100
William Temple, 27, -, 2700, 200, 450

Abraham Poulson, 16 ½, -, 1600, 25, 100
Isaac Davis, 40,-, 4000, 100, 250
John G. Black, 95, 20, 12000, 300, 400
Presley Spruance, 60, -, 5000, 150, 300
Samuel H. Holding, 18, 20, 2000, 100, 120
Ayres Stockley, 30, -, 4000, 200, 250
James R. Clements, 50, 8, 7000, 130, 225
Philip Garry, 70, 2, 3400, 50, 275
Edward Clark, 20, -, 2500, 100, 250
Robert Paterson, 41, -, 3000, 110, 290
John Bacn, 14, -, 1000, 50, 300
Enoch Spruance, 80, 8, 5400, 150, 500
John Raymond, 65, -, 6500, 200, 440
Thomas Crosberry, 40, 20, 800, 35, 230
R. & James Mitchell, 180, 30, 6000, 100, 400
John M. Voshell, 200, 100, 5000, 150, 920
J. & H. Farmer, 100, 5, 2000, 40, 350
Robert Palmatary, 80, 20, 1200, 60, 530
James Hardin, 60, 50, 3000, 84, 250
William Windall, 160, 10, 2000, 25, 250
Johnathan Brown, 230, 57, 8000, 250, 725
James Tomlinson, 17, -, 250, 20, 200
Henry Vining, 17,-, 200, 5, 600
V. & S. Ford, 150, 10, 2500, 40, 425
Caleb Thomas, 100, -, 200, 50,170
Philip Morris, 19, -, 500, 30,70
Albert G. Sutton, 100, -, 2000, 150, 445
Obadiah Voshell, 70, 30, 2000, 30, 210
Robert Miller, 120, 60, 2000, 25, 125
Zadiah Miller, 125, 80, 4000, 100, 475
Enoch Miller, 125, 70, 3000, 50, 225
William E. Boyer, 123, 15, 1500, 30, 175
James F. Boyer, 137, 10, 2500, 30, 268
Henry Eubanks, 54, 6, 1200, 75, 150
Naihaniel Severson, 135, 84, 2500, 125, 310
John VanGesel, 130, 30, 5000, 150, 437
William F. Smith, 90, 20, 2500, 50, 348
Thomas Foster, 125, 55, 4000, 50, 370
William W. Nelson, 170, 30, 4500, 150, 275
William Dawson, 230, 92, 8000, 300, 600
John Young, 145, 47, 5000, 125, 380
Jesse Huffington, 120, 32, 2000, 60, 170
James D. Wilds, 140, 43, 7000, 100, 580
Mason Bailey, 190, 25, 5000, 100, 385
David O. Downs, 100, 15, 3000, 100, 600
John Bailey, 130, 40, 3000, 125, 275
Robert Foreacres, 8, 12, 200, 20, 100
George Buckmaster, 75, 30, 1000, 65, 130
Loadman E. Downs, 65, 35, 1000, 100, 160
John Rash, 30, 30, 600, 50, 136
Jonathan Gordon, 50, 50, 1000, 40, 190
William Taylor, 90, 40, 1500, 80, 215
Joseph Foreacres, 125, 60, 2500, 75, 283
William Clark, 100, 119, 2500, 90, 480
J. W. Scowdrick, 100, 50, 1500, 25, 210

Ann W. Downs, 70, 130, 1800, 35, 250
Joseph C. Downs, 70, 100, 1200, 40, 185
Thomas Copes, 85, 190, 2500, 50, 270
Edward Coppage, 33, 120, 1000, 50, 360
Josiah Wallace, 60, 25, 800, 40, 160
Moses Quillen, 120, 100, 2000, 25, 125
Robert Ringgold, 25, 20, 400, 30, 160
Jacob Jackson, 40, 35, 700, 50, 140
Edward Gibbs, 12, 2, 300, 35, 110
James Blackiston, 12, 10, 300, 25, 60
William Boyer, 25, 5, 300, 20, 140
Albert Layton, 160, 166, 6000, 275, 600
Hiram B. Layton, 200, 40, 3500, 100, 156
Henry Stephens, 100, 40, 1400, 25, 135
Sarah Stephens, 100, 40, 1500, 4, 125
Thomas Hillyard, 60, 5, 1900, 40, 285
William D. Burrows, 120, 80, 4000, 75, 450
James Brister, 15, -, 200, 20, 130
George View, 15, 19, 500, 15, 114
John Coalter, 185, 15, 200, 150, 260
William F. Foreman, 125, 80, 3000, 100, 375
James Cussford, 100, 100, 2000, 125, 380
Armwell S. Durban, 95, 55, 2000, 30, 160
William Husbands, 120, 10, 2500, 125, 430
Pennell Emerson, 100, 10, 3000, 150, 400
Samuel Moore, 120, 15, 8000, 150, 449
John Pleasanton, 80, 10, 1500, 20, 150
William Dill, 75, 15, 3000, 50, 220
Edward Pleasanton, 110, 17, 4300, 75, 370
Clement Nowell, 110, 40, 3000, 55, 361
William Hison (Hirons), 250, 50 6000, 100, 470
Joseph Soward, 170,15, 8000, 7, 701
Samuel C. York, 200, 30, 3000, 50, 399
William Rutledge (Ratledge), 101, 88, 1800, 50, 180
Alfred Bell, 120, 60, 1800, 40, 175
Farris Pobler (Potter), 90, 4, 2500, 225, 394
Stephen Spencer, 50, 20, 800, 20, 125
Clement Spistale, 278, 100, 4000, 150, 845
Robert B. Jump, 220, 67, 9000, 200, 147
Edward E. Palmer, 80, 12, 3500, 100, 290
Thomas Poor, 120, 12, 2000, 50, 245
Edward Burrows, 180, 10, 3000, 50, 300
Ebenezer Burrows, 100, 100, 2000, 40, 220
John A. Jackson, 50, 3, 800, 40, 300
Saml. Jackson, 15, 52, 700, 20, 150
John Baynard, 193, -, 8000, 100, 345
Matthew Hutchenson, 120, 40, 4000, 60, 265
James Hawdain, 175, 25, 10000, 150, 620
David Boggs, 83, 10, 2000, 75, 287
Thomas B. Lockwood, 150, 55, 7000, 200, 508
John Shahan, 170, 30, 2000, 150, 495
William Hillyard, 75, 25, 2000, 45, 200
William Denney, 200, 65, 10000, 200, 950
John C. Raymond, 165, 38, 4800, 150, 465

William Collins, 30, 20, 2000, 100, 350
John Frazier, 232, 75, 12000, 300, 887
George Williams, 250, 80, 12000, 400, 865
James E. B. Clark, 110, 44, 3000, 75, 300
David Smith, 65, 14, 1000, 25, 175
Thomas Lester, 80, 60, 12000, 100, 565
Daniel Cummins, 226, -, 20000, 400, 1065
John Mustard, 30, 3, 3000, 75, 162
John Cummins, 60, 24, 1200, 30, 160
James M. Shaire (Shairs), 80, 25, 1200, 50, 200
John Clow, 100, 70, 2500, 75, 470
Edward C. Phillips, 66, 34, 1500, 50, 310
Edward T. Phillips, 140, 40, 1200, 35, 75
Samuel Riggs, 135, 25, 3000, 100, 530
Peregrine Cork Jr., 26, 10, 500, 25, 75
George Boyce, 200, 50, 4000, 100, 450
William Faerns, 120, 30, 1000, 50, 203
Samuel Daniels, 75, 22, 1000, 75, 135
Hiram Finlow, 58, 2, 500, 75, 485
Benjamin Knotts, 100, 40, 1400, 10, -
William Deen, 66, 30, 1800, 50, 265
William Williams, 125, 25, 4000, 50, 316
Elihu Jefferson, 100, 60, 2500, 50, 350
Henry R. Keith, 150, 50, 4000, 125, 423
William Keith, 150, 50, 3000, 100, 360
David B. Richardson, 200, 20, 2000, 35, 195
William Pruitt, 135, 110, 3000, 50, 330
Man___ Laws, 100, 40, 2000, 75, 375
Joseph Farrow, 100, 40, 1500, 40, 202
Moses Price, 130, 20, 3000, 75, 345
William Surgen, 200, 21, 2500, 200, 340
David R. Moor, 175, 26, 4000, 150, 448
Frederic Digger, 100, 50, 1200, 10, 80
John Jones, 75, 15, 1400, 40, 380
William Boyd, 140, 60, 2000, 50, 485
Rees David, 100, 100, 1200, 40, 325
James Williams, 160, 11, 6000, 250, 400
William Roderfield, 150, 30, 4500, 200, 330
Garnet Forcum, 140, 75, 4000, 75, 402
James English, 130, 60, 3000, 80, 405
James S. Boils, 100, 75, 3000, 50, 230
William Burrows, 100, 75, 300, 10, 100
William Morris, 30, 11, 825, 25, 80
Randall Braman, 27, 3, 500, 30, 100
John Numbers, 60, 28, 1760, 75, 255
William L. Hazel, 150, 32, 3000, 125, 375
Thomas Carney, 75, 105, 1800, 75, 200
Mathew Hazel, 245, 130, 8000, 250, 850
Jasper Boils, 20, 20, 500, 25, 110
James Walls, 30, 55, 800, 25, 120
Risdon Wilkenson, 15, -, 200, 25, 95
Robert Fortner, 40, 30, 700, 25, 125
Peter Short, 15, 57, 400, 20, 160

Loadman Honey, 50, 102, 1500, 25, 125
David Mitchell, 20, 50, 600, 50, 175
Valentine Everett, 5, 48, 1100, 10, 50
Samuel Cuney, 10, 19, 300, 20, 45
Henry Ford, 30, 77, 1000, 20, 75
William H. Conner, 15, 25, 480, 25, 60
Isaac Bery, 80, 25, 1260, 100, 327
William Berry, 125, 70, 2500, 100, 387
Joshua Hays, 7, 43, 500, 60, 100
James Seeney, 12, 12, 360, 20, 70
George Farrell, 15, 65, 500, 25, 90
Samuel Boulden, 25, -, 300, 25, 60
James Ford, 30, 6, 500, 50, 225
Solomon Ford, 100, 50, 1000, 50, 175
Isaac J. Short, 100, 200, 3500, 200, 545
James Short, 40, 32, 800, 10, 100
William S. Short, 40, 32, 800, 10, 100
James Knight, 200, 80, 5500, 250, 625
Jesse Reed, 150, 80, 3500, 50, 200
Isaac Clayton, 45, 12, 80, 75, 225
Dennis Conner, 90, 25, 1500, 100, 250
John Jackson, 163, 100, 1500, 20, 200
William & Philip Clayton, 90, 60, 1500, 50, 160
Elijah Peregrine, 80, 14, 950, 100, 150
John Miller, 120, 60, 2000, -, -
William S. Bishop, 60, 60, 1200, 50, 270
John Catt, 7, 4, 250, 20, 600
Jesse Deen, 12, 3, 350, 20, 100
James Farrow, 100, 61, 2000, 25, 224
James Dodd, 52, 100, 1000, 25, 130
Elisha Durham, 70, 11, 600, 30, 225
Joseph Farmer, 35, 10, 400, 25, 110
Jesse P. Green, 92, 50, 1300, 30, 307
Thomas H. Denney, 185, 40, 3000, 175, 728
John Carney, 100, 45, 1400, 85, 275
Noah Laws, 45, 45, 800, 20, 145
Charles W. Walker, 75, 25, 1000, 50,2 60
Thomas Moor, 60, 48, 1000, 40, 245
Abraham N. Moore, 92, 30, 2000, 125, 340
Dr. William Burton, 200, 100, 8000, 130, 700
Ann T. Green, 8, -, 800, 15, 32
Henry Hudson, 55, 215, 5200, 80, 1500
James R. Mitchell, 90, 70, 4500, 130, 445
John W. Adkins, 3, -, 500, 30, 220
John W. Collins, 80, 60, 4000, 75, 435
James Tumlin, 39, 15, 2800, 75, 730
Jabez H. Cropper, 5, -, 37, 540, 80
Bethuel Watson, 21, -, 1600, 50, 560
Ann P. Hall, 6, 1, 1000, 10, 600
Henderson Collins, 28, -, 2800, 50, 300
William M. W. Dorsey, 10, -, 1000, 40, 135
James P. Lofland, 40, 50, 4500, 25, 300
Peter F. Causey, 145, 50, 6000, 300, 950
John H. Eccleston, 48, -, 1600, 40, 210
George S. Adkins, 12, -, 1200, 30, 190
David H. Jones, 20, 5, 1200, 20, 160
John D. Eubanks, 33, -, 350, 30, 100
Harriet White, 300, 200, 7500, 50, 250
Jesse Sherwood, 30, -, 1500, 35, 145
Elias Primrose, 18, -, 1820, 25, 150
Benjamin Henderson, 100, 200, 5000, 100, 253
Jacob Shockly, 29, -, 630, 20, 24
William Paisly, 25, 150, 600, 20, 160

John C. Smith, 40, 10, 1000, 20, 100
Samuel Paisly, 12, 118, 500, 20, 80
John Redden, 200, 700, 6000, 30, 175
Pevin Lathem, 60, 140, 1000, 30, 83
Hilyard Griffith, 70, 35, 1500, 30, 220
Robert Griffith, 46, -, 1000, 30, 173
David Walls, 70, 30, 1000, 30, 143
William H. Owens, 60, 137, 3000, 15, 160
Collins Stevens, 30, 80, 600, 25, 105
George B. Dennis, 30, 80, 600, 15, 102
James Salmons, 70, 130, 3000, 25, 255
John R. Bennett, 120, 176, 4000, 30, 850
Eli F. Hammond, 200, 60, 2500, 40, 352
James Postles, 120, 47, 3000, 135, 540
William Morgan, 50, 30, 700, 12, 105
Shadrach Postles, 100, 62, 2000, 75, 316
Thomas Lynch, 150, 50, 1500, 15, 212
James Raughley, 60, 60, 3000, 15, 81
John Needles, 40, 20, 500, 20, 100
Samuel Hersey, 75, 70, 1800, 50, 170
Curtis B. Boswick, 225, 225, 2500, 50, 381
Daniel Clifton, 20, 20, 50, 20, 120
John A. Bickel, 50, -, 4000, 10, 60
William Quillen, 60, 60, 1000, 18, 118
William Hall, 100, 200, 4000, 50, 1364
Benjamin Dickerson, 75, 75, 2000, 50, 400
Benjamin B. Ennis, 20, 50, 450, 50, 270
John Maloney, 53, 359, 2500, 50, 514
Solomon Hevaland, 80, 20, 1500, 25, 130
James Wyatt, 30, 270, 2000, 40, 224
James Cullin, 120, 480, 7000, 75, 610
Phillip Torbert, 200, 700, 5000, 50, 485
John N. Primrose, 120, 480, 5000, 125, 570
James Davis, 100, 300, 2000, 75, 448
Mary Whitehead, 40, 20, 1000, 15, 178
Eli R. Wadkins, 50, 30, 1000, 20, 193
George Truitt, 48, 14, 1000, 20, 190
James Hudson, 100, 200, 2000, 50, 632
Edward Mills, 100, 100, 300, 50, 350
William Polk, 300, 200, 12000, 25, 257
Draper Hall, 60, 60, 800, 30, 347
Ezekiel Fitzgerald, 30, 50, 800, 15, 150
Daniel Douglass, 12, 13, 350, 15, 165
Joseph Jester, 10, 6, 200, 15, 115
Charles Jester, 67, 233, 4000, 75, 600
Noah Blades, 200, 400, 7000, 140, 450
John Bennett, 150, 250, 2000, 50, 390
Joshua Bennett, 60, 350, 3000, 50, 390
William H. Richards, 42, 42, 500, 30, 125
Isaac Jester, 15, 28, 500, 20, 100
George Thomas, 15, 35, 600, 75, 150
William Fowler, 30, 30, 600, 20, 156
Albert Jones, 50, 50, 1000, 30, 90
John Richards, 40, 110, 1000, 20, 63
Nathan Smith, 60, -, 400, 20, 158

John Coverdill, 100, 40, 1200 120, 150
John Hollis, 100, 200, 1600, 50, 150
James Tumlin, 60, -, 1000, 50,285
James Vinyard, 150, 250, 3000, 40, 305
Joseph Houston, 60, 51, 1000, 30, 226
James Vaules, 50, 100, 2000, 10, 90
Joseph Quillen, 35, 19, 500, 20, 135
Thomas Mason, 60, 30, 1000, 50, 210
Curtis Dillehan, 50, 30, 1000, 30, 90
John Short, 100, 63, 1600, 5, 241
Benjamin P. Needles, 20, 50, 900, 100, 145
Warner Townsend, 100, 300, 2000, 40, 200
James B. Sipple, 60, 360, 2500, 40, 400
Mathew Milton, 100, 100, 1500, 100, 278
Sylvester Webb, 75, 75, 1000, 50, 242
William Verdin, 25, 75, 1100, 100, 190
William T. Maston, 125, 134, 5500, 50, 465
Joseph French, 100, 150, 1000, 20, 165
David K. Watson, 100, 100, 2500, 40, 250
Reuben Bowman, 100, 200, 5500, 50, 100
Mary Meredith, 75, 75, 700, 20, 140
John Thompson, 80, 190, 2500, 75, 620
William D. Maston, 30, 31, 800, 30, 215
John T. Jester, 25, 42, 500, 40, 125
James Holland, 100, 100, 3000, 20, 275
Dickerson Meredith, 125, 175, 3000, 75, 370
John C. Webb, 100, 200, 3500, 100, 650
William Thompson, 50, 50, 500, 40, 300
Nathaniel Thomas, 100, 100, 2000, 25, 245
John Martin, 36, 18, 2000, 25, 227
Quinton Camper, 200, 150, 4000, 50, 312
Curtis Laws, 100, 20, 1500, 25, 300
George Bateman, 100, 60, 2000, 30, 253
Elias Fowler, 100, 100, 2000, 40, 100
Anna Johnson, 150, 150, 3500, 40, 167
Joseph H. Owens, 80, 20, 1200, 50, 157
Thomas Tharp (Sharp), 135, 65, 3000, 75, 195
John Quillen, 120, 150, 11000, 100, 370
James Sharp, 240, 140, 5000, 50, 300
Daniel Bateman, 120, 40, 1500, 15, 120
Davis H. Mason, 150, 150, 3000, 50, 240
Sarah Alexander, 75, 56, 1400, 30, 162
John H. Alexander, 75, 50, 1000, 30, 162
Thomas Tomlinson, 100, 68, 2000, 200, 230
Jacob Quillen, 105, 45, 3000, 25, 182
Purnel Postles, 160, 80, 2000, 30, 165
George B. Collins, 80, 20, 2000, 15, 275
Mary M. Burton, 150, 250, 4000, 50, 200
George W. Buckmaster, 70, 63, 1500, 60, 205
Lister A. Houston, 177, 177, 4000, 100, 192
Ivan Morgan, 100, 100, 1000, 10, 155

James Smith, 90, 90, 1100, 40, 140
Joseph O. McColley, 150, 150, 10000, 200, 872
Caroy Frazier, 150, 150, 2500, 50, 210
William Tomlinson, 300, 100, 4000, 50, 240
Sarah Smithers, 120, 50, 1200, 30, 410
David Register, 100, 127, 1500, 50, 190
William D. Griffith, 80, 86, 2000, 150, 425
William Tucker, 50, 50, 1000, 20, 140
Clement L. Sharp, 100, 260, 5000, 100, 280
James Johnson, 75, 75, 1000, 30, 155
James J. Wood, 170, 67, 2200, 50, 300
Clement Murphy, 100, 100, 566, 30, 200
Phillip Marvill, 100, 100, 3000, 20, 141
Clarisa Marvill, 100, 100, 600, 20, 79
Collins Tatman, 100, 50, 1200, 30, 191
Outten McColley, 100, 100, 600, 20, 157
Charles W. Morris, 21, -, 500, 15, 61
William L. Jump, 100, 100, 600, 30, 135
Wendall Lynch, 100, 100, 1500, 30, 125
Jesse Walton, 100, 150, 1000, 30, 75
Joshua Hill, 100, 100, 1000, 75, 233
Charles Townsend, 60, 60, 1000, 30, 220
Curtis Vinyard, 95, 95, 1500, 50, 306
Henry Marvill, 70, 140, 1500, 15, 128
George McColley, 40, 200, 1000, 25, 141
Nehemiah Cary, 100, 140, 800, 20, 120
Riley Lynch, 100, 200, 1500, 26, 170
Josiah Marvill, 60, 60, 600, 25, 287
John Booth, 157, 100, 2000, 30, 300
John Hill, 250, 214, 2000, 100, 315
Molton Jacobs, 75, 25, 1200, 25, 190
Alfred Newsom, 70, 60, 1400, 50, 160
Susan Sapp (Sass), 125, 125, 1500, 35, 125
Mitchell Webb, 100, 75, 2500, 50, 286
Robert Powell, 45, 3, 1400, 40, 164
Isaac Harrington, 80, 160, 1000, 30, 190
John Maston, 125, 52, 2000, 50, 270
Daniel Fisher, 50, 190, 500, 20, 170
Nimrod Minner, 90, 86, 2300, 100, 300
John M. Lofland, 100, 150, 3000, 50, 172
Joseph Booth, 150, 250, 2000, 25, 270
Thomas Booth, 70, 45, 1000, 15, 235
John W. Slayton, 50, 100, 500, 20, 128
John Satterfield, 100, 150, 1500, 20, 235
Elias Lofland, 130, 130, 500, 30, 242
Samuel Minner, 75, 25, 500, 20, 151
Matilda Herrington, 150, 100, 3000, 30, 339
Elizabeth Minner, 110, 110, 1000, 30, 213
Alexander Fleming, 50, 106, 1200, 30, 236
John Fitzgerald, 95, 181, 2000, 30, 270
Asa Herrington, 170, 68, 2500, 30, 250
Minas Tatman, 100, 100, 1500, 20, 95
Samuel Short, 80, 170, 2000, 40, 205
Thomas Smith, 80, 186, 2000, 30, 200
Henry Wyatt, 50, 50, 700, 25, 165

Ezekiel Fitzgerald, 75, 75, 1000, 30, 175
Charles Warren, 200, 120, 2500, 125, 376
Jane Minner, 67, 150, 900, 20, 113
Hasty Wyatt, 150, 70, 2000, 25, 200
Elijah Sapp, 75, 75, 700, 30, 13
William Sapp, 125, 125, 2000, 125, 362
Andrew Sapp, 75, 25, 600, 50, 120
Caleb Eliott, 100, 219, 1000, 20, 92
William M. Jester, 166, 166, 2200, 25, 376
John Wyatt, 60, 15, 1000, 40, 280
Cabb Jarvis, 20, -, 300, 15, 66
William Herrington, 150, 150, 1500, 35, 370
Gore Cox, 90, 90, 600, 15, 135
Thomas Jester, 100, 10, 1000, 50, 258
Jacob Whitaker, 100, 100, 1000, 20, 175
Ezekiel Reed, 60, 30, 700, 30, 175
Thomas Henton, 55, 45, 850, 35, 55
William Master, 100, -, 1800, 150, 575
Bias Benson, 40, 60, 800, 20, 171
Henry W. Herrington, 70, 6, 1000, 60, 246
John H. Camper, 24, 33, 700, 30, 206
Thomas Laramore, 15, 15, 500, 10, 208
Eli Laramore, 9, -, 150, 10, 50
Thomas Laramore, 50, 50, 1000, 40, 222
Solomon Conner, 50, 50, 600, 40, 600
Thomas P. Reed, 80, 80, 500, 80, 290
John Baker, 25, 69, 600, 30, 100
Edward Morgan, 20, 30, 950, 50, 260
John Hopkins, 20, 30, 500, 40, 150
Nathaniel Scott, 40, 20, 600, 30, 39
Jon W. Reed, 41, 41, 500, 30, 175
Jacob Hickman, 15, 50, 450, 25, 100
James Camper, 15, 45, 450, 30, 100
Samuel Hughes, 130, 81, 1500, 50, 321
Ezekiel Cooper, 65, 50, 1000, 50, 218
Alexander Daherty, 150, 50, 2000, 50, 255
Eli Daherty, 60, 86, 600, 20, 214
Thomas Baker, 20, 81, 300, 25, 64
John D. Travis, 35, 64, 400, 12, 85
John B. Reed, 100, 125, 1000, 5, 310
John Lister, 300, 200, 5000, 150, 500
James H. Smith, 200, 80, 2500, 100, 342
James Melvin, 90, 40, 1500, 40, 290
Solomon Melvin, 230, 230, 3000, 40, 212
John Daherty, 63, 127, 1000, 20, 177
Wheatly Graham, 80, 20, 1000, 40, 266
William Smith, 100, 200, 1500, 45, 156
Pemberton Clifton, 125, 100, 2500, 45,223
Benjamin Downs, 50, 50, 700, 30, 140
William B. Fitzgerald, 60, 76, 600, 25, 104
Benjamin Thistlewood, 60, 470, 1600, 40, 137
Joshua Hill, 90, 70, 800, 25, 150
Daniel Harrington, 70, 30, 600, 25, 146
Peter Newell, 60, 40, 600, 25, 185
David Harrington, 100, 125, 1000, 26, 125
Nimrod Herrington, 100, 115, 1200, 30, 280
William Herrington, 80, 72, 800, 30, 240
George W. Dornlan, 150, 50, 2000, 30, 288
John T. Sarvas (Parvas), 200, 300, 5000, 35, 230

William Parvas (Sarvas), 250, 250, 2500, 35, 230
Josiah Wolecott, 135, 100, 2500, 50, 300
Daniel Nichols, 50, 50, 1000, 30, 147
Benjamin Herrington, 100, 100, 2000, 100, 207
Benjamin Herrington, 135, 10, 1200, 50, 219
James Herington, 130, 170, 2000, 35, 322
John Meridith, 116, 100, 2500, 35, 400
William Shaw, 100, 113, 3000, 125, 502
Charles Jones, 180, 109, 3000, 100, 264
Hasty Cain, 70, 30, 1000, 100, 275
Curtis Calaway, 43, 48, 500, 23, 310
Benjamin Calaway, 6, -, 1000, 10, 92
Shadrach Raughley, 220, 40, 9000, 400 1176
Luther Swiggett, 50, 53, 2000, 150, 230
Isaac McNatt, 100, 50, 1000, 100, 240
Lemuel Morris, 30, -, 730, 20, 56
William H. Taylor, 30, -, 6000, 100, 86
James Porter, 87, 9, 1500, 35, 180
William Paris, 150, 350, 1200, 25, 190
Waitman Hopkins, 100, 100, 1200, 60, 225
John Cain, 120, 190, 2000, 90, 274
James Thawley, 30, 32, 350, 20, 74
Daniel Antony, 125, 51, 1000, 50, 244
Robert Raughley, 170, 170, 500, 1265, 650
Stephen A. Anderson, 80, 16, 100, 50, 300
Andrew Anderson, 75, 75, 2000, 100, 135
Outten Anderson, 151, 100, 4000, 150, 364
John Cain of Danl., 38, 37, 800, 75, 214
Dennis Minner, -, 72, 3700, 25, 131
Lodewick Cain, 58, 2, 700, 60, 280
John W. Clark, 76, -, 1000, 110, 243
John McNatt, 70, 90, 2000, 25, 180
Nancy McNatt, 20, -, 500, 25, 146
Levi Cain, 87, 145, 2000, 100, 233
Moses Herrington, 125, 125, 2000, 150, 420
Stephen Scott, 15, 19, 250, 25, 130
John Jones, 200, 130, 1500, 100, 361
James B. Ross, 75, 34, 600, 20, 97
William Prettyman, 22, 21, 1500, 20, 225
Jacob F. Lewis, 100, 200, 2500, 140, 370
Clement C. Simpson, 133, 67, 1500, 40, 370
John Porter, 70, 80, 1200, 100, 406
Clement Smith, 30, 23, 300, 20, 137
Greenbury Fountain, 70, 30, 1000, 25, 210
Reuben J. Cavender, 30, 23, 500, 10, 50
Samuel Graham, 200, 64, 10000, 50, 620
Allen Thomas, 120, 80, 1500, 50, 303
Job Willoughby, 100, 50, 1500, 110, 270
Henry Adkison, 14, 26, 500, 30, 73
Stephen Cain, 75, 15, 1500, 50, 274
Robbert Commeen, 75, 75, 900, 20, 130
Eli Calaway, 70, 30, 500, 25, 110
William H. Maston, 100, 223, 2000, 50,3 60
Lewellen Tharp, 125, 175, 3000, 150, 409
William Collison, 75, 32, 1000, 40, 187
Thomas H. Highnut, 50, 50, 1000, 50,110

Peter Calaway, 200, 63, 3000, 125, 400
James S. Currey, 150, 150, 1500, 40, 337
Daniel Wyatt, 125, 125, 1500, 90, 350
John Smith, 120, 130, 1000, 40, 277
Jonathan Herrington, 125, 183, 2500, 100, 278
James Anderson, 166, 84, 3000, 100, 364
Robert Anthony, 80, 10, 700, 30, 98
John Brown, 75, 225, 1000, 75, 250
Robert H. Wicks, 180, 70, 2500, 125, 534
Tilghman Brown, 130, 117, 3500, 200, 608
William Spence, 80, 66, 1200, 35, 122
Henry Carter, 100, 173, 1500, 100, 231
James Hopkins, 40, 10, 500, 25, 275
William Scott, 35, 25, 1000, 100, 128
Levi Sapp, 85, 85, 1600, 100, 235
Wesly Scott, 66, -, 1500, 100, 209
William Simpson, 100, 20, 1400, 40, 154
Robert E. Graham, 75, 20, 1000, 40, 105
Levi Bowen, 40, 26, 500, 10, 65
George Redden, 105, 85, 2000, 110, 264
Noah Cain, 50, 30, 1000, 80, 175
James Calaway, 45, 93, 700, 15, 211
Henry Clark, 90, 60, 800, 20, 171
George Morgan, 70, 20, 700, 25, 231
Major Bowen, 30, 81, 700, 20, 86
John Calaway, 75, 45, 1500, 100, 296
Benjamin Fleming, 100, 100, 2500, 100, 300
Ezekiel Walker, 120, 47, 900, 25, 153
Nathan Fleming, 100, 200, 2500, 80, 350
Robert Lord, 300, 120, 3000, 50, 316
John H. Powell, 100, 125, 1500, 30, 275
London Bradley, 13, 4, 400, 10, 40
George Powell, 175,75, 2000, 100, 310
Nathaniel C. Powell, 33, 67, 2000, 100, 330
William H. Powell, 150, 57, 2000, 125, 762
Jordan Tindle, 35, 35, 400, 25, 48
John Lewis, 70, -, 800, 25, 260
James J. Walker, 75, 36, 600, 15, 109
Joseph W. Willis, 130, 130, 1000, 25, 235
Sarah Herrington, 80, 80, 1500, 100, 230
Richard Adams, 73, 73, 1000, 25, 250
Oliver Hammond, 15, -, 300, 15, 84
Antony Lane, 12, 75, 700, 12, 25
Noah Linch, 30, 81, 700, 10, 91
Lavinia Hollis, 60, 20, 1500, 45, 225
Lydia Sapp, 67, 68, 800, 15, 230
Mary Calaway, 75, 75, 1500, 120, 405
Clement Scott, 50, -, 246, 20, 140
Burton McNatt, 18, 18, 300, 18, 187
Peter L. Cooper, 150, 69, 5000, 100, 260
Jane Bradley, 40, 25, 1500, 100, 305
Benniah Tharp, 120, 60, 4000, 20, 580
Meshack Ducre, 90, 90, 2000, 25, 138
Artemas Smith, 50, 50, 1000, 20,108
Shadrach Collison, 80, 70, 1000, 40, 270
Samuel Anderson, 136, 137, 3000, 80, 238
Isaac Simpson, 120, 150,1800, 120, 230
Harriet Maston, 75, 50, 1000, 50, 187

Thomas Minner, 75, 25, 1200, 30, 205
Major Wyatt, 80, 20, 1500, 50, 309
Richd. J. Harrington, 33, 17, 400, 75, 153
Samuel Harrington, 30, 36, 600, 30, 170
Mathew Fleming, 40, 20, 500, 10, 107
Waitman Hopkins, 100, 160, 1100, 50, 144
Samuel Hopkins, 40, -, 250, 20, 154
Elias Spence, 100, -, 1500, 100, 384
John Bell, 70, 40, 1000, 50, 155
William Vickery, 80, 70, 1500, 50, 230
Nathan Morgan, 35, 23, 600, 30, 351
Thomas Brown, 100, 17, 1500, 100, 274
Samuel Price, 75, 75, 1000, 50, 193
Solomon Barwick, 33, 67, 2000, 80, 190
Ambrose Kinneman, 100, 111, 1200, 25, 144
Jacob Welch, 150, 75, 1200, 50, 208
Alexander Russel, 150, 150, 1500, 25, 190
Riley Melvin, 120, 100, 1200, 100, 400
John W. Vause (Vanse), 150, 130, 1800, 100, 262
Ferdinand Baynard, 160, 140, 2500, 200, 325
William Roe, 60, 40, 1000, 25, 113
John Cahal, 135, 65, 2000, 30, 262
Elisha Maloney, 184, 150, 3000, 120, 415
Isaac M. Fisher, 150, 150, 4000, 140, 450
Major Wyatt, 80, 70, 1500, 50, 416
Jesse Butler, 100, 100, 2000, 40, 180
John Fearns, 90, 30, 800, 20, 168
Henry Calaway, 100, 50, 1200, 20, 185
Waitman Jones, 20, -, 200, 20, 160
James Scott, 100, 30, 780, 30, 40
Reuben Ross, 115, 115, 2000, 75, 214
Sydenham Melvin, 106, 100, 1500, 50, 307
William Jones, 50, 35, 650, 20, 159
John R. Smith, 35, 35,650, 35, 1200
Joseph Ward, 100, 50, 2200, 50, 392
Robert Smith, 100, 65, 2500, 30, 219
John W. Smith, 100, 50, 3000, 150, 525
Stephen Redden, 100, 81, 2000, 50, 257
Gibson Collins, 150, 95, 4500, 100, 300
William H. Andrew, 65, 46, 1500, 20, 157
John W. Redden, 72, 72, 1000, 20, 117
John Hopkins Sr., 300, 245, 6000, 250, 814
Jeremiah P. Cordrey, 150, 150, 1500, 90, 286
Stephen Redden Jr., 80, 64, 1500, 50, 244
David Taylor, 80, 135, 2000, 150, 375
James Dawson, 75, 61, 1000, 40, 195
Isaac Jester, 86, 80, 1000, 50, 190
Mirian Oldfield, 75, 125, 1500, 40, 225
Hooper B. Hopkins, 125, 127, 3500, 125, 408
John W. Stevens, 75, 100, 2000, 40, 102
Eli Wroten, 80, 50, 1500, 110, 343
Charles Wroten, 65, 20, 900, 30, 247
Curtis Hopkins, 100, 68, 1000, 30, 100
Granby Enolds, 65, 65, 800, 20, 170
John Russel, 75, 75, 1600, 35, 206
Henry Q. Herrington, 32, 18, 400, 10, 23
Margaret Salsbury, 150, 90, 1500, 150, 308
Thomas Williams, 70, 90, 1000, 15, 216

James Layton, 75, 75, 1000, 30, 500
Samuel Carlisle, 75, 75, 600, 30, 193
Thomas Layton, 100, 300, 1700, 20, 175
Jonathan Wilson, 100, 65, 800, 20, 68
Priestly Chaffirck, 190, 50, 1400, 85, 449
Nicholas O. Smith, 137, 60, 1200, 70, 463
John Hopkins Jr., 190, 145, 3500, 150, 606
Uriah Meridith, 133, 69, 1000, 50, 222
William Williams, 200, 100, 2500, 250, 584
James Wroten, 75, 35, 600, 51, 56
Henry Spence, 120, 30, 600, 45, 201
William H. Hardesty, 175, 75, 1600, 40, 150
Richard H. Morelan, 200, 100, 2000, 50, 272
Jabez Fisher, 160, 90, 2500, 100, 457
Charles Williamson, 150, 110, 3000, 100, 300
Garrison Salsbury, 75, 75, 700, 25, 244
Thomas Hardesty, 70, 39, 650, 25, 147
Martin Smith, 140, 54, 1000, 30, 247
Ben Smith, 80, 60, 1000, 20, -
William Ward, 75, 75, 700, 15, 85
John Hickman, 60, 44, 820, 20, 184
Robert Adkins, 100, 60, 700, 30, 169
Snowy Jones, 150, 150, 1000, 20, 195
Elizabeth Johnson, 120, 80, 2000, 40, 233
James Rust, 60, 34, 600, 40, 100
Nathan Willis, 45, 17, 400, 10, 50
Jonathan Hamilton, 90, 60, 1000, 20, 214
James B. Hamilton, 60, 37, 500, 15, 81
Elias Booth, 150, 77, 2000, 150, 370
Samuel O. Jones, 125, 83, 1500, 60, 157
William N. Hopkins, 100, 108, 700, 15, 208
George Pratt, 100, 50, 800, 20, 100
Joshua M. Hall, 100, 185, 700, 30, 118
George Collins, 100, 50, 1800, 100, 405
Johnathan Herrington, 100, 147, 1000, 25, 197
Jacob Cordrey, 200, 100, 1200, 100, 398
Burton Prettyman, 90, 60, 2000, 100, 392
Robert King, 65, 47, 550, 15, 16
Robert Ralston, 175, 125, 2500, 150, 297
Alexander Johnson, 350, 350, 14000, 200, 520
Sylvester Smith, 80, 70, 6000, 100, 300
Solomon Murphy, 122, 122, 2000, 125, 265
John Dilleha, 133, 67, 1500, 25, 281
William Dawson, 99, 99, 2000, 30, 286
Robert Johnson, 75, 162, 1712, 80, 224
Miras Sammons, 90, 106, 800, 25, 37
Josiah Dickerson, 100, 140, 1100, 60, 175
Noble Cordrey, 120, 120, 1000, 25, 140
John A. Collins, 250, 250, 2500, 150, 405
James Tatman, 150, 150, 2000, 35, 355
George Shockley, 100, 100, 1000, 30, 194
Nathaniel Riggs, 79, 79, 500, 20, 53
John & Wm. Voshal O, 175, 75, 1000, 50, 200
Thomas Pickering O, 220, 40, 5000, 75, 300

John Truitt T, 130, 30, 1000, 15, 125
Bennett Dyer O, 150, 100, 5000, 50, 275
John R. Norris T, 150, 600, 1000, 30, 250
Wm. Stubbs T, 160, 50, 3000, 60, 250
Daniel McBride O, 100, -, 1200, 75, 200
John Wilson T, 100, 25, 1200, 60, 105
Emanuel Hignutt (Highutt), 65, -, 700, 60, 125
Cornelius _. Coffin O, 65, 30, 800, 25, 60
Woodman Stackley T, 280, 20, 2000, 125, 350
James Ward T, 100, 40, 2500, 100, 200
Elisha Truitt O, 20, -, 600, 25, 125
Wm. Dempsey T, 100, -, 1200, 10, 62
James J. Williams O, 90, 10, 900, 50, 190
James W. Dill O, 50, 40, 2000, 40, 65
Jonathan Jackson O, 45, 12, 600, 5, 18
Thomas R. Fiasthwait O, 37, 25, 1000, 50, 55
Walter Warren O, 26, -, 1500, 250, 225
David Hess T, 15, -, 300, 40, 21
Peter Massey O, 20, -, 1000, 25, 90
George Spurry O, 60, 15, 700, 60, 197
Timothy Terry O, 92, 15, 600, 60, 175
Jane Jester O, 32, -, 600, 35, 100
Wesly Hess O, 29, -, 600, 5,100
Cornelius Dewees O, 170, -, 2500, 50, 225
Charles Conwell O, 130, 70, 2500, 100, 170
Benjamin Gildersleve O, 140, 20, 2000, 70, 200
Wm. Brown O, 100, 20 1500, 70, 150
Benjamin Brown T, 40, 20, 600, 30, 160
John Anderson O, 70, -, 1100, 50, 200
Wm. Spurry O, 45, -, 400, 60, 100
Thomas H. Draper T, 60, -, 700, 10, 175
Samuel M. Carter, 270, 17, 5000, 300, 635
Robert M. Carter O, 124, 6, 2000, 50, 175
Henry Colescott O, 100, 30, 1000, 30, 225
Wm. H. Colescott O, 23, -, 500, 30, 185
Elias Sapp T, 100, -, 300, 15, 35
John Wells T, 60, 44, 2000, 30, 100
James Jackson O, 40, 30, 800, 40, 125
Thomas McIlvaine O, 75, 25, 2400, 50, 125
John R. Jackson T, 800, 300, 10000, 200, 1200
John B. Conner O, 60, 45, 5000, 100, 457
Wingate Massie T, 75, 20, 3000, 50, 235
Alexander Orr O, 75, 20, 3000, 50, 235
John Chambers, 130, 65, 5000, 40, 175
Samuel Chambers O, 250, 50, 3000, 60, 180
McIlroy McIlvaine, 275, -, 7000, 150, 783
James Shaw O, 210, 30, 1000, 35, 130
Wm. S. McIlvaine O, 90, -, 3000, 125, 565
Wm. S. McIlvaine T, 60, 40, 1500, -, -
Wixham McIlvaine O, 100, 16, 3000, 100, 540

John G. Conwell O, 80, 20, 2250, 35, 150
Joshua B. Wharton O, 100, -, 1000, 50, 140
Joseph Burchnal O, 200, 30, 3000, 60, 275
Thomas James O, 300, 170, 6000, 150, 850
James Lindel O, 50, 20, 1000, 50, 40
Solomon Warren O, 139, 21, 5000, 60, 475
Thomas Sipple T, 150, 35, 8000, 100, 500
George R. Warren T, 120, 120, 1200, 40, 250
George Simmons T, 60, -, 600, 5, 50
Samuel R. Mifflin O, 200, 252, 5000, 70, 75
Salsbury Sparkland T, 140, -, 1200, 50, 175
Samuel Cook T, 80, 20, 1200, 30, 125
John Rauley T, 145, 100, 5000, 70, 580
Henry Williams O, 150, 50, 2000, 75, 360
Samuel Warren O, 130, 233, 8000, 200,773
John Warren T, 250, 250, 4000, 50, 350
Nathan Davis T, 100, 100, 1000, 30, 210
Nathaniel L. Warren T, 50, 50, 2500, 50, 60
Peter McLindel T, 100, 40, 300, 50, 200
Thomas Reed T, 160, 210, 400, 50, 750
Peter Lindel O, 130, 40, 3000, 50, 460
John Saxton O, 150, 50, 1700, 40, 235
Wm. Hirons O, 325, 225, 4000, 100, 455
Jonathan Sipple O, 144, 74, 4000, 150, 445
Zekiel Coope T, 80, -, 2500, 50, 175
James Grice O, 160, 40, 3500, 75, 400
George Fowler T, 82, 18, 1500, 25, 175
James Downs T., 100, 100, 2500, 65, 450
Risdon Williams T, 150, 150, 6000, 50, 325
George Smith O, 120, 40, 4000, 75, 340
Jonathan Massey, 170, 70, 3000, 40, 377
Maria Bostic, 90, 20, 700, 10, 80
Wm. Lewis, 130, 106, 3000, 200, 285
Avery Draper, 122, -, 2500, 60, 225
Wm. Lewis, 137, 100, 1300, 40, 230
John Rochester, 50, 6, 380, 60, 125
Jesse Sherwood, 100, 100, 1800, 25, 200
John Sherwood, 100, 100, 2500, 140, 467
Ezekiel B. Clements, 100, 50, 3000, 200, 485
Wm. Stubbs, 70, 50, 1500, 60, 120
Even Lewis, 64, 31, 2000, 75, 215
John Everitt, 300, 60, 5000, 100, 330
James Hawkins, 140, 30, 3500, 75, 307
Henry Seeney, 100, 15, 1000, 58, 175
Dennis Conner, 100, 90, 1800, 75, 205
Thomas Pratt, 75, 50, 1000, 100, 160
Thomas B. Cooper, 95, 25, 1000, 50, 200
James Clark, 60, 30, 1000, 50, 220
Daniel Gooden, 60, 60, 1000, 75, 160
Shadrack Johnson, 11, -, 110, 8, 68
Samuel Herd, 100, 50, 200, 40, 168
John Taylor, 150, 50, 2500, 75, 332
Caleb Clements, 90, 39, 1250, 25, 290

Thomas Summers, 332, 40, 400, 30, 230
Thomas Goodin, 100, 125, 4000, 100, 473
Wm. Gooden, 7, 32, 390, 40, 75
Samuel B. Cooper, 160, 70, 3000, 120, 578
Nehemiah Cohee, 60, 40, 300, 25, 77
Wm. Johnson, 60, 40, 1000, 20, 162
Thomas E. Frazier, 90, 34, 1500, 100, 350
Henry Salsbury, 225, 175, 3700, 50, 280
Joseph Steele, 100, 75, 1500, 75, 302
Andrew Slaughter, 130, 20, 1200, 40, 250
John Bell, 45, 5, 200, 40, 100
Susan Cooper, 85, 65, 1200, 50, 300
John Gooden, 100, 96, 2000, 50, 450
Richard Cooper, 100, 100, 2000, 50, 210
Ebenezer Bennet, 100, -, 500, 10, 30
Thomas Cabbage, 90, 43, 1600, 75, 300
Samuel Dill, 160, 20, 1600, 30, 300
John Goodin, 80, 136, 4000, 148, 309
Lewis Melvin, 127, 20, 1000, 20, 100
Thomas Bell, 15, -, 300, 20, 95
Thomas Vickrey, 200, 25, 5000, 50, 250
Thomas Lockwood, 100, 200, 3000, 80, 390
Joseph J. Lewis, 50, 77, 2000, 40, 230
John Simpson, 120, -, 3000, -, -
Robert & J. Lober, 39, -, 4300, 40, 180
Caleb Smithers, 92, -, 5000, 100, 600
Frusten Mason, 200, 300, 10000, 40, 275
John W. Hall, 100, 20, 3000, 175, 800
John Emery, 90, 60, 3000, 75, 250
Lewis D. Mecmcken, 37, -, 2500, 40, 130
Edward Lord, 22, -, 1200, 25, 200
John Lowden, 8, -, 1000, 20, 105
Samuel Dickson, 11, -, 1200, 20, 50
James Lord, 35, -, 1250, 30, 130
Michel Louber, 85, 15, 4000, 200, 650
Jab__ Jenkins, 18, -, 1080, 75, 100
Isaac Dolby, 14, -, 700, 50, 75
John Corrube, 15, -, 3000, 30, 100
Henry Temple, 150, 70, 6000, 250, 760
Richard Jones, 20, -, 800, 30, 6
Ezekiel Jenkins, 15, -, 800, 75, 180
Joseph Rimmey, 12, -, 600, 25, 85
Robert Davison, 60, -, 2500, 50, 270
John Jackson, 190, 60, 5000, 150, 200
Jesse Thompson, 35, -, 2000 75, 200
John Belville, 140, 24, 2500, 75, 312
Ezekiel Woodall, 160, 70, 3500, 75, 180
Robert Lewis, 120, 50, 3000, 30, 200
Charles Lewis, 25, -, 700, 10, 120
James Anderson, 40, 35, 800, 150, 1100
Joseph Powel, 20, 24, 1500, 75, 340
Joshua Lindel, 20, -, 500, -, 75
James Wise, 75, 25, 1000, 25, 130
Roger Morrison, 75, 15, 1000, 50, 260
Jacob Tinley, 200, 400, 6000, 25, 100
Wm. Hudson, 100, 88, 2500, 50, 150
James Case, 30, 10, 500, 125, 150
Nathan Ribet, 45, 9, 600, 20, 250
Thomas Lindel, 70, 25, 700, 60, 170
Wm. Holstein, 40, 60, 500, 15, 57
John Thomas, 55, 50, 1500, 40, 180
Wm. Simpson, 220, 80, 3000, 40, 400
Robert Catlin, 75, 70, 2500, 40, 135
Henry Hargadine, 120, 30, 3000, 75, 450
John Waller, 70, 30, 1000, 30, 85

Pierson Spence, 70, 30, 1000, 20, 100
Joseph Downham, 60, 15, 1000, 5, 66
James Massey, 135, 30, 2000, 50, 360
James Conner, 80, 20, 2000, 10, 116
Edward Barrows, 100, -, 1000, 50, 140
Caleb Burchenal, 36, 4, 1000, 50, 150
Cornelius Johnson, 100, 50, 900, 20, 160
James Palmer, 50, 48, 1500, 150, 605
Peter L. Bonwell, 258, 40, 5000, 200, 555
Mark Chambers, 200, -, 1400, 75, 175
Joseph Ridgway, 172, 130, 5500, 200, 575
George Bateman, 160, 48, 440, 30, 175
Wm. Herring, 238, 70, 3000, 250, 470
Rubin Johnson, 60, -, 500, 75, 210
Wm. Salisbury, 125, -, 1500, 75, 240
Isaac Godwin, 140, 33, 1800, 75, 340
Wingate Harmon, 100, 60, 500, 25, 30
Wm. Maginis, 200, 50, 1200, 25, 130
Emanual Harmon, 100, 8, 600, 25, 100
Nehemiah Davis, 100, 21, 1500, 35, 80
John Simpson, 100, 20, 4000, 25, 170
Zacharia Young, 20, 2, 300, 15, 150
Horace Green, 75, 25, 700, 25, 75
George Bonwell, 30, -, 2000, 75, 475
Samuel Virden, 278, 170, 7500, 70, 726
John Anderson, 60, 30, 1200, 50, 110
Noah Wheatley, 75, 15, 400, 5, 4
Abner Herring, 35, 15, 600, 40, 15
Thomas Reynolds, 180, 60, 2400, 50, 60
Stansbury Murrey, 30, -, 300, 30, 9
Wm. Billing, 50, -, 500, 30, 40
James Blake, 100, 20, 600, 40, 23
Spencer Perry, 40, 80, 1000, 40, 4
James Downham, 150, 50, 2800, 30, -
James Clymer, 100, -, -, 100, 120
Wm. Brown, 260, 40, 2000, 35, 260
John Emerson, 37, 84, 5000, 200, 379
Nathaniel Young, 37, 80, 5000, 40, 276
Samuel Grace, 30, 10, 2000, 50, 138
Wm. Harrington, 35, 65, 1000, 20, 126
Wm. Saterfield, 240, 60, 3500, 45, 440
Thomas Brown, 200, 60, 1200, 40, 150
Henry Harrington, 180, 120, 1500, 45, 330
Thomas Coursey, 180, 20, 3000, 100, 450
Wm. Roe, 90, 10, 3000, 50, 425
Stephen Fountain, 60, 40, 1500, 25, 100
Andrew Anderson, 90, 10, 200, 5, 50
Wm. Crodier (Cordlie), 80, 20, 1500, 40, 166
John Case, 80, 70, 700, 30, 125
James Bradley, 150, 100, 3000, 50, 325
Noah Holden (Holder), 130, 106, 1000, 50, 125
Benjamin Reese, 275, 125, 6000, 200, 265
Eliza Trovis (Travis), 100, 60, 600, 200, 400
Elias Townsend, 150, 146, 800, 40, 160
John Chambers, 80, 120, 1000, 60, 180
George Whitacre, 18, 18, 500, 40, 90

James Hopkins, 200, 211, 6000, 100, 419
Henry Cooper, 150, 40, 2000, 40, 160
Thomas Moore, 62, 90, 5000, 150, 490
David Needles, 150, 175, 1500 40, 175
Nathaniel Harrington, 109, 100, 1900, 50, 389
John Herd, 140, 20, 800, 50, 235
Stephen Postels (Postles), 175, 15, 550, 150, 555
Joseph Simpson, 150, 100, 1500, 40, 245
Wm. Spencer, 150, 180, 6000, 100, 405
James Cahall, 180, 100, 4000, 75, 300
James Jester, 109, 10, 2000, 50, 225
Wm. Case, 120, 178, 1000, 30, 375
Louden Carter, 200, 130, 1200, 20, 118
John Clifton, 250, 50, 6000, 100, 500
Alexander Fleming, 200, 200, 5000, 30, 212
Thomas Sipple, 100, 100, 1000, 40, 160
Robert Ross, 200, 100, 3000, 20, 205
Jonathan Longfellow, 200, 50, 1500, 30, 325
John Anderson, 300, 30, 4700, 75, 362
Wm. H. Deweese, 175, 125, 2000, 50, 344
John Harrington, 150, 6, 400, 41, 100
John Cullen, 100, 40, 850, 21, 235
George Davis, 100, 150, 3000, 40, 263
Nathaniel Anderson, 25, -, 400, -, 40
Mathew Benson, 35, 8, 300, 35, 200
Thomas Rawley, 85, 20, 700, 50, 270
Wm. Jarrald, 175, 75, 1500, 40, 300
Wm. Graham, 70, 50, 1000, 20, 200
James Roe, 35, 15, 500, 25, 225
Isaac Harris, 50, 60, 400, 15, 60
Alexander Fountain, 30, 16, 400, 20, 180
Jonathan Downham, 75, 45, 600, 40,2 60
James Pritchet, 40, 10, 400, 40, 220
Gediah Beauchamp, 250, 100, 1500, 50, 265
Sylvester Willey (Hilley), 50, 10, 1500, 25, 145
Warner Barcus, 270, 30, 1000, 50, 330
Andrew Calley, 200, 49, 2000, 150, 300
Samuel Gibbs, 100, 40, 700, 40, 90
John Moore, 138, 60, 1000, 20, 200
John Reed, 100, 15, 500, 20, 300
Wm. Cock (Cook), 150, 50, 3000, 100, 400
Willard R. Dawson, 96, 80, 3000, 200, 530
Luke Postles, 25, 9, 200, 30, 115
John Jarrell (Jarreld), 110, 51, 1500, 50, 506
Major Gay, 75, 40, 1000, 40, 239
Lowden Layton, 50, 40, 500, 50, 125
Thomas Saxton, 80, 20, 1000, 40, 246
Jacob Clark, 100, 50, 1000, 20, 225
Alexander Jackson, 100, 100, 3000, 60, 470
Alsolem Gay, 28, -, 200, 20, 30
Richbell Allabane, 140, 60, 1000, 60, 165
Wm. Allabane, 45, 43, 100, 100, 181
Wm. Harris, 90, 10, 1500, 100, 140
John Hawkins, 75, 25, 1200, 20, 80
Thomas Jakes, 60, 30, 5000, 75, 380
Thomas Maddin, 135, 20, 2800, 150, 145
George Moore, 68, 100, 1000, 40, 100
Griffin Moore, 100, 40, 1000, 50, 190
Wm. Wallace, 100, 42, 1500, 40, 295

Samuel Smith, 100, 200, 1000, 40, 166
Samuel Fisher, 225, 90, 1500, 35, 380
Edward Pinder, 90, 70, 1000, 15, 75
John Willaby, 48, 10, 200, 20, 70
Wm. Greenlee, 54, 6, 700, 150, 240
George Reynolds, 84, 40, 500, 35, 112
John Clark, 90, 33, 400, 35, 168
Wm. Dill, 95, 40, 650, 30, 150
John Ervin, 60, 40, 500, 40, 240
Louis Draper, 11, 10, 200, 20, 90
Peter Dill, 109, 20, 600, 30, 192
Lemuel Clark, 215, 115, 3000, 50, 223
John W. Cooper, 250, 250, 3000, 150 400
James Draper, 75, -, 300, 10, 75
Hezekiah Dill, 140, 60, 1000, 100, 275
Philamon Dill, 105, 20, 1000, 25, 201
Jonathan Longfellow, 100, 63, 1500, 90, 410
Joshua Leister, 200, 200, 4000, 70, 551
John Carter, 80, 10, 1000, 60, 302
Neick Sipple, 130, 70, 800, 8, 90
Philomon Carter, 200, 100, 1500, 85, 386
Benson Dill, 100, 120, 1000, 20, 231
Ruben Kean, 100, 30, 1000, 10, 250
Major Kean, 60, 40, 600, 10, 110
James Landers, 80, 120, 100, 20, 100
Alexander Hews, 100, 20, 1500, 40, 140
John Reynolds, 200, 78, 5000, 20, 300
Uriah Sipple, 125, 25, 450, 100, 287
Wm. Milshan, 120, 142, 1500, 24, 230
George French, 150, 100, 2000, 75, 130
Benjamin Dill, 125, 125, 1500, 35, 250
Jonathan Finley, 90, 80, 1500, 75, 300
Mathew Kemp, 32, 7, 390, 15, 75
Thomas Kemp, 17, -, 170, 10, 66
Thomas Davis, 20, 15, 300, 15, 90
Abner Dill, 175, 10, 1350, 50, 202
Wm. Langrell, 120, 55, 450, 40, 162
John Reed, 75, 80, 1380, 50, 210
George McNett, 60, 60, 1800, 50, 275
Solomon Melvin, 90, 48, 1000, 15, 226
Hinson Melvin, 8, 7, 160, 16, 140
James Frazier, 175, 75, 2500, 75, 386
Robert Frazier, 105, 35, 600, 30, 125
Wm. Frazier, 130, 70, 2500, 60, 268
Amos Hinsley, 120, 40, 1000, 40, 200
Joel Clemsuts, 339, 55, 4000, 200, 550
Wm. Sherwood, 190, 40, 1800, 25, 334
Joshua Clemsuts, 125, 120, 2000, 225, 970
Elizabeth Jump, 38, 49, 2000, 30, 300
Joseph Crammer, 100, 100, 1200, 30, 190
Gideon Clark, 230, 200, 4600, 50, 495
Edward Carter, 200, 80, 400, 145, 425
Samuel Cooper, 120, 40, 1500, 100, 330
Mary Gruwell, 145, 5, 1200, 60, 325
James Griffith, 30, 12, 600, 50, 100
Philoman Dill, 100, 50, 1200, 60, 225
Alexander Dill, -, -, -, 40, 200
Mary Meredith, 170, 20, 2000, 30, 250
Thomas D. Cabbage, 200, 60, 2000, 40, 200
John Reed, 300, 20, 4000, 1000, 705

Henry McIlvaine, 50, -, 5000, 150, 390
Thomas Fortner, 50, 72, 825, 30, 100
Sarah Williams, 8, 18, 300, 20, 193
Wm. Meredith, 120, 71, 1800, 130, 480
James Longfellow, 125, 100, 1800, 50, 268
James Herd, 55, 56, 2000, 50, 271
James Hews, 100, 50, 1000, 30, 200
George Doherity, 16, 16, 300, 75, 221
Samuel Dill, 60, 20, 500, 20, 142
John Clark, 105, 185, 1500, 20, 285
Samuel Conner, 70, 130, 1000, 50, 239
John Cooper, 110, 65, 1000, 40, 231
Nathaniel Clark, 140, 15, 1000, 100, 225
James Bostic, 145, 55, 1000, 35, 146
James Fortner, 275, 30, 1000, 30, 195
Philomon Edwards, 260, 40, 1000, 25, 156
Eunice Sylvester, 150, 50, 1000, 30, 259
John Cooper, 142, 10, 80, 20, 194
Joseph Clark, 40, -, 200, 20, 75
Wm. Herd, 84, -, 600, 30, 234
Thomas Herd, 60, 144, 900, 40, 277
Wm. Fortner, 10, 22, 200, 5, 70
James Longfellow, 150, 50, 1500, 150, 378
Wm. Weeks, 150, 100, 1500, 100, 320
Robert Greenlee, 70, 80, 1500, 30, 180
Wm. Edwards, 120, 195, 1500, 30, 398
Ann Cooper, 60, 30, 700, 20, 168
Levin Blades, 400, 325, 4500, 100, 352
Thomas Herd, 60, 68, 511, 35, 220
Aaron Wyatt, -, -, 400, 25, 180
James Killen, 60, 20, 650, 40, 55
Wm. Slaughter, 70, 40, 500, 20, 185
David Sylvester, 100, 35, 1200, 20, 137
John Sipple, 130, 130, -, 25, 161
John Pinson, 35, 61, 60, 10, 75
Richard Barrows, 23, 10, 300, 12, 50
Charles Hobden, 125, 125, 1200, 25, 180
Thomas Kellen, 75, 75, 700, 25, 200
Benjamin Goodin, 100, 50, 1500, 45, 270
Gluge Kemp, 20, 60, 200, 35, 85
Jacob Kemp, 40, 20, 350, 4, 45
Louis Simmons, 60, 40, 1000, 50, 216
John Downham, 40, 30, 365, 45, 65
Wm. Kenton, 80, 32, 1000, 50, 62
David Perry, 70, 65, 1000, 40, 225
Joseph Downham, 150, 210, 3000, 60, 437
Jame Billeas, 150, 100, 1500, 40, 219
Ann Cohee, 80, 100, 1440, 40, 261
James Godwin, 50, 50, 700, 40, 180
John Grenwell, 150, 100, 2300, 100, 351
Rizla Grenwell, 20, 10, 300, 20, 55
James Downham, 120, 80, 2000, 100, 185
Wm. Bell, 80, 125, 1000, 20, 150
Thomas Downham, 194, 40, 1000, 12, 85
James Craig, 15, -, 180, 20, 139
Major Minor, 5, -, 150, 10, 125
James Anderson, 100, 80, 1500, 20, 335
Wm. Market, 8, 15, 230, 5, 157
Henry Cowgill, 300, 300, 6000, 205, 575
Peregrine Baker 140, 10, 2500, 100, 400
Caleb Berry, -, -, -, 30, 250
James Cohee, 110, 50, 1800, 50, 346
Alexander Frazier, 120, 80, 3000, 100, 534

New Castle County Delaware
1850 Agricultural Census

This agricultural census was filmed from original records in the Delaware State Archives in Wilmington Delaware by the Delaware State Archives Microfilm office. It is negative film.

There are some forty-eight columns of information on each individual. Only the head of household is addressed. I have chosen to use only six columns of the information because I feel that this information best illustrates the wealth of the individuals. These are shown below:

1. Name of Owner
2. Acres of Improved Land
3. Acres of Unimproved Land
4. Cash Value of the Farm
5. Value of Farm Implements and Machinery
13. Value of Livestock

Thus, the numbers following the names represent columns 2, 3, 4, 5, 13.

The following symbol is used to maintain spacing where information in a column is left blank (-). This symbol is used where letters, names or numbers are not legible (_).

Page 159 and page 169 are missing and page 185 is too dark to read.

Levi Cannon, 150, -, 4000, 50, 150
Daniel Corbit, 45, -, 3000, -, -
John Mathews, 300, 150, 20000, 500, 500
Gabriel Nickerson, 470, 150, 9500, 400, 600
James R. Collins, 220, 150, 16000, 400, 900
John Appleton, 250, 38, 10000, 1100, 500
Enoch Leatherman, 300, -, 15000, 500, 1000
Edward Thomas, 400, 50, 20000, 200, 700
Robert M. Latimer, 235, 10, 15000, 207, 800
W. M. Vandergrift, 220, 80, 15000, 250, 700
J. W. Vandergrift, 55, -, 3000, 50, 300
John Smith, 208, -, 7000, -, 400
Tilghman Foxwell, 500, 1000, 25000, 250, 1200
Nicholas Vandyke, 210, 30, 1000, 200, 600
Edward Webb, 150, -, 6500, 100, 125
Wesley Jefferson, 80, 90, 4000, 100, 210
William Jones, 100, 200, 7000, 100, 300
Veronica Deangne (Vandyke), 100, 12, 5000, 200, 400
Annanias Enos, 140, 450, 15000, 150, 500
Isaac Staats, 150, 300, 10000, 200, 600

Samuel Armstrong, 400, 70, 20000, 350, 900
Archy McLane, 200, -, 6000, 100, 500
Robert Derrick, 800, 130, 150, 5000, 200, 500
Benjamin David, 175, 40, 5000, 200, 600
Jacob Staats, 100, 105, 3500, 150, 400
John Jarrald, 200, -, 4000, 200, 500
George Collins, 300, 25, 6000, -, 300
Isaac Archer, 30, -, 600, 25, 100
Jeremiah Cole, 50, -, 1000, 25, 150
Thmas Bratten, 75, 23, 3000, 50, 100
Peter S. Deakgne (Deakyne, Vandyke), 200, 40, 7000, 150, 100
George Vandyke, 70, 60, 25000, 100, 700
A. S. Vandyke, 80, 100, 4000, 100, 600
John Cooper, 59, -, 1800, 100, 150
G. A. Vandyke, 45, -, 1500, 95, 150
Jacob Vandyke, 160, 180, 6700, 200, 50
Henry Walker, 160, 30, 8000, 600, 700
Josiah Flemming, 100, 120, 5000, 100, 600
Aaron Reynolds, 110, 10, 3000, 100, 400
Abraham Hayden, 10, 20, 3000, 100, 400
Jacob Harris, 200, 100, 4000, 150, 300
John Walker, 100, -, 3000, 150, 500
Henry Keen, 200, -, 6000, 100, 400
Thomas Jones, 100, -, 3000, 50, 300
Joseph Thompson, 90, -, 3000, 75, 200
Job Townsend, 130, 100, 5000, 100, 250
John Riley, 300, 100, 15000, 100, 700
Morris Collins, 20, 12, 1600, 50, 150
John Lynam, 160, 40, 6000, 100, 400
Alexander Lee, 45, 40, 1200, 100, 300
John Barnett, 85, 25, 2000, 700, 450
William Alston, 130, 70, 4000, 150, 400
John Pierson, 200, -, 6000, 100, 250
James Reynolds, 30, -, 900, 25, 18
James Brockson, 270, 200, 10000, 200, 350
John McMurphy, 60, 60, 1000, 50, 100
Benjamin Smith, 24, -, 500, 25, 100
Mathew Durham, 40, 50, 1200, 50, 75
Perry Hamilton, 12, 3, 400, 25, 75
James Nolan, 40, 29, 1200, 25, 100
Daniel Bartlet, 45, -, 1000, 25, 100
George _. Buckannon, 175, 85, 8000, 200, 200
Jabes Gooding, 28, 7, 500, 15, 75
James Brockson Jr., 275, 125, 9000, 200, 500
Benjamin Abbot, 50, -, 1000, 25, 350
Jacob Daniel, 50, 40, 1200, 25, 400
William Wright, 25, -, 500, 10, 100
Edward Price, 140, 28, 3000, 100, 350
Eli Welsh, 90, 60, 3000, 50, 250
Nemiah Crowder, 80, 20, 2000, 100, 250
James Cara (Card), 65, 65, 1500, 25, 300
Lynam Sweatman, 140, 25, 1650, 100, 200
Daniel Sweatman, 100, 50, 1500, 25, 75
Jacob Clifton, 30, -, 300, 25, 250
John Rash, 25, -, 250, 25, 150
Lambert Melvin, 70, 60, 1500, 50, 200
Benjamin McKay, 150, 120, 2000, 50, 250
William Clayton, 50, 62, 1500, 50, 150
William Budd, 18, -, 500, 25, 150

Richard Ponza, 100, 38, 15000, 50, 300
James Harris, 250, 50, 3700, 100, 300
Ezekiel Anderson, 160, 50, 5000, 300, 500
William Scaggs, 170, 160, 3000, 100, 300
Henry Webster, 120, 240, 4000, 300, 250
John Skaggs, 50, 41, 1000, 50, 200
John Reynolds, 80, 70, 1500, 50, 300
William Sweatman, 50, 50, 1000, 50, 200
Benjamin Garner, 75, 25, 2000, 125, 200
Thomas Marin, 150, 150, 2500, 75, 1100
Nahan Dilsaver, 30, 270, 2000, 25, 200
Robert Boots, 125, -, 2000, 50, 1100
John McKay, 140, -, 2500, 25, 150
Timothy Claytor (Clayton), 130, -, 2500, 50, 200
John Moffitt, 150, 50, 2000, 100, 500
James Prior, 150, -, 3000, 50, 450
John Prior, 75, 25, 2000, 50, 450
Lambert Simmons, 50, 40, 1000, 50, 100
Benjamin Chivins, 65, -, 1500, 50, 250
William Ford, 20, -, 600, 25, 150
Bassett Fergurson, 200, 100, 6000, 125, 300
John Gott, 100, 27, 3000, 100, 250
John Hutchinson, 16, 24, 1000, 50, 200
Moses Marshall, 80, 100, 2000, 50, 175
William Ellas, 65, 20, 2000, 50, 200
Henry Lightcap (Lightcup), 45, 20, 1200, 25, 200
M. Hutchenson, 180, 120, 4000, 200, 400
M. Connolin, -, -, -, -, 200
Robert Ratledge, -, -, -, -, 150
Ward Deakyne (Vandyke), 30, 45, 1200, 50, 100
Thomas Riley, 80, 24, 2000, 75, 175
John Nandain (Naudain), 120, 80, 4000, 200, 450
Rayworth Welden (Wilder), 110, 155, 3300, 125, 400
Stringer Tingley, 60, 70, 2000, 35, 300
Henry Lent, 150, -, 2000, 50, 175
Wiliam Watts, 316, 80, 3000, 200, 250
Joshua Fennimore, 170, 25, 20000, 800, 1000
Robert A. Cochran, 170, 30, 7000, 600, 1000
James Roberts, 170, 50, 6500, 300, 600
James Kanely, 200, 70, 8000, 350, 700
John McCrone, 300, 113, 9000, 250, 600
James Williams, 240, 40, 8000, 150, 450
Richard Hamilton, 160, 110, 6000, 150, 500
William Wilson, 430, 150, 13000, 650, 900
Nemiah Davis, 360, 140, 10000, 100, 500
William Francis, 150, 70, 4500, 100, 350
William Rothwell, 386, 170, 14000, 500, 1200
Joshua Bennett, 225, 50, 7000, 100, 525
John Bedwell, 100, -, 3000, 50, 300
James Tush, 80, 20, 2500, 75, 350
Zachariah Robert, 205, 75, 7500, 200, 500
William Grim (Ginn), 140, 60, 4500, 150, 550
Benjamin Money, 200, 65, 8000, 150, 600
Josiah Taylor, 185, -, 5000, 100, 300

William Daniels, 128, -, 3500, 100, 600
Christopher Sweatman, 200, 60, 6000, 200, 600
Isaac Gibbs, 250, -, 10000, 500, 600
James S. Crawford, 350, 10, 14000, 400, 900
Benjamin Gibbs, 205, -, 8000, 200, 600
Peregrine Hendrickson, 145, 141, 5000, 100, 250
Thomas Jones, -, -, -, 150, 600
Edward Silcox, 125, -, 4500, 200, 300
Nathaniel Williams, 150, 7, 5500, 200, 250
Manlove Wilson, 190, 48, 7000, 200, 600
John McCoy, 250, 150, 10000, 250, 650
Joseph Hutchinson, 70, 24, 3000, 100, 400
Charles Rhody, 25, 26, 600, 25, 300
Thomas Deakyne (Vandyke), 150, 145, 5000, 100, 150
William Johnston, 75, 25, 2000, 100, 150
Mary Chambers, 200, 100, 6000, 300, 200
Joseph Beck, 50, 100, 1500, 75, 150
Peter Burchard, 150, -, 3000, 100, 500
Bayman Collins, 150, 50, 4000, 150, 300
William Durham, 160, 100, 4000, 150, 300
Abraham Brown, 250, 150, 6000, 200, 400
James M. Derickson, 150, -, 4000, 150, 300
John Scotton, 190, 10, 11000, 100, 300
Peter Warner, 125, 30, 3000, 50, 200
John Bennett, 85, 13, 2000, 100, 400
Isaac Walker, 75, 15, 2000, 100, 300
Jonathan Jester, 100, 50, 2500, 100, 400
Gedeon E. Rothwell, 250, 117, 8000, 175, 800
Wesley Stanley, 150, 102, 4000, 100, 500
William Naylor, 200, 75, 2000, 150, 400
Robert W. Wright, 100, 100, 1000, 75, 200
William C. Clayton (Claytor), 60, 15, 600, 75, 250
Thomas Thornton, 140, 10, 2000, 100, 225
William Thornton, 20, 25, 500, 25, 350
Abraham Pickard, 50, 50, 500, 25, 200
James Powers (Powel), 20, 90, 500, 50, 150
John Thomas, 100, 100, 1200, 75, 250
Arther Alston, 90, 19, 1000, 50, 300
Benjamine Shime, 225, 54, 4500, 100, 400
Samuel Roberts, 100, 20, 3000, 200, 300
Ezekiel View, 40, 23, 1000, 50, 100
William M. Johnston, 100, 97, 2500, 75, 250
Jacob Ryland, 150, 20, 3000, 100, 200
William E. Riggs, 190, 76, 5000, 100, 500
Jacob Harris, 175, 45, 4500, 150, 500
William Vandyke, 60, 40, 1200, 100, 250
Cusby Wright, 50, 25, 600, 25, 100
V. P___ Gott, 15, 25, 300, 50, 100
Joseph Rash, 140, 560, 3000, 75, 300
John Flaerharity, 75, 25, 1500, 50, 200
Isaac P. Walker, 200, 50, 3500, 200, 500

James Reynolds, 70, 30, 2000, 50, 400
Sarah Reynolds, 36, 9, 800, 50, 100
Isaac Price, 13, -, 1000, 50, 500
William Budd, 80, 54, 1500, 25, 150
Robert Jones, 160, 164, 2000, 75, 400
John W. Thomas, 54, 14, 1000, 50, 200
Jesse Callen, 40, 25, 1200, 50, 300
Robert Jones, 160, 164, 2000, 75, 400
William C. Horner, 200, 120, 6000, 100, 250
Robert Alexander, 100, 50, 2000, 75, 250
William Smith, 150, 225, 3000, 75, 200
Jeremiah Allen, 60, 24, 2000, 75, 175
Samuel Missig, 50, 75, 2000, 50, 150
Philine Dill, 30, 50, 1000, 25, 75
John Brister, 80, 28, 1500, 850, 300
John Howell, 125, 105, 3500, 50, 150
Joseph Donohoe, 120, 70, 4000, 50, 500
Thomas Middleton, 50, 100, 1500, 75, 350
Daniel Walls, 200, 100, 6000, 150, 400
Sarah Hill, 80, 100, 2500, 50, 350
James V. Moore, 130, 44, 10000, 800, 200
James N. Mathews, 200, 140, 10000, 300, 500
Thomas Scott, 120, 40, 600, 250, 400
Mark Davis, 220, 60, 8000, 100, 300
Isaac Ratliff, 70, 50, 3000, 300, 300
Peter Jester, 200, 90, 6000, 75, 250
James Young, 120, 60, 5000, 100, 500
Robert Miller, 60, -, 3000, 75, 250
Abraham Moore, 240, 60, 10000, 100, 450
James Mannon, 70, 600, 8000, 75, 500
Joseph Green, 180, -, 4000, 75, 500
William E. Hyatt, 120, 130, 4000, 300, 450
William W. Armstrong, 150, 150, 5000, 150, 250
John Briscoe, 62, 12, 2500, 75, 150
Isaac Gibbons, 30, 23, 2000, 75, 250
William Allfrey, 90, -, 2500, 50, 250
Benjamin Johnson, 85, 15, 2000, 50, 275
Benjamin Denny, 120, 40, 4000, 100, 250
Samuel B. Johnston, 50, 30, 2000, 75, 400
William Driggers, 40, 10, 1000, 50, 250
Louis Cornal, 40, 10, 1000, 50, 250
William Hyatt, 30, 30, 1000, 50, 150
James Dean, 30, 20, 1000, 25, 250
Samuel Slater, 30, 10, 1200, 75, 400
George Colbert, 160, 20, 6000, 100, 400
John L. Durham, 100, -, 4000, 100, 150
Ann Fennimore, 12, 12, 600, 50, 211
Andrew Donohoe, 90, 57, 1500, 75, 200
Henrietta Hartcup (Hartsup), 100, -, 5000, 75, 300
Delight Gardner, 90, 28, 6000, 75, 150
Joseph D. Parker, 100, 150, 4000, 100, 400
Thomas Lyons, 14, -, 500, 50, 200
Jethro Thompson, 70, 30, 2000, 75, 150
John Thompson, 70, 30, 2000, 75, 150
John S. Townsend, 50, -, 1500, 100, 200
Thomas Davis, 130, 20, 4500, 100, 400
Manlove Davis, 220, 10, 5500, 200, 600

Henry Davis, 200, -, 5000, 300, 500
Joseph Wirt, 160, -, 6000, 300, 500
Elias M. Staats, 150, -, 5000, 200, 250
Laurence Nandain (Naudine), 100, 25, 4000, 300, 200
John M. Nandain (Naudain), 790, 22, 6000, 300, 500
Garrit Ohoson, 260, 40, 11000, 300, 500
John C. Shaw, 320, 30, 15000, 400, 800
James Clayton, 67, 20, 2700, 50, 100
Henry Watson, 30, 50, 1000, 25, 125
____ Mattecks, 85, 62, 2000, 50, 50
William Smith, 80, 60, 2000, 75, 100
Jacob V. Nandain, 45, 110, 2000, 100, 150
Experience Loire, 100, 196, 6000, 100, 250
Lewis W. Stidham, 180, 83, 4000, 150, 400
Frederick Corch, 60, 40, 1500, 50, 100
Gideon E. Barlow, 70, 95, 3000, 50, 250
William Lambert, 60, 30, 2000, 25, 250
William Webster, 80, 70, 500, 250, 200
John Townsend, 300, 80, 12000, 700, 800
Henry Burgess, 120, 30, 4000, 50, 200
Jacob Tush (Lusk), 100, -, 3000, 50, 400
Nicholas Johnson, 100, 50, 2000, 25, 250
Isaac Caulks, -, -, -, 25, 300
Samuel Bartley, 25, 17, 600, 20, 250
James Colgate, 70, -, 800, 20, 200
Mary Vandyke, 80, 20, 3000, 100, 350
William S. Vandyke, 110, 78, 3000, 100, 400
John R. Lattimer, 140, -, 30000, 300, 1600
Richard Hartest (Hardesty), 80, 60, 10000, 150, 450
Stephen Woodruff, 17, -, 4000, 50, 300
John A. Brown, 80, -, 1300, 800, 900
Robert P. Robinson, 60, 26, 9000, 250, 600
George Cleland, 90, 10, 12000, 250, 1100
James J. Bird, 60, 10, 1000, 500, 400
George Simpson, 87, 20, 7000, 200, 200
Ellen Brynberg, 50, 50, 7000, 200, 300
John W. Tatem, 120, 33, 10000, 350, 800
Thomas Helm, 30, 15, 4000, 100, 200
Col. Sam. B. Davis, 90, 8, 25000, 1000, 1200
John Campbell, 44, 8, 15000, 300, 400
James H. Warren, 1112, -, 15000, 200, 250
Alexander Moore, 150, -, 15000, 1000, 2000
William Webb, 100, 8, 10000, 600, 500
John Platt, 32, -, 9000, 200, 200
James McGar___, 23, -, 4000, 150, 220
Joseph Lloyd, 135, 45, 20000, 200, 350
Emet M. Robinson, 7, -, 3000, 800, 120
John Glendening, 20, 8, 2000, -, 35
Isaac Ford, 22, -, 2000, 200, 227
Edward Doherty, 38, -, 3500, 200, 380
Bridget McCaw, 17, -, 3000, -, 60
John R. Brinkle, 78, -, 25000, 500, 500
Jeremiah A. Bois, 60, 7, 6500, 200, 775

Thomas Fitzgerald, 140, 9, 11000, 250, 800
John Perry, 15, 1, 1500, 50, 10
John Egbert, 12, 7, 1400, 25, 25
Daniel Cafferty, 16, -, 2500, 100, 250
Joseph Blackburn, 150, 50, 10000, 250, 445
Henry Holmes, 25, 6, 3000, 150, 760
Oliver P. Ely, 30, -, 3000, 150, 175
Jesse Hallowell, 110, 11, 10000, 500, 500
John Vail, 50, 22, 2000, 150, 150
James J. Brimley, 125, 55, 8000, 200, 500
David Taylor, 53, -, 3000, 30, 200
Ben. Waller, 28, -, 7000, 600, 590
Milton H. P. Nichols, 90, 10, 8000, 200, 500
John Cooper, 36, -, 7000, 100, 230
Nathaniel Brown, 60, -, 6000, 100, 450
Joseph Bancroft, 30, 30, 1500, 50, 300
William P. Law, 45, 25, 3000, 100, 400
Thomas Smith, 20, 140, 7000, 75, 630
John Thomson, 30, 30, 3000, 100, 245
William Beck, 50, 10, 7000, 400, 350
Henry DuPont, 300, 130, 40000, 800, 4600
Jesse Grigg, 70, 30, 15000, 200, 600
Henry Faulk, 19, -, 2000, 100, 150
Peter W. Grigg, 100, 20, 10000, 500, 1000
James Milligan, 75, 10, 6500, 100, 325
Joseph G. Anderson, 150, 40, 15000, 600, 200
Michael Clahan, 75, 9, 7000, 300, 480
William Williamson, 45, 63, 5000, 200, 375
Robert Husband, 130, 30, 9000, 500, 1040
William B. Harvey, 80, 46, 6000, 200, 670
Joseph Crapgraw, 15, 15, 200, -, 120
Thomas Wilson, 73, 15, 5500, 75, 540
Herbert Wilson, 25, 5, 1800, 25, 110
William Huey, 190, 60, 15000, 400, -
Marshall _. Chandler, 10, 30, 7800, 50, 1240
Ann Heeth (Keeth), 63, 32, 6000, 40, 490
Robert McCullough, 96, 15, 6000, 150, 516
Hugh Bogan (Hogan), 40, 40, 4000, 40, 224
Mary K. Swayne, 60, 20, 4000, 40, 240
Joseph H. Turner, 101, 125, 85, 300, 700
John Ruthever, 150, 30, 12000, 200, 640
Thos. P. Chandler, 120, 30, 10000, 500, 1000
Jane Wilson, 70, 38, 8000, 75, 413
Benj. Chandler, 130, 70, 12000, 500, 1377
Joseph Clowd, 50, 15, 4000, 100, 255
Titus Mosely, 60, 10, 4260, 100, 380
Joseph Gould, 22, 4, 1500, 60, 227
Heckland Gould, 30, 70, 4000, 100, 132
Joseph Oaks, 40, 4, 1800, 25, 150
Jesse Graves, 60, 40, 3000, 70, 270
Joseph S. Pierce, 35, 10, 2500, 75, 67
Charles _. Crapgraw, 42, -, 2000, 75, 240
Wm. G. Wilson, 20, 15, 1800, 20, 100
James Scott, 12, 14, 1350, -, 135
William Armstrong, 78, 12, 5500, 100, 500

Lewis Stern, 110, 26, 6000, 100, 650
Margt. & Ann Hutchenson, 100, 50, 7000, 50, 250
David Taylor, 100, 65, 12000, 300, 448
Ellis Nichols, 70, 25, 5000, 150, 350
Daniel Nichols, 100, 50, 7500, 100, 600
John Cummins, 85, 20, 7000, 300, 800
Alexander H. Dixon, 100, 38, 9350, 400, 1815
Moses Southens, 48, 3, 2500, 200, 580
John T. Banning, 60, 60, 6000, 80, 200
Joseph Hendrickson, 65, 60, 6000, 50, 610
William P. Robinson, 100, 40, 5000, 200, 410
Thomas Lynam 40, 44, 4000, 100, 268
Samuel Canby, 150, 40, 20000, 500, 1500
John Nixon, 35, 15, 4000, 30, 130
Richard Robinson, 50, 10, 3000, 200, 730
William Morrow, 70, 750, 5000, 100, 467
John Power, 70, 14, 7000, 200, 600
Edward Doherty, 16, 40, 2500, 100, 185
Melton Neals, 14, 12, 1200, 30, 110
Marice H. Fredd, 78, 12, 400, 200, 404
Benjamin Hartley, 15, 11, 1950, 50, 85
Obed. Eachers, 55, 15, 3500, 100, 295
Isaac Fredd, 155, 45, 16000, 600, 1670
Rob. McLaughlin, 30, 90, 1070, 80, 210
Joseph Gert, 35, 8, 400, 40, 150
Jacob Way, 51, 14, 4000, 50, 435
Samuel Osborn, 30, 5, 1800, 40, 160
Lewis Graves, 93, 24, 7000, 250, 718
George Stern, 90, 25, 6900, 60, 580
David Campbell, 25, 15, 2000, 75, 110
William G. Brown, 29, 6, 1400, 40, 100
Caleb P. May (Way), 60, 17, 4300, 200, 410
Thomas H. Way, 90, 32, 7000, 50, 532
Ann W. Brown, 80, 21, 5600, 150, 735
John Braman, 55, 15, 2448, 50, 200
Patrick Doherty, 25, 25, 1500, 25, 125
Eli Call, 9, 3, 800, 30, 30
Miller Speakman, 20, 20, 4000, 200, 350
James Mastin, 125, 25, 10000, 500, 1174
Levi Springer, 60, 40, 7000, 300, 655
Stephen Scott, 75, 22, 7500, 150, 137
Jeremiah H. Dowelsworth, 12, -, 2000, 50, 50
George M. Lodge, 90, 20, 5000, 200, 4000
Marice Green, 70, 8, 7000, 100, 378
Enos Walter, 40, 5, 3000, 150, 430
Benjamin Bartram, 70, 15, 7000, 150, 560
Thomas Neal, 35, 5, 3000, 100, 121
Benjamin Beaty, 30, 64, 2500, 50, 232
James Campbell, 45, -, 4500, 150, 390
James White, 15, 29, 3000, 200, 255
James A. Bideman, 120, 130, 16000, 650, 1150
Ashton Barlow, 60, 20, 5000, 150, 331
Eli Nichols, 30, 30, 3000, 100, 318
James T. Carpenter, 125, 25, 10000, 150, 1035

Levi Walker, 40, 8, 300, 50, 182
Joseph P. Chandler, 27, -, 4000, 100, 374
James Delaplain, 100, 40, 14000, 300, 970
John M. Callogh(McCallogh), 28, 75, 2000, 100, 106
John Steel, 13, 2, 2500, 75, 180
Israel R. Bensel, 16, 10, 5000, 55, 200
Richard Clement, 100, 40, 10000, 200, 850
Chaulkey May, 75, 7, 6000, 150, 575
Lewis Phillips, 60, 12, 4000, 50, 300
Jacob Sanborne, 100, 45, 5500, 200, 520
Harlan Clowd, 100, 20, 10000, 40, 4340
Thomas Pandover, 25,-,1000,40, 115
Jesse Klaire, 79, 15, 4650, 100, 626
Enos Harper, 33, 17, 2000, 100, 280
Samuel Jackson, 90, 26, 7500, 200, 1110
Thos. Ralph, 175, 55, 21200, 400, 2063
James Chandler, 35, 5, 1800, 50, 195
Samuel M. Green, 60, 23, 6000, 200, 560
Harman Davis, 32,10,1800, 100, 150
Evan Phillips, 110, 30, 11000, 200, 710
Caleb Harless, 50, 10, 4000, 200, 310
Harvey Phillips, 85, 15, 7500, 750, 1300
Reece Pyle, 29, 7, 1800, 60, 150
John Armstrong, 82, 18, 6000, 100, 370
Samuel Armstrong, 79, 38, 6000, 75, 360
John W. Quincy, 131, 17, 1150, 210, 588
Thomas Derrickson, 50, 1, 12000, 50, 200
Henry Watkinson, 140, 45, 13000, 400, 1100
Ashton Richardson, 53, 68, 115000, 100, 262
John R. Lynam, 68, -, 4000, 75, 389
Thomas Lampleigh, 35, -, 3000, 100, 335
William R. Flinn, 85, 15, 7000, 100, 520
Joseph Killgore, 25, 4, 1500, 60, 185
Joseph Fitssimmonds, 20, -, 1500, 30, 100
John Clark, 6, -, 1000, -, 75
George M. Bramble, 45, 50, 10000, 150, 490
Jones McGee, 25, -, 2000, 200, 155
William Boutelier, 50, 8, 4300, 150, 200
Jacob Lynam, 124, 6, 11000, 350, 1135
William McElwee, 100, 95, 9000, 150, 220
Samuel G. Chandler, 40, 20, 2000, 100, 310
John B. Marshell, 25, 38, 1500, 115, 104
John P. Armstrong, 80, 40, 10000, 400, 855
George S. Grubb, 100, 20, 10000, 400, 600
Thomas Foot, 80, 32, 9000, 300, 380
John Stewart, 45, 13, 4500, 150, 285
John Folk, 43, 10, 2000, 50, 300
Eli Evans, 42, 12, 4000, 50, 225
Francis Sowder, 115, 12, 9725, 400, 6353
John J. Flinn, 80, -, 4000, 100, 500
William Armstrong, 80, 32, 8400, 25, 395
Joseph Woodward, 100, 45, 100000, 400, 740
Joseph E. Mattieu, 90, 20, 10000, 300, 660
James Armstrong, 70, 50, 10000, 300, 575
Joshua Strang, 2, 50, 10000, 100, 370

Elizabeth Armstrong, 100, 40, 10000, 100, 682
James H. Hoffecker, 95, 5, 10000, 200, 380
John T. Brown, 60, 15, 5000, 100, 190
James Armor, 65, 1, 5000, 150, 270
George F. Jameson, 39, 3, 2500, 50, 120
Moses Jowney, 100, 60, 8000, 200, 230
John B. Justice, 100, 11, 9800, 250, 570
William Z. Derrickson, 60, 55, 3500, 200, 284
Rebecca Miller, 50, 14, 3200, 100, 350
Gilpin Stedham, 80, 10, 7000, 200, 410
James Sowder, 60, 7, 5000, 200, 735
W. P. & J. Richardson, 190, 64, 18000, 600, 2000
George Barrett, 18, 2, 500, 20, 170
Thomas Lynam, 80, 20, 6000, 100, 460
Albert J. Lynam, 80, 20, 5500, 150, 460
Joseph S. Lynam, 700, 10, 5500, 150, 345
Elias S. Nandine, 166, -, 15000, 300, 700
William H. Writtenhouse, 28, 2, 1500, 10, 100
Emma Whiteman, 130, 76, 6500, 30, 265
Lewis C. Flinn, 100, 40, 8000, 150, 850
Joseph Higgans, 170, 60, 8000, 100, 620
Jacob Rothwell, 25, 11, 1800, 30, 235
Franklin _. Flinn, 120, 30, 10000, 250, 540
Francis Sowder Jr., 140, 85, 9000, 150, 350
William Claver (Clover) Jr., 100, 20, 10000, 50, 1000
Joseph Claver, 248, 185, 20000, 200, 2000
John L. Price, 100, -, 7500, 50, 600
George W. Rassner, 260, 20, 20000, 800, 2500
William F. Wilson, 21, 12, 800, 25, 600
William M. Mullen, 140, 10, 9000, 125, 750
James C. Howes, 50, 6, 5600, 50, 300
Thomas Edenfield, 275, 25, 15000, 150, 480
John W. Osborn, 340, 60, 20000, 200, 1000
Curtis Bourman, 100, -, 6000, 50, 650
Jones Young, 211, 25, 13000, 75, 870
James Hessey, 82, 25, 4000, 50, 354
John Eagles, 80, 25, 4500, 75, 500
George T. Clover Jr., 70, 10, 5000, 50, 500
Christopher Vandergrift, 100, 40, 6000, 100, 600
P___ Clover (Claver), 150, 10, 9000, 100, 420
Jacob Grose, 150, 10, 7500, 100, 450
Mary V. Leslie, 90, 20, 45000, 75, 300
William W. Leslie, 85, 25, 6500, 75, 280
William Piguett, 230, 40, 18000, 75, 1675
John T. Robinson, 32, 110, 12000, 75, -
Fredan P. Vanaklen, 300, 60, 10000, 35, 450
George H. Aydelott, 230, 60, 4000, 100, 300
Louis Green, 120, 60, 4000, 50, 500
George O. Burgess, 126, 160, 8000, 75, 350

Michael Robson, 150, 100, 5000, 150, 115
Thomas Bayards, 100, 45, 4000, 75, 400
Jacob Vandergrift, 290, 30, 15000, 150, 900
Edward Townsend, 240, 125, 14000, 200, 1000
Legarde G. Vandergrift, 125, 12, 6000, 100, 750
William Reybold, 375, -, 37500, 2150, 4000
Samuel Riley, 125, 50, 8000, 75, 1260
James Carpenter, 210, -, 12000, 200, 2100
Eliza Stuart, 105, -, 6000, 100, 470
John D. Dilworth, 240, 60, 12000, 150, 500
John Chaver, 75, 75, 9000, 150, 560
Andrew Biddle, 114, 55, 6000, 50, 500
Phillip Grosey, 20, 10, 6000, 100, 300
William Kennedy, 220, 60, 14500, 150, 500
Thomas T. Cooper (Coopes), 105, 35, 7000, 100, 300
Joseph Diehl, 60, 40, 6000, 100, 600
Joseph Diehl, 100, 40, 5000, 100, -
Johnathan P. Smith, 150, 60, 10000, 100, 1600
Edwin A. Smith, 40, 44, 2000, 50, 300
Joseph Aiken, 60, 40, 5000, 50, 150
James B. Alricks, 120, 120, 15000, 1000, 1000
John C. Vail, 152, -, 10000, 150, 400
Samuel McVay, 55, -, 2500, 50, 150
John R. Moody, 150, 36, 11000, 100, 300
Thomas Nandain, 150, 40, 8000, 200, 350
Louis Mills, 380, -, 12000, 100, 375
Elias A. Nandain, 100, 40, 4000, 100, 300
John Houston, 375, 25, 15000, 200, 600
Benjamin V. Armstrong, 70, -, 3500, 50, 250
Michael Daugherty, 250, 150, 20000, 100, 400
Joshua Clayton Sq., 151, 21, 9000, 100, 400
Samuel T. Price, 100, -, 6500, 100, 300
John Jones, 374, -, 22500, 1200, 1000
John Hanson, 300, -, 15000, 150, 575
Ed. Joshua Clayton, 90, -, 8000, 150, 650
Andrew Beaston, 400, 100, 24000, 150, 900
Joseph Williams, 221, 30, 10000, 150, 1125
James Statts, 400, 50, 20000, 200, 885
Henry _. Hoffecker, 158 ½, -, 10000, 200, 1100
Thomas Murphy 158 ½, -, 10000, 200, 700
Zachariah Jones, 260, 40, 10000, 200, 800
Richardson Armstrong, 300, 100, 20000, 200, 500
William C. Parker, 140, 60, 10000, 200, 600
William T. Stroope, 250, -, 10000, 150, 300
Hudson Clifton, 140, 17, 7500, 100, 400
John T. Gears, 140, 10, 7500, 75, 350
Henry Hogan, 160, 20, 6500, 50, 300
Thomas Eavender (Cavender), 500, 18, 15000, 400, 800
John P. Cochran, 560, -, 30000, 100, 1310
John P. Cochran, 300, -, 15000, -, 260

James T. Band (Bond), 200, -, 8000, 1000, 300
Benjamin T. Hanson, 160, -, 800, 100, 450
Thomas P. Hanton (Houston), 120, -, 6000, 100, 450
George Reynolds, -, -, -, -, -
Benjamin Gill, 200, -, 16000, 150, 600
A. P. Reading, 240, -, 10000, 400, 800
Amos T. Lynch, 200, -, 10000, 300, 650
Spencer Holton, 165, -, 6000, 150, 350
Horation N. Willets, 315, -, 20000, 1000, 600
John Hunn, 167, -, 9500, 150, 350
R. W. Cochran, 150, -, 9500, 400, 650
W. A. Cochran, 380, 60, 16000, 1000, 650
Joel Pierce, 396, -, 15000, 500, 200
Thomas Hayes, 250, -, 10000, 200, 300
Nicholas D. Appleby, 175, -, 10000, 500, 400
John McWhorter, 300, -, 15000, 800, 1000
David Appleby, 200, -, 1000, 150, 600
Adam Diehl, 100, -, 7500, 100, 450
James Gray, 140, -, 7000, 100, 450
George W. Townsend, 80, 20, 6000, 100, 260
Thomas McWhorter, 185, -, 7000, 500, 450
Cyrus Polks (Folks), 250, -, 125000, 300, 900
N. McWhorter, 190, -, 10000, 500, 600
John P. Hudson, 150, -, 10000, 200, 700
James Pogue, 107, -, 7000, 150, 250
William L. Sparks, 150, -, 7000, 200, 300
Louis Vandergrift, 300, 50, 10000, 200, 600
James M. Vandergrift, 200, 46, 12000, 100, 600
Eli Biddle, 256, -, 10000, 300, 600
James Pont, 102, -, 3500, 150, 250
A. J. Vandergrift, 240, -, 12000, 500, 700
Lawrence Aspril, 150, -, 4500, 200, 300
William Diehl, 250, -, 1000, 200, 400
John Boyd, 150, -, 9000, 50, 400
William Bowman, 190, -, 15000, 1000, 800
Robert Huggins, 220, 10, 12000, 150, 600
Daniel Stevens, 250, -, 10000, 200, 600
Thomas Ford, 455, 170, 25000, 250, 1200
John Janvier, 160, -, 11000, 600, 800
Daniel S. Craven, 130, 30, 7500, 200, 500
Samuel Higgins, 535, -, 15000, 100, 800
Purnell Jefferson, 200, -, 10000, 100, 800
John Aspril, 100, -, 5000, 75, 500
Thomas Henson, 120, -, 5000, 50, 400
Thomas Jamison, 240, -, 12000, 800, 800
James Money, 400, -, 16000, 300, 750
Purnell J. Lynch, 225, -, 10000, 100, 500
William Loire, 220, -, 15000, 100, 750
Joshua B. Pierce, 200, -, 1000, 100, 500
James L. Compt, 135, 15, 6500, 100, 400
James McKay, 280, 56, 20000, 150, 600
John Bradly, 150, -, 7500, 25, 500

Joseph Thompson, 300, -, 18000, 300, 600
James M. Ellison, 140, 30, 15000, 150, 100
Cyrus Bell, 135, -, 3500, 50, 250
John Hayes, 100, -, 6500, 100, 140
David McRee (McKee), 300, -, 2000, 175, 500
Thomas S. Merrett, 260, -, 20000, 500, 1000
John Alston, 153, -, 12000, 100, 400
Alexander Vail, 41, -, 4000, 75, 300
William Cochran, 175, -, 13000, 300, 800
Cyrus Polk, 200, 100, 12000, 300, 900
Henry Whitlocks, 300, 26, 18000, 200, 600
Sereck Shapeross, 460, 19, 35000, 320, 1000
Wesly Irons, 150, -, 11000, 150, 600
John H. Cannon, 80, 20, 6000, 100, 400
Isaac Cannon, 85, 20, 6000, 100, 250
James B. Moore, 43, -, 3000, 50, 200
John Doolan, 52, -, 3500, 75, 300
Isaac H. Vandergrift, 610, 24, 3400, 300, 1200
Samuel Pennington, 200, 120, 15000, 300, 650
W. H. Crawford, 100, -, 10000, 600, 650
Joseph Gary, 240, 40, 16000, 200, 500
Nemiah Barrons, 250, 60, 14000, 200, 700
J. A. Vandergrift, 85, -, 10000, 100, 300
James H. Burnham, 145, -, 11000, 100, 500
Robert T. Cochran, 510, 20, 37000, 600, 1200
Joseph T. Hanson, 380, 50, 20000, 250, 1000
Alexander Porter, 40, -, 2500, 25, 150
William Cavender, 270, -, 10000, 150, 650
Samuel H. Rothwell, 220, 13, 15000, 200, 800
George W. Prettyman, 200, 36, 7000, 50, 300
James Sillcox, 108, 38, 5000, 50, 300
Charles Polk, 400, 50, 20000, 500, 500
Charles Polk, 80, -, 8000, -, 200
Daniel Corbit, 74, -, 7400, 300, 1000
Ephraim Be-zton, 270, 30, 12000, 300, 700
Sarah Taylor, 155, 20, 7000, 100, 400
John Hosuton, 55, -, 2500, 50, 200
Henry Templeman, 185, -, 8000, 50, 500
Samuel Pennington, 100, -, 8000, 150, 200
Richard Lockwood, 160, -, 9000, 40, 600
Joseph Smithers, 200, 13, 10000, 50, 200
John Whitby, 70, -, 7000, 50, 200
George Templeman, 80, -, 4000, 50, 200
Samuel Peterson, 60, -, 3000, 50, 150
L. George Vandergrift, 100, -, 4000, 75, 200
Abraham Vandergrift, 150, 40, 10000, 410, 500
Gasway Watkins, 130, -, 1000, 150, 500
John Appleton, 325, 325, 32000, 1600, 1200
John W. Gallahan (Callahan), 40, -, 1000, 25, 150
George Derrickson, 100, -, 8000, 100, 300
Samuel T. Crawford, 150, 10, 7000, 100, 400
Samuel Moody, 400, 96, 15000, 350, 800

William Davis, 80, -, 6000, 75, 350
Samuel Jefferson, 60, 24, 6000, 200, 300
Isaac. Woods, 150, 20, 7500, 700, 400
John Flemming, 20, 20, -, 25, 200
John M. Woods, 200, -, 10000, 200,800
James Price, 94, -, 16000, 150, 600
Wm. D. Clark, 200, 65, 26500, 450, 1160
John Exton, 230, 95, 24400, 312, 1966
Levi H. Clark, 190, 20, 15000, 200, 770
Anthony Reybold, 218, 160, 20000, 660, 1800
Philip Reybold, 30, -, 10000, 25, 868
David Martin, 200, -, 12000, 300, 1400
Joseph Martin, 104, 36, 15000, 350, 440
Grantham Reynolds, 265, -, 21000, 400, 680
Anthony M. Higgins, 370, 100, 25400, 500, 940
Thomas M. Rodney, 22, 8, 2200, 50, 300
Wm. W. Stubert (Hubert), 285, 50, 9950, 350, 695
Joseph E. Capelle, 97, -, 6790, 200, 256
Sarah Bowers, 70, -, 6000, 150, 235
Robt. Ocheltree, 215, 25, 17000, 600, 1072
John L. Clark, 800, 100, 72000, 200, 4020
Isaac Woods, 126, 120, 12000, 300, 785
Richard Eaton, 126, 120, 12000, 250, 785
Alex. Biddle, 110, 40, 8000, 50, 571
Philip Reynolds, 260, -, 20800, 1040, 1732
Thomas Bird, 170, 30, 15000, 500, 960
John Tumlin, 420, -, 29400, 750, 3000
Charles Deputy, 110, -, 7700, 100, 300
Asbury Pennington, 125, 25, 10500, 200, 505
Barney Reybold, 290, -, 29000, 920, 1071
Obediah Clark, 92, -, 6000, 200, 318
John L. Deputy, 317, -, 15800, 300, 120
John W. Calhoun, 110, -, 3000, 150, 295
Asher Ougston, 125, -, 5000, 250, 380
Wm. Casperson, 25, -, 5000, 50, 380
Michael Dunning, 5, -, 300, -, 50
George Toppin, 6, 6, 600, 70, 90
Alfred Caulk, 190, -, 10000, 200, 420
Thomas Holmes, -, -, -, -, 120
Maria Harris, 30, -, 500, -, -
James Shuster, 235, -, 14100, 1000, 1214
Saml. Draper, 274, -, 10000, 200, 400
John W. Corey, 161, 20, 9660, 300, 1082
Thomas Higgins, -, -, -, -, 105
Stephen Lecates, 227, -, 12000, 200, 850
Wm. J. Hurlock, 258, -, 25000, 200, 1614
Sam. Boggs, 70, -, 10200, 200, 560
John P. Belville, 190, -, 15000, 300, 975
Albert E. Newton, 227, -, 13620, 300, 1000
Abner Allston, 190, 30, 10000, 150, 428
Thomas Longfellow, 200, -, 12000, 250, 685
Thomas Clark, 200, 33, 18640, 250, 790
John Vandergrift, -, -, -, -, 50
Wm. Vail, 175, 12, 10000, 300, 460

Enoch Guy, -, -, -, 25, 65
Elias Lofland, 22, 2, 13320, 250, 860
Amos Carson, 250, -, 15000, 250, 1040
Henry Massey, 200, -, 14000, 250, 1000
Thomas W. Belville, 120, -, 7200, 300, 534
Clement Reeves, 180, 60, 17000, 300, 520
Daniel Wolf, -, -, -, -, 175
Wm. Sutton, -, -, -, -, 50
John Sutton, -, -, -, -, 120
James Jackson, -, -, -, -, 100
Gatlebbs(Gotlibbs) Stringer, -, -, -, -, 30
James N. Sutton, -, -, -, -, 205
Isaac V. Clark, -, -, -, -, 180
John Sutton Jr., -, -, -, -, 356
Saml. B. Sutton (Suttle), 200, 52, 19000, 600, 1925
John J. Henry, 475, -, 22000, 200, 3145
Wm. C. Thomas, 165, -, 16500, 300, 335
Clayton Reybold, 450, -, 36000, 1000, 6240
James McMullen, -, 100, 2500, -, 100
James B. Henry, -, -, -, -, 4500
Jonathan Hodgson, 150, 100, 12000, 200, 546
John Reybold, 700, 150, 70000, 2000, 4250
T. Jefferson Clark, 60, 30, 12000, 200, 275
James L. Veazy, 110, 115, 5700, 250, 345
Samuel Dale, 100, -, 2000, 50, 286
James C. Derricks, 100, 52, 10000, 200, 560
John Gooding, 100, -, 6000, 100, 200
William B. Calhoun, 132, 30, 6000, 100, 290
William W. Stewart, 102, 60, 8000, 200, 482
Jacob Ferris, 200, 60, 12000, 20, 805
Francis A. Roop, 50, 35, 25000, 100, 78
David McAllister, 130, 10, 6000, 50, 155
Joseph Roop, 50, 33, 4050, 100, 296
Edward G. Jamerson, 170, 50, 11000, 200, 665
Jacob Harmer, 70, 30, 500, 100, 240
Abraham S. Moore, 80, 20, 11000, 200, 325
John Clark, 120, 100, 11000, 200, 325
Robert Greins, 90, 30, 3000, 100, 202
Frederick Rozine, 30, 2, 1300, 50, 1300
John G.Rozine, 40, 2, 1800, 60, 117
Azariah F. Smith, 70, 30, 7000, 150, 510
John Mausby (Mausley), 100, 1903, 3500, 100, 320
Stephen Cunningham, 80, 40, 3500, 100, 320
James W. Alexander, 60, 40, 1700, 100, 200
Adam Hyatt, 100, 50, 5000, 150, 194
Andrew Polce Powell, 160, 30, 8000, 1000, 188
John Ded Cracker (Cracken), 210, 20, 12000, 500, 1000
Manlove Jester, 180, 20, 17000, 300, 890
James Taylor, 150, 10, 4000, 600, 570
James Nicholson, 100, -, 3000, 100, 465
Lewis Ellison, 210, 80, 15000, 150, 373
Richard T. Cann, 344, 60, 20000, 300, 1500
James Cann, 230, 50, 14000, 300, 714
William Conallin, 190, 10, 10000, 300, 591

Andrew Ellison, 160, 20, 11000, 700, 1180
Benjamin T. Biggs, 236, 60, 12000, 125, 682
John Biggs, 125, 106, 7000, 200, 375
Thomas L. Biddle, 329, 40, 10000, 300, 820
John Kane, 66, 20, 5000, 200, 284
Jonathan Ellison, 380, 40, 21000, 600, 1285
George Herbert, 318, 12, 20000, 500, 800
Sampson Law, 208, 16, 11000, 150, 560
John R. Price, 196, 6, 16000, 200, 810
John S. Warren, 146, 3, 5000, 150, 250
John Davidson, 200, 10, 14000, 400, 240
John Cazier, 195, 20, 13000, 400, 729
Lewis P. Ellison, 300, 60, 21000, 400, 1128
William Dashavi, 160, 20, 9000, 200, 511
Sylvester Lawrensend(Lawrense), 140, 50, 9000, 500, 632
John Wiley, 12, 18, 750, 60, 85
Peter Wright, 20, 1, 630, 50, 90
Stephen Wiley, 30, 10, 1200, 50, 163
James Rider, 10, 5, 900, 40, 600
_. Henry Cazier, 250, 150, 20000, 600, 885
James McCoy, 100, 95, 5000, 100, 360
Sylvester Clement, 50, 18, 4680, 200, 445
C. Henry Gray, 100, 87, 8000, 100, 350
Rylands (Highlands) Graves, 115, 40, 7500, 200, 425
James A. Short, 180, 190, 11000, 150, 546
James Louver, 10, -, 400, 50, 100
Charle Boulden, 145, 50, 9250, 100, 480
Curtis B. Allison, 140, 209000, 400, 635
George Jester, 90, 70, 2000, 95, 100
James Boulden, 100, 50, 7000, 100, 380
Isaac Boulden, 100, 80, 7200, 200, 368
Nathan Boulden, 200, 100, 15000, 400, 985
Michael Greer, 100, 40, 4000, 100, 345
Reuben Roy(Ray), 20, -, 800, 20, 110
Marcellus Price, 200, 100, 12000, 200, 785
David Stevenson, 19, 4, 800, 100, 165
Isaac Nichols, 8, 3, 800, 75, 135
David J. Shaw, 40, 40, 2000, 70, 147
Thomas Frazier, 30, 70, 2000, 100, 103
George Boulden, 400, 200, 15000, 500, 745
Charles H. Warren, 126, 60, 8680, 300, 920
David Cann, 95, 50, 4000, 100, 500
Isiah Stanton, 68, 12, 3200, 175, 416
George Walker, 50,50,2000,100, 360
Robert Price, 20 ,-, 200, 40, 70
Nathaniel Jaquette, 63, 63, 3780, 150, 312
James Cavender, 60, 90, 5000, 50, 295
Joseph Lindle, 100, 43, 5000, 135, 296
Samuel Frazier, 168, 30,10000, 95, 405
John Alricks, 85, 85, 8000, 150, 425
Andrew Harman, 80, 120, 4000, 200, 276
William Nickerson, 10, -, 1000, 65, 152
Thomas Truitt, 100, 31, 4300, 100, 288

Pery Sanders, 25, 5, 500, 20,80
William Vansant, 85, 55, 5000, 75, 205
Willian Vinser, 80, 90, 7000, 100, 418
William Curlett (Caslett), 70, 29, 4000, 100, 397
William McMancel, 75, 55, 6000, 200, 360
Samuel Wright, 105, 20, 6000, 400, 457
Jacob Carrinder, 142, 32, 8500, 250, 499
Arthur Booth, 40, -, 2500, 100, 157
George Baynards, 70, 50, 6250, 250, 492
Thomas Bradly, 160, 60, 10000, 150, 183
Andrew Graves, 130, 17, 7000, 150, 405
Samuel Comely, 94, 100, 4200, 100, 331
William Cooch, 40, 20, 4000, 200, 350
Levi Champion, 20, 9, 1300, 20, 162
James Griffith, 86, 40, 300, 400, 620
Andrew H. Fisher, 9, 12, 3593, 150, 1345
Hannah Davis, 15, -, 700, -, 30
Solomon Sanders, 9, -, 450, 60, 75
James Lindsey, 138, -, 8280, 400, 527
William M. Shakespear, 40, 65, 1200, 200, 755
L. D. Hamel, 190, 80, 16000, 150, 320
Thomas C. Bradly, 100, 23, 6150, 300, 320
John W. Evans, 180, 40, 12200, 350, 4700
Theodore Warren, 61, 50, 6000, 150, 239
Peter Springer, 141, 20, 9660, 300, 342
Joseph Griffith, 130, 5, 8600, 350, 411
Edward Armstrong, 190, 20, 12600, 250, 590
James Frazer, 75, 30, 5150, 150, 400
William Frazer, 70, 36, 5300, 200, 391
J. M. McConaughey, 260, 319, 15000, 350, 799
William L. Mote, 140, 60, 8000, 100, 314
John S. McCraw, 50, 35, 2200, 100, 305
William G. Powell, 180, 210, 6200, 70, 195
Jacob Stott, 60, 50, 2200, 120, 261
Seth Stewart, 10, 15, 900, 50, 89
George Carter, 40, 12, 1500, 50, 149
William Slacke, 130, 52, 4100, 100, 492
Robert Melvin, 45, 25, 2000, 50, 132
Charles Commpton, 60, 41, 1200, 25, 56
Fergus Maney, 78, 61, 1500, 50, 180
Samuel Green, 30, 10, 1000, 30, 182
Samuel Rambo, 75, 20, 4000, 100, 265
Samuel Lorphus, 50, 10, 1200, 30, 135
Michael Nealy, 18, 17, 300, 20, 35
Charles Herd, 40, 10, 1000, 25, 70
Andrew Fisher, 65, 5, 1400, 30, 215
Alexander Colter, 60, 20, 4000, 150, 360
Samuel Clendenning, 50, 80, 1700, 60, 199
Benjamin R. Eustic, 140, 60, 5000, 125, 276
Robert King, 14, -, 400, 50, 134
Henry Pratt, 110, 40, 3000, 100, 250
James Livingston, 86, 36, 3000, 100, 200
Alexander Simpson, 80, 24, 3000, 75, 220
James Livingston Jr., 50, -, 1000, 50, 100
Levi Green, 24, -, 600, 50, 199
John Lewis, 60, 10, 2000, 100, 428

Joseph Pennington, 40, 10, 1800, 60, 132
George Lewis, 40, 10, 1000, 60, 225
Benjamin F. Wilson, 50, 10, 1200, 100, 192
Levi G. Cooch, 340, 210, 22500, 900, 788
Cantwell M. Clark, 200, 500, 18000, 250, 2184
Jones Coverdale, 300, -, 12000, 400, 840
Stephen Townsend, 140, 207, 7000, 100, 358
William Irons, 11, 1, 250, 25, 75
Valentine Skimether, 72, 30, 1800, 60, 158
James Stewart, 60, 70, 1950, 60, 240
George Veach, 70, 38, 2200, 150, 205
Joseph Crawford, 8, 8, 1000, 25, 130
Joseph Veach, 16, 1, 594, 100, 125
John Veach, 18, 5, 690, -, -
John Frazer, 100, 15, 7380, 150, 374
Samuel McIntyre, 100, 57, 5750, 150, 482
James McIntyre, 90, 60, 6000, 150, 320
James Raymond, 150, 50, 8000,150, 371
Robert M. Black, 200, 50, 2400, 350, 765
John M. Clark, 90, 50, 5000, 100, 194
John Price, 10, -, 510, 30, 169
Benjamin Reed, 105, 35, 3900, 100, 176
Hannah Adair, 140, 16, 1160, -, 70
David Ford, 150, 50, 9000, 200, 634
John W. Gillinghouse, 270, 110, 26000, 800, 3149
Isaac J. Lane, 200, 53, 17700, 400, 942
Henry Ross, 120, 30, 3000, 150, 395
Samuel Burnham, 120, 44, 10000, 200, 570
Robert Graham, 8, -, 1200, 50, 200
Solomon Townsend, 140, 20, 8000, 100, 310
William Silvers, 85, 15, 6000, 150, 570
Peregrine Kemp, 11, -, 1000, 50, 88
Samuel McMullen, 60, 36, 4000, 200, 360
Joseph Moore, 80, 45, 8000, 100, 380
Archibald Enos, 51, 55, 4300, 75, 150
Edwards Streets (Struts), 90, 8, 6000, 100, 290
James Sharp, 115, 69, 11000, 60, 173
William Gray, 152, -, 8000, 100, 212
William Day, 110, 95, 7000, 100, 355
Jonathan Hainer, 25, 5, 3400, 70, 208
John S. Turner, 177, 41, 13000, 310, 545
John Diehl, 235, 18, 20000, 500, 1744
James Strayse, 81, 32, 5000, 100, 230
George Boal, 90, 22, 7000, 150, 580
Charles Gooding, 150, 80, 14000, 302, 955
George Watson, 95, -, 2000, 100, 180
David W. Gerennell, 327, 20, 25000, 600, 2795
James Gray, 160, 55, 14000, 200, 655
Stephen S. Stradl, 140, 50, 10000, 100, 740
Charles S. Compt, 50, 6, 2000, 30, 182
David Biddle, 250, 50, 16000, 50, 658
John Deputy, 222, 150, 18000, 100, 143
Samuel Holland, 70, 15, 2000, 50, 164
John Caulk, 170, -, 14000, 400, 1185

Mary McMullen, 190, -, 15000, 488, 860
William B. McCrone, 120, -, 9000, 200, 623
George R. Townsend, 90, -, 6600, 200, 580
James F. Clayton, 353, 44, 3600, 1000, 2895
Joseph Barnaby, 140, -, 9000, 175, 561
John Marsh, 140, 31, 11510, 175, 271
Daniel Diehl, 94, 6, 8000, 320, 510
Matthew Rothwood, 159, 14, 9000, 250, 440
Thomas Sampson, 180, -, 10000, 6, 4263
James Biggs (Riggs), 25, 5, 2400, 100, 213
Ferdinand Janvier, 97, 50, 6000, 200, 610
John Pugh, 300, 100, 2000, 350, 604
John T. Simmons, 185, 15, 13000, 300, 550
Lewis Ashton, 115, -, 9000, 150, 622
Jesse Tatman, 100, -, 5000, 50, 162
Josiah C. Linden, 30, 130, 13000, 250, 625
George Allen, 35, -, 2500, 200, 320
William R. Strope, 100, 90, 3000, 175, 447
William N. Brooks, 175, 285, 15000, 400, 1038
Louisa Marley, 103, 30, 7200, 150, 478
Peter Riley, 100, 6, 7000, 100, 285
James Morrison, 96, -, 10000, 100, 375
Thomas Booth, 100, 7, 6000, 150, 405
Linwood Pennington, 200, 100, 12000, 200, 644
David Laville, 100, 10, 700, -, 486
David Morgan, 97, 20, 6000, 150, 213
Lewis Davis, 270, 22, 5000, 150, 1415
William McAllister, 115, 5, 9000, 200, 772
Aron Stoopes, 101, -, 8000, 300, 391
Thomas Smith, 145, 10, 6000, 150, 415
Adam Turnbull, 130, -, 7000, 200, 735
George Turnbull, 112, 12, 9760, 100, 386
Thomas Whiton, 200 50, 15000, 125, 531
Micajah Churchman, 145, 4, 8000, 100, 1769
Richard Jackson, 170, 30, 12000, 200, 1370
John Pugh, 130, 2, 11000, 350, 629
James Demphsy, 8, -, 900, 60, 55
James C. Cadson, 154, -, 17000, 400, 914
Patrick Donolly, 75, 1, 5000, 150, 340
James Ogram, 130, -, 12000, 200, 684
Isaac Stanton, 150, 40, 10000, 150, 598
Thomas M. Pennington, 125, 10, 13000, 360, 883
Isaac H. Kirby, 385, 15, 18000, 250, 1149
Thomas J. Snyder, 60, 9, 4000, 100, 307
John Paoch, 87, 57, 7000, 150, 415
Joshua Haynes, 220, 70, 18000, 200, 825
Bryan Jackson, 220, -, 22000, 600, 1410
Francis Dinsworth, 255, 120, 20000, 200, 1251
Thomas Appleby, 100, -, 8000, 300, 406
Jonathan E. George, 163, 10, 14000, 300, 700
Henry L. Churchman, 290, 10, 20000, 700, 4163

James W. Johns, 150, 60, 20000, 410, 1025
Marcellus Mathes, 95, 3, 6000, 100, 275
Joseph J. Taggert, 128, 15, 11440, 300, 935
David G. Nirvin, 160, 49, 11000, 500, 657
Jesse H. Alexander, 10, -, 5000, 110, 153
James Bennett, 10, -, 5000, 150, 125
Taylor Townsend, 71, -, 12000, 400, 905
John Ridgeway, 126, -, 14000, 300, 1619
George Alexander, 52, -, 6000, 150, 343
Jacob Brown, 70, -, 7000, 100, 545
Francis Hobson, 150, -, 15000, 100, 574
Jacob Walton, 216, -, 16000, 250, 2450
Peter S. Alricks, 125, 40, 12000, 600, 975
Clayton Platt, 130, 7, 12000, 500, 1055
William Wetherill, 125, -, 10000, 260, 790
John Lesterson, 185, 9, 13000, 400, 1115
Giles Lambson, 156, -, 15000, 700, 1475
Benjamin Elliot, 207, 31, 20000, 300, 1595
John Smith, 170, -, 12000, 400, 494
Lewis Reynolds, 80, 12, 5000, 100, 230
James Caulk, 125, 10, 9450, 150, 414
Barrett Stoopes, 100, 40, 7000, 220, 549
George McCrone, 140, 23, 20000, 500, 1000
Thomas Hill, 17, -, 5000, 200, 130
Mary F. Reynolds, 150, -, 12000, 400, 868
Joseph Bartlet, 92, 12 6000, 200, 405
William Banks, 100, 40, 11000, 300, 570
Robert Moore, 192, 30, 18000, 400, 401
Jonathan Stepson, 95, -, 9000, 300, 388
Moses Lambson, 205, 30, 15000, 1100, 815
John Eckles, 120,70, 21000, 150, 388
William Motherlin, 170, 4, 10000, 500, 895
John Patterson, 82, -, 7000, 300, 560
C. P. Holcomb, 450, 50, 4000, 900, 2495
John Morrison, 250, 50, 18000, 600, 1253
John Rattlewood (Kettlewood), 140, 50, 6000, 200, 500
William Booth, 160, -, 13000, 100, 440
Edwin M. Nivin, 60, 42, 6000, 150, 407
John W. Burton, 70, -, 700, 20, 325
John Dorman, 125, 5, 11000, 200, 566
John Speaksman, 80, -, 8000, 200, 1145
Robert E. Enos, 180, 26, 16000, 150, 543
Ann Edmonston, 160, -, 15000, 300, 703
Benjamin T. Booth, 150, -, 12000, 300, 795
Robert T. Johnston, 19, -, 4000, -, 175
John Broadbelt, 280, 40, 13000, 200, 845
Isaac Sutton, 215, 10, 12000, 200, 983
George Hanson, 65, 15, 7000, 200, 42
John Sample, 162, 8, 12000, 400, 815

Thomas Sowder, 15, 10, 2000, 200, 225
Ephraim Sutton, 10, -, 900, 75, 154
William White, 400, -, 25000, 200, 860
John W. Andrews, -, -, -, 300, 545
Samuel Rambo, 118, 5, 1200, 150, 831
John McFarland, 118, 15, 11000, 400, 710
Stephen Biddle, 90, 70, 4000, 150, 550
John Newlove, 140, 20, 13000, 400, 1185
Henry C. Veazy, 80, 45, 4000, 150, 290
George Bartholomew, 90, 30, 8000, 200, 444
Jesse Z. Tybut, 370, 10, 30000, 800, 2342
James B. Rogers, 125, 40, 15000, 500, 739
Catharine Lass (Larr), 7, 7, 700, -, 36
Hezekiah Tally, 30, 17, 1800, 100, 227
John E. Grubb, 5, 6, 500, -, 25
Isaac Smith, 6, 5, 600, 50, 500
John McReeve, 30, 31, 2400, 50, 200
Jehu (John) Tally, 45, 7, 2500, 100, 230
Eli B. Tally, 120, 20, 6000, 300, 1200
Joseph D. Tally, 27, -, 2000, 50, 60
Michael McGuire, 70, 10, 4000, 50, 75
Isabella Butler, 7, -, 700, -, 6
Isabella Jordan, 7, -, 600, -, -
Newlin Arment, 50,8, 2500, 100, 210
Joseph Hanby, 70, 32, 6000, 250, 200
John A. Jordan, 30, 10, 900, 25, 25
John Saville, 55, 40, 5000, 150, 100
Amos Chandler, 70, 20, 6000, 100, 400
Samuel Tally, 3, -, 900, -, 100
James H. Field, 8, -, 1500, 50, 125
John M. Bullock, 60, 40, 400, 40, 270
Daniel McGuire, 35, 7, 150, 50, 230
Cornelius Sweeney, 20, 5, 800, -, 45
Parker Green, 20, 10, 1200, 5, 45
William Smith, 90, 40, 8000, 300, 800
Isaac Smith, 25, -, 3000, 300, 635
Elihu Tally, 100, 30, 8000, 300, 462
John Chandler, 45, 6, 3500, 150, 450
Blythe Butler, 80, 12, 3000, 80, 250
William M. Day, 30, 32, 1750, 70, 168
Benj. Landeman, 10, 6, 1200, 30, 100
John Day, 55, 17, 3000, 100, 350
Thomas B. Day, 45, 30, 2000, -, 80
John Mancell, 13, 10, 600, -, 70
Francis Icendell, 18, 10, 2000, 30, 185
William Huntsman, 80, 40, 600, 100, 560
William Tallgsey, 80, 50, 6000, 200, 400
Elizabeth Keneday, 15, 10, 1000, -, 65
Caleb Taylor, 90, 40, 4000, 300, 462
Elias Pierce, 40, 10, 2000, 100, 225
James Righter, 25, 45, 3000, 100, 145
Christopher Landeman, 16, 10, 1500, 25, 220
James A. B. Smith, 80, 20, 5000, 200, 450
John Phillips, 30, 50, 2000, -, 150
Thomas L. Tally, 90, 8, 5000, 200, 520
Curtis Mousley, 50, 12, 4000, 100, 715
Samuel M. Tally, 40, 12, 2000, 100, 310
Robert Johnson, 40, 20, 2500, 80, 170
Milton S. Barlow, 15, -, 1000, -, 50
Solomon Beeson, 60, 12, 3000, 100, 280

Joseph Pierce, 30, 10, 2000, 40, 200
Robert Morrison, 30, 18, 3000, 60, 156
John Friel, 60, 30, 3600, 257, 230
Joseph Pierce, 40, 35, 3000, 150, 170
John H. Tally, 35, 10, 3000, 60, 160
Rev. Jno. Tally, 24, 6, 3000, 50, 130
Frederic Hay, 5, -, 700, 60, -
Alexander Hand, 35, 5, 800, 40, 100
George Dougherty, 40, 60, 2000, 50, 240
Lewis Bird, 25, 3, 2000, 80, 145
Valentine Feorwood (Fearwood,Fernwood), 40, 30, 3500, 150, 278
Clark Webster, 70, 25, 4000, 250, 362
James M. Cannon, 35, 23, 2000, 45, 200
Martin Petre, 20, 11, 1400, 175, 230
Joseph L. Derrickson, 180, 36, 4000, 250, 498
Reece Baldwin, 43, 5, 2000, 130, 225
Isaac Nees (Hess), 18, 2, 15000, 35, 125
Charles Willbank (Hillbank), 30, 18, 1800,5 0, 270
Uriel Pierce, 26, 10, 100, -, 80
Adam Pierce, 90, 40, 6000, 400, 350
John Foulke, 100, 30, 7000, 200, 385
Joshua Pugh, 55, 8, 5000, 200, 638
Thomas _. Zebley, 16, 2, 1100, -, 100
Owen Zebley, 23, 5, 1500, 15, 160
William Hanby, 53, 25, 3200, 300, 435
Jesse Ford, 125, 25, 8000, 300, 860
Thomas King, 20, 10, 1500, 55, 130
George Mousley, 35, 21, 2500, 20, 90
Robert Casey, 70, 10, 2000, 50, 270
Robert Miller, 70, 40, 3000, 40, 305
Lewis Tally, 40, 30, 3000, 100, 230
Miller Tally, 22, 3, 2000, 100, 200
Puscella Tally, 40, 35, 3400, -, -
Charles Bowlen, 34, 6, 4000, 100, 250
Adam Grubb, 100, 40, 7000, 300, 620
Samuel L. Grubb, 35, 30, 3000, 35, 80
Byard Grubb, 30, 4, 3000, 150, 280
Samuel Hanby, 40, 10, 2000, 150, 240
James G. Hanby, 30, 10, 3000, 150, 250
Richard Hanby, 32, 11, 2800, 300, 190
Walter Pierce, 70, 12, 5000, 150, 455
Thomas Rambo, 30, 35, 3000, 100, 245
John Casady, 20, 10, 1200, 60, 150
Peter Tally, 20, 10, 1500, 75, 180
James Sleeper, 70, 10, 2000, 100, 200
Thomas Tally, 40, 20, 3000, 100, 245
Samuel Ferowood, 16, 4, 1400, 80, 80
Penrose R. Tally, 85, 45, 7000, 250, 689
George Horan, 125, 50, 7000, 300, 675
John L. Pennell, 50, 20, 4000, 130, 330
Joel Robinson, 35, 25, 1500, 20, 100
Amor Robinson, 35, 5, 1500, 30, 150
Adam Prince, 65, 15, 4000, 275, 400
Jacob Carpenter, 80, 20, 4500, 40, 250
Lot Cloud, 80, 80, 7500, 50, 600
William Grantland, 50, -, 300, 100, 100
Charles Lefferts, 100, 25, 11000, 200, 740
George V. Churchman, 65, -, 7000, 150, 1130
Francis Ennis, 20, -, 1000, 20, 60
Abner Vernice 60, 11, 5000, 40, 200
Thomas Clyde, 140, 10, 12000, 200, 790

Jonas Goodley, 50, 13, 4000, 50, 350
Elizabeth Goodley, 35, 10, 3500, 30, 200
John _. Dailey, 18, 2, 3000, 25, 175
William Gray, 108, -, 8000, 150, 780
William C. Dodge, 300, 135, 20000, 300, 1800
George Dodge, 120, 10, 15000, 200, 1000
William L. Suddards, 30, 12, 8000, 150, 510
John Bird, 50, 7, 8000, 50, 135
William Lawsin, 40, -, 3500, 50, 50
Curtis Mousley, 22, 8, 2000, 25, 330
George Valentine, 100, -, 10000, 100, 735
Lewis B. Harvy, 75, 25, 6000, 300, 300
Maria F. Jordan, 40, 12, 2600, 200, 2500
Benjamin Day, 40, 10, 3000, 200, 280
Curtis Grubb, 15, 15, 1100, 40, 75
John Greenwood, 18, 12, 1000, 50, 106
Rachel Wilson, 30, 13, 11200, 100, 270
James McLaughlin, 25, 5, 2500, 60, 55
Jacob Sharpley, 35, 15, 2000, 100, 200
Mary Pierce, 60, 40, 4000,125, 278
Sebastian Plank, 20, 8, 1800, 150, 310
Thomas Green, 20, 30, 2000, 200, 140
Lewis Zebley, 16, -, 1800, 100, 200
____ Phillips, 90, 70, 5500, 100, 215
John Gorrie, 17, 60, 3800, -, 168
George C. Frank, 135, 40, 8500, 20, 165
William Bird, 72, 12, 3500, 100, 260
George Veal, 80, 20, 10000, 500, 950
Thomas Goodley, 80, 80, 7000, 50, 400
George Williamson, 75, 25, 6000, 150, 470
George Valentine, 40, 20, 2500, -, 75
James Grubb, 57, -, 7000, 150, 295
Hany Williamson, 33, 3, 3000, 100, 200
John Lodge, 90, 20, 5000, 300, 750
Joseph Lloyd, 28, -, 3000, 100, 300
Joseph Grubb, 50, 17, 5500, 200, 350
Thomas Elbert, 80, 15, 10000, 200, 450
Christian Perkins, 50, -, 5000, 50, 200
Joseph Perkins, 75, 50, 8000, 200, 415
William Gardner, 43, -, 2500, 50, 150
Joseph Guesy, 35, -, 2500, 100, 300
Francis M. Lordly (Sordly, Sandley), 10, -, 1200, 75, 200
Daniel Perkins, 16, 9, 1200, 75, 50
Hiram H. Lodge, 69, 15, 6000, 200, 575
John Edwards, 190, 30, 12000, 50, 638
Harriett Fernwood (Ferowood), 50, 10, 5000, 50, 195
John Feorwood, 70, 50, 5000, 100, 480
Joseph Reece, 15, -, 1400, 50, 100
Isaac N. Lodge, 150, 75, 10000, 500, 1250
Amos Perkins, 30, 15, 2500, 150, 260
John Campbell, 50, 50, 8000, 100, 375
Thomas Cardwell, 30, 15, 2500, 50, 225
John Bayliss, 40, 12, 2000, 100, 180
Joseph Orr, 60, -, 3000, 100, 300
Jesse Rindle (Kindle), 40, 3, 3000, 100, 260
Edward Beeson, 100, 35, 8500, 250, 621
John Beeson, 90, 30, 8000, 250, 480

William R. Welden, 54, -, 5000, 300, 1100
George W. Tally, 30, 20, 2000, 400, 1565
Y. W. Tally, 50, -, 2500, -, -
Lewis Welden, 50, 14, 12000, 500, 1225,
William Hobson, 90, -, 7000, 150, 1185
Isaac Landeman, 68, 7, 4000, 100, 395
Theopholis Mange, 100, 40, 6000, 80, 362
John Weldon, 40, 4, 5000, 100, 390
Jacob Zebley, 90, 42, 2000, 400, 678
Stephen C. Blackwell, 155, 25, 20000, 500, 100
Joseph Saving, 113, -, 10000, 200, 420
Ja__ S. Beeson, 90, 30, 8000, 200, 800
David H. Beeson, 65, 3, 5000, 100, 800
Joseph Carr, 250, 83, 20000, 500, 3400
Richard Rowland, 40, 40, 2000, 20, 75
Amos Jordan, 76, 35, 1100, 50, 190
James Ferowood (Foxwood), 40, 18, 4000, 25, 250
Henry J. Paschall, 73, 7, 1800, 200, 435
Jacob Morris, 83, 30, 3000, 50, 182
Josiah Miller, 56, 10, 3000, 250, 500
John Elliott, 75, 25, 7000, 300, 600
Martin Miller, 50, 15, 3500, 200, 270
George Weldon, 40, 30, 3500, 100, 387
John Bradford, 150, 8, 6000, 400, 925
Esau Sharpley, 100, 40, 5000, 300, 500
Jacob Weldon, 40, 20, 3500, 200, 365
Joseph Langley, 200, 57, 12000, 300,753
Jacob Backhouse, 15, 25, 2000, 100, 180
James Whyte, 25, 16, 3000, 30, 110
Samuel Henderson, 10, 7, 1200, 60, 137
William Husbands, 180, 60, 14000, 600, 1470
Daniel Frazier, 35, 8, 1450, 20, 98
Michael Mellon, 40, 37, 2500, 60, 185
Peter Springer, 10, -, 2000, 100, 300
George D. Clark, 100, 55, 6500, 6, 70
Thomas McBride, 80, 15, 5000, 70, 310
Charles Stephens, 45, 20, 3000, 90, 150
John Haganson, 35, 15, 2500, 200, 190
John Alemond, 35, 6, 2000, 60, 170
Joseph H. Weldon, 14, 5, 1500, 10, 145
David Forwood, 36, 30, 3500, 150, 630
George Gasford, 9, 1, 1500, 40, 30
Henry Webster, 75, 75, 6000, 10, 350
William Robinson, 134, -, 12000, 400, 2160
Francis R. Garden, 134, 4, 30000, 400, 1900
Joseph Jefferis, 35, 15, 2600, 200, 265
Thomas Tally, 30, 30, 4000, 150, 275
John Lynch, 73, 10, 6000, 100, 220
Stephen M. Hapler (Stapler), 50, -, 5000, 175, 130
Allison Palmer, 20, 2, 2100, 100, 130
Joseph Mendenhall, 78, -, 6500, 70, 810
James V. Jefferis, 100, 5, 10500, 500, 1490
Eli Wilson, 103, 5, 10000, 350, 1400

Andrew McKee, 70, 30, 6000, 150, 460
Mary Stidham, 65, 15, 5000, 150, 250
Thomas P. Mcclanahan, 160, 90, 10000, 400, 945
Charles John Post (Pugh), 40, 50, 10000, 300, 500
Thomas Husbands, 50, 70, 7000, 200, 600
Peter Collins, 15, 22, 2400, -, 145
Sarah Arbuckle, 175, 6, 6500, 200, 198
Alfred D. Murphy, 40, 20, 4000, 200, 275
G. M. Murphey, 12, 3, 1500, -, 30
William Sharply, 40, -, 2000, 150,275
Isaac B. Elliott, 175, 25, 13500, 600, 400
John C. Elliott, 85, 4, 9000, 200, 725
Augustine E. Jessup, 30, 26, 4500, 75, 180
Setee Bowman, 50, -, 8000,125, 520
Levi Weldon, 30, 30, 3000, 125, 200
John Beeson, 25, 25, 2000, 40, 137
Eduard Beeson Jr., 30, 50, 4000, 100, 375
William Todd, 133, -, 12000, 500, 435
John Tally, 15, 25, 12000, 30, 75
Benj. Brown, 16, -, 1000, 20,70
Edward T. Bellah, 195, -, 17000, 600, 2615
Edward Tatnall, 7, -, 1500, -, 310
Joseph Tatnall, 25, -, 4000, 50, 300
William Lea, 20, -, 4000, 75, 236
Francis P. A. Ademany, 70, 38, 6500, 250, 784
James Bauman, 50, 25, 4500, 100, 242
Irwin W. Pierce, 20, -, 1800, 50, 105
Benj. Elliott, 80, 15, 10000, 400, 622
Wm. F. Husbands, 25, 23, 3000, 100, 285
Abraham Husbands, 18, 12, 2000, 50, 155
Henry duPont, 70, -, 7000, -, 1120
George Miller, 47, 20, 3500, 200, 305
Isaac S. Elliott, 75, 75, 4500, 200, 415
Edward Tatnall, 90, 17, 8000, 100, 215
Edward Tatnall Sr., 113, 100, 30000, 500, 1140
William Bird, 15, -, 1200, 40, 105
John Beeson, 70, 8, 5000, -, 25
John Armstrong, 80, 78, 4500, 100, 485
Jacob Chandler, 32, 30,5000, 140, 427
Jesse Bishop, 75, 25, 4000, 100, 406
Jackson Holmns, 40, 9, 1500, 125, 172
Robert Pool, 125, 25, 11000, 125, 792
James Willson, 110, 12, 4810, 80, 307
William Little, 130, 20, 9700, 40, 489
Caleb Extrican, 95, 12, 5510, 90, 503
Thomas McFarlin, 94, 20, 9280, 45, 935
Thomas Little, 75, 20, 5000, 80, 543
Caleb Heald, 78, 20, 7000, 100, 800
Jacob Heald, 131, 25, 10000, 150, 1165
James Dixon, 138, 12, 8500, 200, 570
William Pearson, 45, 3, 2500, 100, 176
Jeshur H. Dixon, 78, 20, 8000, 200, 460
Thomas Pearson, 80, 10, 4000, 150, 390
Leban Pearson, 24, 2, 1800, 25, 178
John Ochletree, 51, 10, 3000, 60, 197
Robert Walker, 80, 20, 6000, 120, 485

Thomas Walker, 70, 30, 5000, 200, 327
John Walker, 65, 35, 7000, 150, 493
John Thompson, 140, 40, 12880, 300, 515
John Springer, 80,8, 6100, 175, 489
Aquilla Lambson, 100, 25, 8750, 300, 332
Matthew Bunting, 40, -, 5000, 100, 295
Sarah Willson, 132, 8, 11000, 260, 719
George Springer, 65, 10, 6000, 125, 425
Thomas Baldwin, 64, 3, 3500, 80, 453
Joseph Mitchell, 170, 25, 12000, 250, 990
Thomas Mitchell, 83, 18, 6000, 200, 407
Abner Woodward, 111, 15, 9800, 150, 925
Amos Eastborn, 100, 42, 6000, 120, 445
Robert McCabe, 84, 20, 6000, 200, 555
John Beeson, 170, 40, 16000, 400, 2217
Robert T. Weed, 80, 25, 5000, 145, 732
Matthew Lockhard, 80, 17, 5000, 22, 1501
Benjamin Moore, 100, 25, 5000, 75, 160
William Carlisle, 57, 3, 2200, 153, 295
John Little, 165, 25, 7600, 150, 438
Samuel Lloyd, 58, 2, 4000, 75, 240
George Ball (Bell), 100, 60, 11400, 100, 525
Westly Flinn, 75, -, 6500, 100, 317
John Crossin, 60, 7, 2000, 75, 108
Spencer Chandler, 107, 20, 8255, 295, 867
Wayne Jackson, 125, 12, 8220, 200, 1055
David Willson, 140, 40, 12600, 200, 1075
Stephen Willson, 70, 6, 3420, -, 418
James Mendenhall, 75, 20, 4950, 150, 519
Gibbons Hendle, 36, 6, 3000, 50, 165
John Morrison, 105, 15, 7000, 35, 518
Jesse Taylor, 113, 30, 8000, 140, 594
Walter Craig, 41, 4, 2200, 100, 171
James Aikin, 45, 17, 1800, 30, 103
Samuel Tyson, 170, 50, 15000, 500, 1225
Samuel Russel, 24, 6, 2000, 50, 112
Thomas Rankin, 108, 40, 9500, 150, 786
Joseph Rankin, 90, 60, 9500, 100, 714
Jacob Hustis, 135, 25, 5000, 150,388
George Jacobs, 75, 22, 2500, 20, 100
John Cloud, 95, 35, 4300, 150, 334
William Cloud, 195, 6, 4000, 150, 498
John Thompson, 44, 8, 2800, 200, 172
Chandler Lambson, 93, 50, 8580, 250, 275
Mary Armstrong, 86, 12, 5000, 125, 338
William Rankin, 38, 2, 6000, 100, 215
Thomas Hoops, 41, 7, 2580, 75, 112
William Mendenhall, 95, 20, 6780, 75, 272
Rachel Grimes, 45, 5, 2200, 180, 222
John Grimes, 60, 5, 3400, 100, 167
John C. Parat (Sarat), 35, 13, 4500, 125, 275
Maxwell Drake, 79, 25, 4000, 50, 215
Benjamin F. Gibhart, 50, 14, 2500, 50, 209
Johnathan Hoops, 100, 13, 5057, 50, 300
James Walker, 71, 9, 4000, 100, 310

Joseph Lindsey, 150, 50, 12000, 400, 1250
William Brackin Sr., 31, 2, 3000, 100, 123
Rebecca Walker, 104, 6, 6600, 15, 1445
Thomas Hannah, 50, 10, 3000, 100, 253
Maxwell B. Ochletree, 100, 12, 8000, 300, 790
Samuel Taylor, 57, 3, 3250, 100, 187
Stephen Mitchell, 65, 15, 5500, 75, 500
John McDaniel, 140, 20, 9000, 400, 1367
Joseph Eastborn, 429, -, 30000, 500, 461
Samuel Hannah, 80, 20, 5000, 50, 294
James Stinson, 22, 2, 2000, 100, 120
Elizabeth Davis, 80, 10, 4000, 50, 195
Levi B. Moore, 100, 20, 5500, 150, 500
Joseph Pyle, 59, 15, 4700, 120, 262
Elie Hyfeald, 70, 12, 2800, 75, 130
Benjamin Armstrong, 66, 12, 5000, 100, 225
George Montgomery, 24, 4, 2380, 100, 164
Joseph Redman, 88, 16, 6825, 500, 360
John Faust, 51, 12, 4000, 200, 310
Joseph Neis, 50, 6, 4000, 200, 372
Samuel Graves, 38, 6, 3220, 200, 381
Robert Morrison, 97, 20, 8170, 200, 456
Elie Crossin, 43, -, 2500, 150, 298
Isaac Crossin, 94, 16, 8000, 220, 718
James Baily, 57, 8, 4875, 120, 459
James Springer, 80, 20, 6000, 175, 404
Stephen Springer, 80, 20, 6000, 200, 622
Louis Fell, 162, 30, 11000, 400, 1288
Jonathan Mayson, 109 21, 6500, 75, 655
James Griffin, 90, 30, 6500, 125, 516
Samuel Evason, 95, 10, 6480, 250, 845
Thomas Love, 42, 8, 3500, 50, 264
Thomas Vandiver, 190, 18, 5000, 50, 316
William Foot, 77, 30, 5350, 75, 338
Isaac Mendenhall, 70, 20, 4500, 60, 310
David Chambers, 50, 5, 3000, 80, 344
James Woods, 23, -, 1000, 15, 120
Samuel Taylor, 48, -, 2600, 150, 212
Adam Flair, 105, 20, 6250, 200, 674
Robert Walker, 90, 40, 7000, 150, 595
John H. Whitman, 70, 30, 2500, 125, 291
John Springer, 83, 12, 4000, 200, 344
Joshua B. Baker, 53, 12, 4875, 100, 330
James Leach, 21, 5, 1500, 100, 972
W. De__ese Wood, 50, 23, 6000, 150, 830
Richard Smithers, 300, 75, 25000, 800, 2160
John Mitchell, 128, 16, 7500, 100, 741
James McCloy, 200, 50, 9000, 100, 650
Thomas Chandler, 65, 11, 3800, 35, 386
Philip C. Willson, 72, 8, 6000, 95, 344
Thomas Yeatman, 30, 18, 4000, 75, 235
Benson (Beason) Yeatman, 26, 5, 2000, 100, 297
George Thompson, 88, 40, 8960, 125, 795

James McDowel, 81, 25, 5000, 70, 305
Thomas J. Moore, 93, 10, 6180, 150, 340
Albin Buckingbe (Buckingham), 38, 8, 3150, 175, 115
Samuel Dennison, 84, 12, 5000, 100, 240
John _. Jackson 85, -, 7000, 200, 1170
William Philips, 75, -, 3500, 75, 360
Jacob Whiteman, 112, 13, 8000, 350, 642
Eli Moat, 64, 40, 3500, 100, 353
Samuel Harkness, 43, -, 1720, 150, 208
Alexander Guthrie, 43, 7, 2000, 25, 106
George W. Whiteman, 27, 7, 1200, 50, 70
Henry Whiteman, 82, 20, 4500, 75, 268
Smith Moat, 130, 70, 6000, 100, 349
Oliver Eastborn, 85, 15, 5000, 150, 537
William Moat, 85, 15, 5000, 150, 537
William Ayars, 50, 20, 3100, 150, 214
Jacob Finley, 70, 30, 5000, 150, 476
Samuel Little, 96, 6, 500, 200, 510
Elvi Davis, 80, 20, 4500, 150, 249
William McElwee, 54, 30, 3500, 50, 115
John Craig, 120, 25, 8700, 200, 715
Beason Craig, 30, 5, 2100, 100, 204
Martin Bowsinger, 12, -, 1500, 100, 100
John Sinix, 29, -, 1450, 100, 220
Henry Clark, 50, -, 4000, 100, 400
John Baily, 50, -, 4000, 100, 400
Johnathan Fell, 77, 80, 6000, 150, 736
Daniel Chaplin, 75, 10, 6000, 125, 377
James Clarrinax, 80, 23, 5000, 75, 415
Nathan Yearsly, 217, 10, 3000, 125, 230
Daniel Lynan, 90, 10, 5000, 200, 500
Robert Justis, 110, 10, 12700, 200, 740
Caleb Martial, 40, -, 4000, 200, 1066
James Cranson, 100, 25, 6950, 200, 772
Cornelius Derrickson, 94, 4, 6000, 150, 355
Benjamin Cranson, 140, 10, 20000, 300, 1000
Bengamine W. Duncan, 135, 30, 8200, 75, 588
Bengamine Jarret, 25,-, 2500, 200, 277
Levie Workman, 11, -, 1800, 15, 116
Thomas Smith, 77, 3, 1200, 30, 220
Joseph Cranson, 160, 40, 12000, 90, 888
Springer McDaniel, 14, 1, 1500, 60, 235
Jacob Reubincome, 81, 12, 6000, 150, 410
Samuel Baily, 60, -, 9500, 80, 680
James Russell, 245, 25, 13500, 400, 1330
John Foot, 95, 17, 6720, 80, 456
John & J. Harlin, 40, 5, 2250, 50, 140
Zachariah Derrickson, 169, 30, 7645, 150, 395
Thomas Higgins, 122, 8, 6500, 150, 520
Joseph Worrell, 57, -, 2600, 125, 215
William Baldwin, 110, 2, 6000, 200, 750
James Donal, 110, 10, 4800, 50, 260
John Stilwell, 130, 30, 6400, 150, 589
Henry Rusner, 24, 3, 1500, 25, 76
John Tweed, 90, -, 4500, 200, 445
James H. Mitchell, 75, 10, 4000, 200, 310

Nathaniel Richards, 65, 5, 2500, 50, 103
Stephen Broadbent, 45, 6, 3500, 100, 822
William Little, 195, 15, 1200, 200, 758
Isaac Vanzant, 110, 25, 3500, 170, 268
William Thompson, 50, 10, 3000, 50, 186
Robert Fitzsimmons, 56, 28, 1600, 100, 228
Thomas Fitzsimmons, 64, 6, 1400, 100, 157
William Johnson, 80, 46, 5000, 150, 334
Catherine Gairy, 84, 10, 4500, 150, 372
Thomas Mitchell 2nd, 50, 12, 1922, 75, 235
Bengamine Calk, 110, 20, 6000, 200, 535
John Peach, 95, 21, 6960, 150, 462
John A. Reynolds, 277, 30, 14000, 400, 1000
Curtis V. Ruthers, 20, -, 1000, 120, 337
Joseph Prichard, 163, 30, 10000, 300, 762
Uriah Drake, 130, 30, 8000, 150, 305
John Greenwall, 320, 1, 1500, 200, 637
Thomas Jagger, 200, 86, 12000, 150, 535
Isaiah Eastborn, 40, 20, 2500, 50, 143
Alexander Guthrie, 160, 40, 10000, 300, 763
Thomas Reynolds, 135, 20, 8000, 100, 282
John P. Derrickson, 110, 40, 9000, 100, 240
Henry Saunders, 34, 6, 2460, 50, 183
Joseph Greenwall, 48, 2, 2500, 160, 382
Lewis Kennack, 100, 15, 4000, 125, 300
Samuel Montgomery, 72, 8, 2500, 100, 375
Edward Wingate, 125, 25, 8000, 110, 456
Joseph Finster, 40, 18, 2500, 100, 180
William Kelly, 127, 20, 600, 200, 490
Samuel Anthony, 50, -, 3750, 100, 531
Westly Robinson, 50, 15, 4000, 125, 225
David Graves, 73, -, 5475, 90, 515
Calvin Philips, 25, -, 2500, 58, 358
John McNight, 118, 15 8280, 80, 820
Barthmew. Bartlett, 80, 24, 5000, 40, 270
William Rice, 146, 6, 6000, 75, 255
James Denny, 97, 3, 6000, 120, 454
Edwin Janvier, 270, 30, 20000, 400, 975
Robert Lyman, 92, 4, 5000, 200, 210
James Foot, 50, 15, 4550, 150, 275
Bengamine Saunders, 130, 20, 7500, 135, 500
Joseph Woolaston, 90, 10, 5000, 200, 374
George H. Alcorn, 14, -, 1200, 100, 100
John Ball, 190, 40, 14000, 260, 1140
George Flair, 100, 35, 8100, 250, 506
John Sowers, 120, 23, 9000, 100, 475
Louis H. Crossin, 80, 30, 3850, 100, 475
Aquilla Derrickson, 84, 12, 600, 30, 430
John Reubencome, 70, 54, 4000, 83, 285
David Eastborn, 128, 2, 10000, 100, 1015
Fredus Pennington, 155, 20, 9000, 150, 72

Thomas Worrell, 10, 4, 1400, 25, 135
Arnold Nandane, 149, 20, 10800, 200, 800
Alexander Graig, 80, 45, 3750, 100, 215
Amos Sharpless, 75, 25, 6000, 150, 675
Elisha Brown, 100, 47, 9000, 200, 614
James C. Jackson, 135, 15, 12000, 250, 700
Caleb Sharpless, 116, 60, 10000, 102, 629
Louis Thompson, 95, 15, 8000, 125, 655
Andrew H. Heeb, 100, 10, 7700, 130, 490
Joseph Leach, 100, 30, 6500, 200, 368
John McCormick, 90, 16, 6000, 50, 370
Louis McLewee, 95, 5, 6000, 500, 627
Abner Hollingsworth, 62, 5, 3500, 100, 250
John Chambers, 60, -, 4000, 150, 508
Samuel P. Dixon, 100, 44, 8600, 175, 1170
Archibald Armstrong, 70, 30, 4200, 100, 390
John Jourden, 80, 20, 5000, 200, 250
Samuel Whiteman, 76, 14, 4500, 150, 280
John H. Prill, 190, -, 23750, 1500, 2425
John Sayers, 4, -, 2000, 100, 30
Albert Lewis, 105, 20, 10000, 220, 479
Joseph Pritchard, 210, 40, 20000, 150, 465
John Millur, 55, -, 5000, 166, 275
Eliza Holtzbecker, 300, 140, 16000, 75, 4075
James L. Miles, 100, -, 6000, 200, 730
Elizabeth Chamberlin, 95, -, 5000, 125, 320
John Whann, 10, -, 1000, 30, 95
Daniel Fields, 53, 10, 5000, 25, 544
Frances P. S. Blandy, 43, 3, 1000, 50, 90
George A. Casko, 80, 20, 4000, 100, 310
Thomas Holland, 100, 50, 5000, 100, 300
James McCuen, 38, -, 1200, 100, 225
Peter Stewart, 30, -, 800, 50, 118
Thomas Holland Jr., 50, 40, 2000, 40, 350
John E. Maybin, 80,, 25, 5000, 150, 559
James Croes, 30, 30, 4200, 50, 120
George S. Croes, 10, 29, 800, 30, 90
Thomas Steel, 100, 24, 5000, 150, 400
James H. Ray, 150, 38, 9000, 215, 727
William H. Pa__inson, 130, 20, 10000, 200, 460
Samuel Bell, 90, 10, 5000, 100, 210
Richard Chamlee, 120, 30, 10000, 100, 1300
Daniel Thomsin, 150, 50, 16000, 250, 2174
Lamborn Pyel, 15, 5, 1000, 50, 2000
Isaac Moore, 63, 9, 5000, 150, 625
Leedom Moore, 85, 15, 400, 95, 130
Jasoh Beltg (Beltz), 24, -, 1200, 100, 25
George Terrel, 21, 12, 1500, 50, 80
John McLoughlin, 18, 8, 2000, 20, 210
Joseph James, 75, 20, 4000, 200, 425
John Alrich, 200, 86, 14000, 200, 560
Andrew Kerr, 100, 40, 9500, 250, 692
Joseph Hosinger, 120, 40, 12000, 200, 190

Benjamin Priginham, 180, 57, 11000, 250, 1480
Joseph Taylor, 64, 6, 3500, 100, 310
John Concannon, 55, 15, 3000, 100, 300
Hosea T. Riddle, 50, 37, 3000, 50, 450
Aaron Baker, 80, 20, 3000, 54, 325
Harry Warren, 145, 30, 8000, 500, 700
George P. Price, 50, 20, 3000, 50, 130
Benjamin G. Ogle, 23, -, 1500, 50, 110
Thomas Reese, 55, -, 4000, 50, 229
Abraham Cannon, 6, -, 1800, 40, 100
James Culling, 24, -, 900, 20, 45
James Pugh, 50, 5, 2000, 75, 140
Stephen Y. Townsend, 80, 15, 1800, 100, 208
George Johnson, 35, 27, 1800, 60, 133
Benjamin Groves (Graves), 60, -, 1300, 100, 265
John Wright, 100, 65, 6600, 50, 129
William Jones, 80, 24, 3000, 50, 210
Henry Smally, 75, 5, 6000, 150, 1687
William Reynolds, 10, -, 500, 50, 130
Ward Vandergrif, 89, -, 3600, 95, 250
John Carender (Cavender), 12, 12, 600, 20, 80
Dennis Oneal, 20, 90, 6200, 30, 188
James Lee, 20, 16, 1200, 100, 200
Johnathan Groves, 80, 10, 3500, 100, 390
William R. Lynum, 150, 50, 6000, 153, 575
Thomas Broocks, 110, 40, 4000, 160, 535
John Haman, 150, 110, 12000, 70, 620
Edward McFarland, 100, 30, 4500, 100, 340
Robert McFarland, 123, 30, 8000, 200, 450
Benjamin Shakespear, 50, 20, 3000, 60, 189
Ephraim Betts, 90, 92, 7000, 300, 425
George Morton, 26, 5, 800, 50, 170
Joseph Ryder, 30, 45, 1500, 20, 125
Samuel Kelly, 120, 60, 6000, 100, 574
Abraham Lawton, 135, 5, 5000, 200, 360
John T. Long, 109, 14, 5000, 125, 615
Samuel P. Johnson, 75, 25, 2500, 50, 155
Henry Rowan, 250, 250, 12000, 125, 2327
John Hawthorn, 91, 20, 3500, 600, 224
William Haman, 150, 50, 12000, 172, 575
Robert McCoy, 7, -, 700, 20, 100
Robert Hawthorn, 100, 30, 3000, 100, 275
Thomas Layman, 45, 5, 5000, 50, 260
Arnold Kirby, 100, 205, 13000, 100, 140
William Robinson, 220, 80, 12000, 200, 1064
William Cook, 10, -, 5000, 20, 75
Elizabeth Moree, 15, -, 400, -, 50
John Moree, 50, 20, 2000, 40, 340
Samuel Stroud, 90, 30, 6000, 125, 300
Frederick Carender, 80, 120, 2500, 30, 209
Samuel Allen, 250, 140, 20000, 300, 1250
David Morrison, 40, 20, 2000, 200, 615
Joseph Spencer, 110, -, 8000, 200, 1076
Henry Byard, 42, 20, 1000, 10, 120

Samuel Roberts, 110,142, 7000, 125, 375
Richard Carty, 50, 40, 3000, 100, 200
Washington Calmary, 56, 44, 5000, 125, 233
Robert Ferguson, 120, 20, 8000, 200, 550
Lofley Griffin, 100, 100, 6000, 50, 250
James Groves, 60, -, 1500, 100, 410
James Morrison, 120, 45, 10000, 400, 1200
William Morrison, 120, 45, 10000, 200, 500
William Hawthorn, 55, 13, 2500, 200, 170
Samuel Scott, 20, 18, 1500, 25, 100
William Ruth, 95, 25, 4000, 100, 285
Mathew Nandane, 60, 60, 4000, 250, 355
Wesly Morris, 182, 30, 12000, 175, 400
David Madeill (Modeill), 40, 48, 3000, 75, 300
William Dunlap, 30, -, 2000, 25, 102
George T. Hope, 14, -, 1800, 100, 175
Benjamin Peters, 42, 138, 4000, 200, 354
John Sears, 10, -, 900, 25, 90
Levi Ruth (Rutts), 50, 30, 2000, 50, 276
Jesse Moor, 125, 10, 700, 150, 270
George Janvier, 105, 25, 9000, 400, 600
Andrew Rambo, 80, 33, 7000, 150, 360
Robert Armstrong, 145, 20, 13000, 150, 625
Robert Cling, 130, 30, 9000, 150, 435
James Adams, 52, -, 3000, 75, 250
Ezahial Barns (Banus), 53, -, 2500, 25, 253
Christopher Broock, 120, 15, 10000, 45, 438
John Moor, 75, 15, 4000, 100, 330
James Robinson, 10, -, 700, 25, 112
Alexander Crawford, 9, -, 1000, 25, 60
Richard McW__arty, 180, 20, 7000, 275, 387
Rathmel Wilson, 100, 20, 5000, 200, 445
William Smith, 100, 25, 8000, 150, 1200

Sussex County Delaware
1850 Agricultural Census

This agricultural census was filmed from original records in the Delaware State Archives in Wilmington Delaware by the Delaware State Archives Microfilm office. It is negative film.

There are some forty-eight columns of information on each individual. Only the head of household is addressed. I have chosen to use only six columns of the information because I feel that this information best illustrates the wealth of the individuals. These are shown below:

1. Name of Owner
2. Acres of Improved Land
3. Acres of Unimproved Land
4. Cash Value of the Farm
5. Value of Farm Implements and Machinery
13. Value of Livestock

Thus, the numbers following the names represent columns 2, 3, 4, 5, 13.

The following symbol is used to maintain spacing where information in a column is left blank (-). This symbol is used where letters, names or numbers are not legible (_).

Wm. Harris, 125, 25, 200, 50, 200
Ezekiel Reed, 50, 12, 800, 25, 12
Wm. D. Pentin, 175, 125, 2000, 75, 140
Gillis Ellis, 100, 182, 5000, 100, 226
Wm. J. Lloyd, 50, 170, 2000, 25, 110
Alexander Oneal, 75, 31, 1200, 40, 150
Allen Hains, 200, 200, 8000, 200, 350
Oakly L. Eskridge, 65, 75, 700, 25, 75
Green Flower, 25, 151, 500, 20, 100
Wm. Ellis, 80, 235, 5200, 60, 200
James Dawber, 45, 10, 1500, 30, 100
John L. Colbersun, 40, 12, 900, 100, 125
Wm. Morgan, 8, -, 600, 23, 125
Mager W. Allen, 9, -, 1200, 100, 325
Rhodes Haggan, 9, -, 560, 50, 200
Mary Cannon, 12, 60, 5000, 50, 200
Hugh Martin, 160, 115, 10000, 300, 400
Jacob W. Prettyman, 12, -, 1000, 30, 150
J. & H. Shepley, 70, 45, 5000, 75, 400
John Heuston (Huston), 150, 150, 5000, 40, 400
Robt. Tull, 150, 150, 5000, 20, 175
Wm. H. M. Dawson, 135, 175, 4000, 40, 250
H. H. Martin, 200, 140, 6000, 40, 100
Edward Tull, 60, 30, 600, 40, 150
John A. Spain, 250, 150, 3500, 35, 200
Cyrus S. Phillips, 500, 580, 8000, 150, 750
Littleton Jackson, 200, 500, 2000, 40, 150

Elijah Colbourn, 100, 100, 1500, 20, 200
Liven Allen, 15, 172, 500, 20, 150
Minas Conaway, 100, 75, 1000, 25, 150
Jelenuah Coles, 130, 37, 1100, 20, 150
Eben Jackson, 110, 55, 1000, 30, 200
Jno. M. Wainwright, 125, 81, 2000, 20, 100
Wm. D. Conway, 200, 140, 5000, 50, 300
Irvin LeKate, 150, 75, 1500, 15, 150
Percy (Perry) L. Darby, 130, 75, 1500, 50, 200
Peter Morris, 100, 60, 700, 50, 150
Newton Willias, 100, 125, 5250, 50, 150
Saml. Eskridge, 75, 50, 1000, 100, 156
John Williams, 75, 75, 1000, 50, 100
Wm. Fleetwood, 100, 75, 1000, 15, 150
R. C. Allen, 150, 50, 1500, 30, 250
L. D. Morris, 100, 75, 750, 10, 75
Penna (Purna) T. Fleetwood, 100, 50, 1200, 15, 125
Levi W. Delany, 175, 182, 8500, 300, 550
Isaac Olin, 100, 100, 1500, 50, 500
Henry Beason, 40, 18, 600, 10, 50
Henry Wallace, 115, 100, 1600, 15, 200
Joshua Olin, 150, 150, 2500, 100, 250
James Wright, 550, 50, 2000, 15, 200
J. P. Cannen, 150, 50, 1000, 25, 500
Jno. J. Sain, 175, 141, 2500, 80, 250
Jno. H. Twifold, 175, 75, 2000, 150, 500
M. G. Davis, 125, 80, 2000, 40, 100
S. L. Nobb, 100, 50, 1500, 50, 500
D. B. Kinder, 100, 100, 1500, 40, 200
Isaac Kinder, 125, 75, 2000, 100, 200
Owen Kinder, 370, 160, 4800, 200, 500
Wm. Neal Sr., 250, 250, 5000, 20, 500
Chas. Miller, 100, 100, 1000, 100, 100
Catharine Neal, 230, 200, 5000, 20, 500
Jos. Neal, 150, 150, 2000, 100, 300
Wm. W. Spice, 200, 100, 1800, 25, 150
Arther Neal, 110, 60, 1000, 50, 200
J. R. Cannen (Carmen, Conner), 100, 44, 1200, 100, 200
Jesse Layton, 200, 100, 5000, 50, 500
Pinnal Layton, 100, 100, 2000, 40, 150
B. W. Hurby (Henby, Handy, Hurly), 200, 100, 2250, 50, 50
Isaac Conner, 150, 52, 1600, 50, 500
Levi Carmean (Cannen), 175, 85, 5000, 25, 550
Alfred Wright, 150, 75, 1000, 75, 50
Jesse Brown, 150, 100, 2500, 30, 250
W. T. Cannen, 125, 27, 2000, 35, 150
Augh Jones, 200, 75, 2000, 40, 200
Wm. E. Cannen, 200, 150, 5000, 75, 200
Hugh Brown, 209, 209, 4000, 50, 400
Wm. Allen, 200, 111, 2800, 50, 300
Wm. H. Ross, 400, 480, 10000, 650, 1000
Wm. Parker, 125, 50, 1000, 15, 100
Elizabeth Hooper, 250, 50, 2000, 20, 150
John G. Collins, 133, 67, 1100, 20, 175
Caleb D. Kinder, 120, 90, 1200, 25, 125
Henry Little, 150, 120, 5000, 20, 140

Jos. C. Allen, 153, 150, 3000, 80, 300
Wm. B. Adams, 100, 65, 1000, 50, 300
Jas. B. Layton, 200, 100, 1500, 25, 200
Warsen Kinder, 250, 220, 5000, 100, 400
Edward Davis, 100, 90, 1900, 20, 150
Thos. Jacob, 275, 151, 4460, 65, 200
Nath. Hasey (Henry), 100, 44, 5000, 100, 200
WM. N. Cannen (Canner), 167, 67, 2000, 60, 200
Jesse Willias, 157, 100, 1200, 20, 200
Solomon Sumpter, 95, 31, 1200, 25, 200
Stephen Cerban, 100, 200, 1500, 100, 200
Robt. Richard, 150, 150, 1500, 25, 200
Chas. Noble, 150, 50, 200, 95, 500
Joshua Crasm, 150, 150, 2050, 25, 250
Plymouth Ca_um, 70, 5, 700, 10, 100
Steers Jacobs, 500, 200, 5000, 50, 500
John Canen (Carom), 50, 50, 800, 12, 1000
Wm. K. Jacobs, 140, 30, 2000, 50, 500
Elijah Moore, 50, 50, 800, 25, 100
Wm. Edgen (Edger), 125, 75, 1500, 25, 200
Thos. Collins, 150, 50, 1000, 35, 125
Sam. Cabbage (Cubbage), 150, 210, 1700, 20, 240
Wm. Ross, 250, 200, 2250, 75, 300
Wm. H. Stafford, 128, 100, 1200, 100, 250
Archd. Evans, 150, 29, 700, 10, _
Adam Richards, 10, 70, 400, 20, 500
John W. Higman, 120, 80, 700, 50, 150
Wm. Richards, 100, 100, 1000, 20, 100
Noah Lednew, 170, 170, 2500, 100, 400
W. N. Cook, 100, 50, 2000, 30, 500
John Higman, 100, 50, 800, 50, 150
Jacob Kinder, 500, 140, 300, 150, 350
Wm. Turner, 88, 87, 1000, 15, 50
Lewis N. Wright, 175, 175, 5000, 150, 500
John Kinder, 210, 210, 4000, 50, 320
John Shepherd, 25, -, 150, 8, 25
Revil Hersey, 350, 450, 6000, 125, 440
W. Swain, 135, 42, 1500, 75, 225
Saml. Justin, 175, 195, 1200, 15, 225
Thos. J. Cannen, 225, 180, 3000, 100, 500
Jos. Sullivan, 143, 147, 1000, 25, 200
Salisbury Hobbs, 74, 74, 700, 60, 125
Enoch Hollish, 150, 62, 1200, 50, 150
Jason Vyland, 75, 40, 500, 60, 100
Geo. Leompt (Lecompt), 100, 75, 500, 20, 150
Henrietta Morris, 125, 125, 1500, 25, 100
Benton Hobbs, 120, 120, 850, 50, 250
Danl. Reed, 160, 120, 960, 50, 150
Andrew Vyland, 100, 67, 800, 15, 150
Elijah Hignut, 85, 85, 200, 15, 100
Stephen Morris, 75, 75, 600, 10, 100
Thos. T. Brown, 100, 57, 1000, 25, 125
Wm. Adams, 150, 150, 1200, 25, 100
John Atkinson, 82, 82, 1000, 40, 175
Philip Noble, 100, 100, 800, 25, 150
Wm. Jones, 136, 100, 1000, 25, 200

Levin Todd, 150, 53, 2000, 100, 2000
Thos. R Bullock, 70, -, 500, 10, 150
Gore Atkinson, 150, 62, 1275, 40, 260
Benjamin Simpson, 200, 100, 800, 5, 75
John Jones, 124, 124, 1200, 50, 500
N. O. Smith, 125, 75, 700, 50, 150
Henry W. Reed, 150, 100, 700, 50, 100
Peter Cannen, 200, 100, 800, 100, 100
Dr. J. R. Sadler, 280, 200, 7500, 400, 1163
Benj. Lord, 150, 157, 1200, 100, 200
Lonely Wills, 150, 150, 1500, 50, 500
Jno. C. Prettyman, 225, 75, 1500, 10, 100
John Wills, 200, 100, 1800, 60, 150
Thos. W. Dawson, 100, 100, 2000, 50, 225
Wm. E. Rogers, 140, 54, 4000, 75, 225
L. R. Jacobs, 200, 100, 1500, 50, 200
Archd. Satterfield, 90, 30, 800, 25, 200
Nehemiah Wills, 211, 100, 2500, 75, 250
Jos. Records, 180, 50, 150, 75, 500
Jas. M. Goslen, 150, 170, 1000, 200, 180
Alexander Jones, 152, 82, 1500, 75, 550
Wm. Quick, 100, 75, 1500, 75, 500
Jno. R. Adams, 90, 66, 700, 5, 100
James Nutter, 100, 50, 600, 40, 125
Leonard Cooper, 150, 50, 2000, 50, 250
N. M. Maylin, 175, 75, 4000, 25,250
Curtis Scott, 75, 21, 400, 20, 100
Jno. M. Collins, 160, 60, 1500, 50, 400
John Cade, 175, 50, 1500, 75, 500
Nathan J. Barker, 225, 100, 1000, 75, 150
L. C. Wadman, 60, -, 800, 50, 125
Wm. Carmen, 90, -, 3000, 100, 1000
M. A. Jones, 80, -, 500, 80, 400
Jas. Cury, 125, 72, 1000, 25, 50
Henry Smith, 100, 100, 1200, 20, 100
Clement Hitch, 250, 86, 1200, 50, 200
Hudson D. Swain, 100, 80, 1200, 50, 50
Jno. Ledquick, 100, 75, 1500, 50, 125
Waitman Jones, 100, 101, 1500, 125, 300
Whit__ Smallen, 100,90, 600, 50, 1250
Chas. Scheck, 1000, 100, 2000, 50, 250
Sol. Stephens, 200, 100, 1500, 40, 125
Simon Pennwell, 400, 300, 4500, 150, 250
Jas. Barwick, 200, 125, 2000, 50, 200
Wm. P. A. Joswaters, 200, 125, 1400, 15, 150
Ed. Morris, 100, 60, 1000, 50, 125
David Linch, 125, 75, 1000, 20, 75
Henry Pratt, 100, 100, 1000, 40, 150
Hezekiah Morris, 50, 50, 500, 25, 200
A. G. Ladd, 100, 150, 1200,75, 250
Jno. Morris, 125, 175, 1500, 50, 200
Geo. P. White, 40, -, 2000, 150, 500
Wm. A. Jacobs, 150, 110, 1200, 25, 150
Wm. M. Simpson, 70, 60, 700, 30, 125
St__ Reed, 1250, 50, 1200, 25, 200
Jas. Morgan, 150, 175, 2000, 75, 200
Jas. Lawless, 60, 22, 550, 15, 150
Wm. Brown, 150, 150, 1800, 20, 200
Jas. Downey, 150, 100, 1500, 25, 50
Wm. Allen, 100, 50, 700, 25, 100

Theodore Wilson, 160, 140, 1800, 10, 200
Revil Bozman (Boyman), 125, 125, 1500, 40, 350
Dr. Wm. W. Stuart, 80, 14, 2000, 100, 500
Jas. H. Prettyman, 50, 50, 800, 30, 125
D. B. Short, 140, 100, 1000, 30, 200
Jno. Dutton, 100, 67, 1200, 50, 140
Heston Gorlin, 225, 175, 5000, 50, 250
Chas. Wright, 200, 150, 10000, 200, 575
Jno. Simpson, 250, 50, 300, 50, 150
Geo. H. Adams, 150, 100, 2000, 75, 200
Henry Messick, 100, -, 1000, 40, 200
Alfred Messick, 200, 100, 3000, 15, 150
Jas. B. Cannen, 150, 9, 1200, 20, 100
Mager Smith, 165, 114, 1680, -, -
Danl. Cannen, 170, 114, 5000, 200, 4000
Gevin (Levin) Phillips, 150, 50, 1500, 25, 100
Dewitt Harting (Hasting), 100, 65, 900, 25, 75
Jones Carroll, 150, 150, 2000, 50, 200
D. Ellensworth, 150, 150, 2500, 50, 200
Wm. C. Polk, 220, 200, 5000, 300, 800
Wm. Ladd, 100, 100, 150, 50, 100
Wesley Jackson, 100, 75, 1500, 60, 100
James Vincent, 175, 125, 3000, 50, 200
Thos. M. Colburn, 150, 150, 1500, 50, 50
Thos. Brown 150, 100, 500, 50, 150
Clement Jones, 260, 100, 2000, 50, 250
Moses Dawson, 100, 25, 1000, 75, 60
Martin Haman, 100, 100, 1500, 25, 100
Saml. Nooks, 100, 50, 700, 50, 200
Jas. Swain, 243, 210, 3500, 110, 450
Eph. Moore, 200, 100, 1500, 210, 200
Abrm. Nicholls, 50, -, 400, 30, 150
Levi Neal, 154, 100, 1000, 50, 50
Danl. Brown, 200, 100, 1500, 25, 150
Isaiah Neal, 150, 100, 1500, 50, 200
Burtin Layton, 150, 75, 1200, 25, 200
Wm. Robinson, 120, 60, 1000, 55, 175
David Smith, 80, 40, 500, 50, 125
Henry Smith, 200, 250, 1800, 100, 200
John Richards, 400, 450, 3000, 75, 625
Eli Nelson, 65, 65, 650, 12, 40
Roger Adams, 300, 500, 6000, 500, 1000
Henry Todd, 35, 15, 240, 15, 50
Ed _. Jones, 275, 125, 2000, 50, 150
W. Clifton, 100, -, 500, 6, 35
Vincent Perkins, 9, 110, 100, 25, 175
Jno. M. Rankin, 65, 65, 1000, 50, 175
Jas. Stuart, 100, 150, 2500, 50, 150
Garretson Adams, 150, 200, 1500, 15,150
Wm. H. Hatfield, 140, 77, 1500, 100, 250
Westly Smith, 210, 110, 2000, 50, 250
Twiford Collison, 200, 150, 800, 50, 200
Stansbury Collison, 108, 100, 800, 15, 250
Ca__ Willy, 200, 400, 5000,70, 50
Thos. Gray, 150, 100, 1200, 75, 550
Elias Morris, 100, 100, 1200, 75, 200
Julius Smith, 125, 150, 1200, 50, 225
Alley Smith, 75, 80, 800, 25, 100

William Harting (Hasting), 350, 150, 5000, 20, 200
Levi E. Hitch, 200, 100, 200, 100, -
Levi Callaway, 100, 150, 1200, 40, 200
Wm. Cannen, 100, 400, 5000, 50, 200
Matthew C. Wright, 200, 100, 1000, 50, 100
John Baker, 54, 67, 600, 50, 75
James Ward, 100, 100, 1600, 10, 500
Alfred Adams, 100, 60, 1500, 10, 200
Thos. N. Sullivan, 50, 16, 500, 20, 50
Jno. R. Morris, 100, 50, 1000, 20, 800
Levin LeKate, 150, 100, 3000, 100, 250
Elijah Hitchens, 75, 75, 1000, 10, 30
Wm. T. Hall, 100, 50, 1000, 55, 100
John T. Moore, 275, 100, 5000, 25, 100
Geo. Polk, 50, -, 230, 40, 75
Josiah A. Moore, 75, 25, 800, 50, 25
Joseph Smith, 40, -, 250, 50, 70
C. A. Waddington, 150, 130, 1600, 100, 360
Jonathan A. Hearn, 90, 70, 4000, 65, 300
Wm. H. Anderson, 100, 90, 4000, 100, 250
IsaacL. Whaly, 50, 50, 600, 10, 125
John G. Gann, 100, 100, 800, 60, 200
Isaac Hearn, 65, 65, 1000, 200, 300
Geo. M. Wootten, 20, -, 500, 60, 250
N. Hearn, 60, -, 900, 75, 200
Jas. Wainwright, 100, -, 100, 8, 50
W. James, 120, -, 2000, 50, 150
Ridgway Vincent, 160, 40, 5000, 150, 200
Robt. Elzey, 300, 100, 5000, 100, 400
Nathl. Hersey, 50, 100, 2500, 100, 300
Mrs. Ralph, 160, 40, 2400, 150, 290
Mrs. Harting, 35, 58, 726, 50, 10
Isaac Carmean (Cannen), 100, 75, 2000, 25, 175
Noah Carmean, 175, 55, 2500, 25, 125
Henry Sharp, 200, 100, 3200, 28, 425
Wm. Culon, 100, 44, 1084, 20, 100
Thos. Ralph, 57, 28, 200, 15, 150
Henry Bacon, 120, 80, 1200, 50, 50
Geo. Vincent, 56, 57, 500, 25, 450
Benjamin Dennis, 100, 70, 2000, 50, 400
Kendol B. Hearn, 140, 40, 1000, 40, 200
Jonathan Beach, 228, 100, 1500, 2, 520
Jacob Wootten, 150, 100, 2000, 100, 150
John Elliott, 150, 150, 2000, 40, 200
Levin Calloway, 225, 75, 2000, 20, 150
Jonathan Waller, 250, 65, 1500, 25, 200
Elijah Williams, 100, 200, 2500,75, 500
Isaac Wootten, 150, 150, 3000, 100, 400
Saml. Kenny, 150, 150, 1200, 50, 175
N. Jacobs, 160, 150, 1500, 75, 200
James Spencer, 80, 40, 2000, 50, 100
Thos. Perdue, 75, -, 500, 25, 200
Danl. Boyce, 80, 120, 4000, 50, 200
Jno. M. Phillips, 80, 120, 4000, 50, 200
G. W. Henry, 60, 40, 700, 20, 200
Thos. J. Phillips, 120, 250, 5550, 82, 265
Wm. T. Phillips, 80, 90, 4500, 25, 275
Thos. H. Jackson, 5, 16, 6500, 20, 65
Betsy Davis, 10, -, 120, -, 6
John Rhodes, 16, -, 130, -, -
Henry Spencer, 1 ½, -, 200, -, 22

Wm. T. Collins, 20, 80, 200, -, -
Geo. Moore, 25, 40, 400, -, 100
Jane Pennwell, 20, 30, 600, -, 12
Josiah Pennwell, 75, 75, 2500, -, 175
Levin A. Collins, 70, 40, 1400, 5, 110
Thos. Dunn, 10, -, 125, -, 37
Hiram M. Collins, 80, 20, 10000, 40, 194
Jacob E. Phillips, 50, 25, 800, -, 15
Elijah Pennwell, 100, 50, 1000, 50, 411
Nelly Waller, 20,, 15, 150, 5, 5
Henry Sharp, 30, 20, 200, 5, 150
James Moser (Moren), 70, 110, 2000, -, 50
Henry Culver, 50, 50, 400, -, 8
Ann Elliott, 20, 60, 600, -, 25
James Rhodes, 10, 40, 300, -, 6
Geo. Melner, 30, 40, 200, -, 55
Wm. Culver, 35, 20, 200, 10, 65
Levi Collins, 40, 30, 500, -, -
James Collins, 20, 100, 2000, 25, 275
Robt. J. Venibles, 60, 500, 600, 15, 125
Perry J. Kennekin, 80, 30, 500, -, 4
Arena Keenedy, 100, 100, 600, -, 5
Brozant Briley (Baily), 50, 50 500, 5, 100
Wm. Kucobs (Jacobs), 50, 50, 500, -, 22
Whitfield Moore, 6, 20, 200, 30, 160
Mary Adams, 500, 380, 9000, 100, 700
Jeremiah Adams, 100, 50, 2300, 52, 140
James Lowe, 50, 40, 400, 10, 95
Jeremiah Hitch, 60, 30, 500, -, 25
Hester Knowles, 50, 125, 1000, -, 56
Purnal Jones, 3, 22, 500, 16, 60
E. Colbourn, 5, -, 100, -, 15
Isaac Phillips, 400, 300,800, 75, 250
Nancy Phillips, 16, -, 200, -, -
David M. Knowles, 90, 40, 1500, 25, 12
Danl. F. Wilson, 20, 8, 500, -, 5
Barth. Woolford, 50, 50, 800, 25, 2
Thomas Woolford, 60, 24, 500, 1, 4
Isaac R. Hearn, 10, 10, 200, 10, 6
Saml. Spencer, 50, 50, 600, 50, 6
Elijah Phillips, 65, 35, 6500, 50, 11
Theodore Phillips, 125, 35, 1200, 100, 6
John Knowles, 45, 5, 500, -, 5
Wm. B. Kinnekin, 16, 6, 400, 10, 10
Gold__ W. Phillips, 18, 9, 500, 5, 5
Matthew Kinnekin, 30, 20, 500, -, 2
Thos. Bradly, 100, 50, 1500, 50, 9
Waitman Kenny, 90, 70, 1000, 6, 8
Stephen Ellis, 150, 150, 2000, 60, 250
James Parmer, 100, 100, 1800, 50, 150
Maria Clark, 10, -, 150, -, 10
Stephen Ellis, 500, 50, 7000, -, -
Danl. Boyce, 150, 150, 2500, 15, 200
Isaac Hasting, 125, 50, 1000, -, 60
Spencer Cordrey, 75, 50, 1000, -, 75
William Cordrey, 10, -, 180, 10, 50
Isaac Ricards, 70, 50, 600, -, 5
Joshua Ricards, 6, -, 130, -, 10
Thos. Phillips, 160, 400, 2000, 50, 200
Griffith Mucian 125, 75, 2000, 100, 200
Burton Dunn, 150, 50, 200, 50, 200
William Dunn, 25, 15, 500, -, 20
W. Henderson, 50, 50, 1000, -, 25
Leo S. Collins, 500, 400, 7000, 80, 250
Jno. M. Husten, 8, -, 120, -, 20
Geo. Collins, 125, 75, 1600, -, 70
Thos. Morris, 90, 40, 1000, 20, 120
Doughty Collins, 100, 61, 2415, 50, 500
John B. Collins, 75, 25, 800, 50, 100
Henry Collins, 8, -, 100, -, 500
Oah Carnean (Carmean), 100, 100, 2000, -, 40
Wm. Workman, 150, 110, 1700, 20, 150

Elijah Hutchins, 120, 100, 3000, 100, 250
Benj. Hitch, 120, 120, 2000, 75, 250
Levin A. Collins, 80, 80, 1000, 5, 200
John Mills, 50, 50, 1200, 25, 100
Thos. W. Records, 200, 100, 3000, 100, 400
Elizabeth C. Kenny, 75, 75, 1200, -, -
Mary Jones, 75, -, 175, -, 8
Eli Harting (Hasting), 150, 75, 1500, 20, 100
Columbus Henry, 50, 38, 800, 20, 120
Jones Harting, 50, 20, 900, 20, 75
Sally Hastings, -, 20, -, 20, 80
John Kennekin, 200, 150, 2000, 25, 225
George Wolston, 80, 82, 1700, 60, 325
Joshua Kenney, 50, 19, 544, 30, 225
Jas. Finley, 100, 50, 1500, 80, 300
Isaac Henderson, 60, -, 300, 5, 80
Levin Twilly, 75, 50, 1500, 50, 600
Harnutt English, 30, 30, 400, -, 30
Aaron Owens, 160, 150, 5000, 40, 200
Isaac Baily, 60, 60, 500, -, 50
G. W. Bennett, 20, 70, 600, -, 10
Ephraim Kurals (Rurals), 30, 30, 500, -, 15
W.. Rurals, 140, 100 720, 20, 46
Robt. Twilly, 160, 80, 2500, 50, 200
Cannon Knowles, 50, 60, 1000, 20, 175
Elijah Goslin, 40, 30, 4000, -, 34
Richard Bradly, 100, 80, 1000, 50, 80
Phebe Bradly, 90, 60, 1000, -, 60
Johnson Elliott, 60, 90, 1000, 15, 90
James Bradly, 70, 20, 1500, 20, 300
Jesse A. D. Bradly, 70, 80, 2500, 90, 625
Huffington, 90, 90, 1000, 30, 100
Mary Whatly, 5, -, 150, -, 3
Morgan & Cooper, 100, 100, 500, 60, 200
Josiah Baily, 100, 75, 1200, 35, 200
Clayton Owens, 170, 60, 1000, -, 25
Charles Wright, 40, 50, 700, 5, 20
Ricahrd Stephens, 70, 50, 1000, 8, 130
Darcus M. Phillips, 75, 90, 1200, 30, 250
Robt. Hitchens, 80, 30, 700, -, 7
Geo. Ellis, 75, 75, 1200, 35, 235
Stephen Ellis, 10, -, 120, -, 25
James Marshal, 60, 30, 600, -, 28
Eben Waller, 250, 150, 3500, 100, 400
Isaac Giles, 200, 250, 6000, 200, 330
Stephen Kinnikin, 10, -, 120, -, 12
Benj. Waller, 100, 50, 2000, 25, 120
W. Lloyd, 108, 14, 1800, 20, 90
John Hill, 75, 75, 2000, -, 20
Wm. Ricards, 150, 100, 2700, 15, 200
James Wooten, 10, -, 100, -, -
Charles N. Moore, 50, 50, 2000, 40, 200
Azarias Phillips, 60, 75, 2006, 35, 140
Sarah S. Ralph, 150, 75, 2008, -, 70
Huey P. Cullen, 140, 75, 5000, 20, 400
Jonathan Baily, 150, 100, 3000, 200, 800
Joseph Phillips, 200, 200, 5000, 200, 800
Hiram Phillips, 100, 40, 100, 50, 150
Danl. Phillips, 100, 75, 1500, 56, 200
Henry Douglas, 100, 95, 1500, 8, 100
Jas Pritchett, 80, 90, 1000, 75, 175
Saml. Phillips, 120, 120, 2500, 12, 550
Elias Calvin, 30, 20, 175, -, 7
Noah Cooper, 75, 75, 1500, 70, 200
Roger Phillips, 70, 60, 1000, 20, 100

Stephen T. Baily, 50, 50, 800, 20, 120
Soveriegh Baily, 60, 40, 800, 20, 120
Thos. Gray, 150, 50, 1400, 5, 50
Mathias D. Wilson, 76, 76, 900, 12, 825
John S. Harny, 62, 62, 800, 30, 100
Wm. Cooper, 75, 100, 1000, 75, 150
Joseph Ellis, 70, 80, 600, 10,75
Isaac Peppin, 8, -, 125, -, 2
James Elias, 50, 70, 600, 50, 200
W. A. R. Phillips, 70, 100, 2000, 60, 250
Rachel Waller, 100, 20, 1500, 30, 3150
James Howard, 40, 40, 500, -, 12
Isaac Henry, 75, 75, 1000, 20, 100
Burton Hearn, 50, 86, 900, 5, 100
Geo. Phillips, 60, 40, 400, 10, 125
Wm. Lowe, 1100, 100, 2100, 50, 180
Jos. Ellis, 125, 175, 2500, 150, 300
Dinah Lowe, 100, 100, 800, 15, 50
Thos. Ralph, 85, 85, 1000, 50, 150
Thomas Ralph, 125, 125, 250, 95, 150
Arnold Elzey, 66, 25, 800, 6, 40
Martin M. Elzey, 100, 75, 1200, 20, 150
James Mills, 10, -, 175, -, 15
Hudson L. Mills, 12, -, 180, -, 5
Wm. Hearen, 100, 50, 800, 20, 150
Moses Harting, 200, 100, 150, 80, 250
Beaucamp Harting, 8, -, 150, -, -
Thos Harting, 10, -, 120, -, 12
Jas. E. Ellis, 200, 150, 2600, 50, 200
Littleton Ellis, 20, 20, 500, -, 20
Peter White, 80, 60, 1100, -, 30
Danl. Hasting (Harting), 50, 10, 400, -, 50
Danl. Calvin, 100, 50, 1000, 30, 150
Nancy Collins, 20, 20, 200, -, 14
Warden Harting, 125, 127, 1500, 50, 320
Elihu Harting, 75, 25, 800, 20, 100
John Calvin, 30, 10, 200, -, 100
Ialathiel Calvin, 50, 25, 400, 18, 100
Jonathan _. B. Tooks, 100, 100, 1200, 100, 150
Isaac Hearn of S, 100, 100, 1500, 50, 200
John S. Calvin, 75, 25, 1000, 20, 120
Hultz (Hatty) L. Katt, 75, 75, 1200, 30, 125
Sarah Elliott, 70, 50, 1500, -, 30
James Gann, 8, -, 100, -, 5
Mary Ellis, 8, -, 110, -, 2
Chandler Gann, 75, 60 1000, 30, 75
Winder Harting, 70, 50, 1200, 25, 200
Sylvester Wootten, 100, 50, 1500, 25, 150
James R. Serman, 150, 200, 2500, 65, 150
John H. Gordy, 100, 100, 1200, 65, 150
Nancy Weredson, 125, 25, 1500, 30, 100
Wm. Cordrey, 90, 98, 2500, 50, 150
Thmas Adams, 54, 2, 600, 5, -
Nancy Adams, 12, -, 200, -, 5
Spencer Cordrey, 90, 15, 1400, 20, 150
Catman Miller, 50, 50, 500, -, 5
Wm. Hill, 100, 75, 2000, 50, 160
Joshua Hitch, 6, -, 110, -, 6
Wm. P. Marvel, 150, 150, 3000, 20, 100
Sarah A. Marvel, 150, 100, 1500, 20, 100
John Harting, 75, 15, 5000, 15, 6
Wash. _. Cole, 50, 50, 500, -, 4
Eben Calloway, 80, 90, 1400, 50, 200
Elihu Calloway, 90, 40, 1200, 20, 100
Philis Ralph, 50, 50, 600, 20, 200
Mary Kinny, 50, 50, 600, -, 150
Eleanor Kinny, 60, 40, 600, -, 125
Hardy Beach (Brash), 50, 50, 1000, 20, 250
Wm. Hearn, 150, 100, 1500, 20, 200

Charles Calvin, 225, 40, 2000, 50, 200
Headwell Hartin, 50, 50, 500, 25, 15
Sarah Calloway, 75, 25, 700, 20, 80
K. B. Hearn, 40, 57, 1000, 10, 100
Peter Calvin, 10, -, 120, -, 15
Eben LeKate, 100, 75, 1000, 50, 200
Margaret LeKate, 20, 10, 200, -, 15
Elijah Oliver, 20, 5, 500, 10, 100
John E. Low, 20, 10, 200, -, 4
Washington Calloway, 60, 60, 1500, 40, 250
James LeKate, 80, 20, 2000, 20, 60
John Serman, 35, 25, 900, 50, 200
Jacob Messick, 8, -, 120, 5, 4
Danl. Harting, 175, 150, 300, 12, 200
Asa Terpin, 50, 50, 500, 40, 75
Josiah Cordrey, 125, 25, 1500, 2, 200
Job Macklin, 25, 25, 200, 50, 6
Joshua Cannon, 125, 125, 1500, 18, 225
Saml. Ward, 25, 80, 500, 7, 125
Luther W. Collins, 100, 50, 600, 40, 100
Philip West, 50, 40,700, -, 150
Josiah Collins, 80, 20, 40, 20, 40
Purnel Baker, 30, 70, 800, 5, 100
Saml. Elliott, 50, 50, 500, 25, 75
Narnett (Harnett) Hearn, 100, 100, 800, 100, 60
Jos. J. Hearn, 150, 100, 3000, -, 300
Geo. Selby, 10, -, 110, 75, 108
Geo. W. C. Hearn, 100 65, 1500, 5, 250
Wm. Workman, 20, 40, 600, 15, 80
Jos. Palmer, 60, 75, 800, -, 80
William Vincent, 8, -, 110, 30, 100
Freeborn G. Wells, 50, 50, 1500, 28, 150
John Parsons, 40, 25, 800, 8, 125
Arther J. Jonsin, 40, 60, 500, 25, 75
Jas. B. Harkey, 60, 50, 500, 10, 125
John Workman, 60, 75, 800, -, 60
Elijah Parker, 10,75, 200, 35, 30
Wm. Melson, 50, 250, 1500, 5, 125
K. Brittingham, 60, 5, 800, 5, 50
Jos. J. White, 100, 75, 1000, 20, 50
Thos. Sullivan, 80, 90, 1000, 7, 130
Elihu Hitchens, 60, 40, 600, 50, 30
Levin Sullivan, 70, 25, 800, 15, 200
Noah Hearn, 60, 54, 800, 10, 60
Sarah Hearn, 70, 30, 1000, 40, 160
Thos. W. Sullivan, 30, 100, 300, 60, 150
James Ward, 20, 18, 175, 4, 70
Leonard Ward, 40, 100, 1000, 20, 70
Jacob W. Elliott, 70, 115, 700, 25, 160
Huet Elliott, 50, 25, 500, -, 15
Mary Elliott, 10, 10, 200, -, 26
Stokly Elliott, 66, 66, 1000, 40, 150
Burton Elliott, 50, 50, 600, 20, 75
Elias Elliott, 44, 22, 500, 30, 120
Josiah Collins, 75, 34, 500, -, 25
George W. Hearn, 150, 170, 300, 50, 125
Thomas N. Hearn, 125, 160, 3500, 50, 200
William Elliott, 40, 50, 700, 10, 50
Benjamin West, 150, 150, 1800, 100, 300
Benjamin Carman, 75, 80, 1000, 20, 60
Josiah Downs, 30, 20, 500, 15, 100
John C. Gordy, 40, 100, 900, -, 200
Benton H. Gordy, 200, 175, 3000, 100, 200
Philemon Carmosne, 50, 100, 400, 25, 170
Charles B. Green, 8, -, 180, 60, 500
William W. Hall, 175, 82, 2000, 50, 200
Phillip Truitt, 8, -, 100, 50, 150
William Colbourn, 15, 30, 20, -, 10
Charles I. Hitchens, 50, 40, 700, 30, 12
Isaac Vinson, 100, 50, 500, 50, 125
Sarah Wootten, 100, 60, 1500, 50, 150
Joshua S. Elliott, 50, 50, 1000, 20, 200

Isaac Adams, 20, -, 100, -, 75
William Carmean (Cameron), 75, 68, 750, 30, 16
Ishea Hasting, 50, 75, 1200, 5, 150
Joseph G. White, 10, -, 150, -, 50
Henry Eason, 500, 150, 3500, 200, 18
John Hosea, 50, 50, 500, 36, 600
Cyrus Carmean, 50, 50, 1200, 20, 75
Shedrack Hill, 80, 20, 1000, 20, 125
Marshel Smith, 100, 50, 800, 70, 150
Elijah Hitchens, 75, 75, 800, 35, 125
Elizabeth Smith, 90, 44, 900, 15, 130
William Smith, 75, 60, 800, 20, 150
Joseph Hearn, 160, 20, 1800, 60, 75
Jacob Hearn, 50, 50, 500, 15, 225
Augustus Davis, 30, 50, 500, 15, 75
Leonard Hasting (Harting), 50, 50, 800, 60, 250
Benjamin Hearn, 70, 40, 800, 40, 175
John Morris, 50, 90, 700, 150, 350
Andrew Elliott, 80, 59, 450, -, 45
Lewis W. Elliott, 65, 50, 500, -, -
Wm. H. Sermon, 50, 70, 1000, 15, 100
Wm. Sermon of L, 73, 75, 1500, 200, 250
Hannah Morris, 75, 100, 1500, 80, 500
John Elliott, 60, 60, 500, 20, 150
Benjamin Elliott, 75, 60, 500, 15, 250
William Elliott, 500, 50, 500, -, 50
James Morris, 75, 35, 1100, 20, 85
Luther Williams, 150, 100, 2000, 150, 500
Benjamin Cordrey, 100, 125, 2000, 4, 75
Nathan J. Elliott, 70, 50, 700, 25, 70
Elizabeth James, 100, 80, 1500, 55, 140
Cannen Lauk, 90, 25, 600, 12, 100
John Linch, 150, 53, 2500, 38, 250
Jane Cooper, 200, 200, 6000, 60, 500
James Husey, 60, 40, 600, 20, 100
Job Polk, 100, -, 1 20, -, 16
Julia A. Magee, 100, 50, 750, -, 25
John Magee, 175, 56, 1075, 25, 200
Lucy Bradly, 100, 68, 1500, 70, 80
Wm. B. Oliphant, 60, 60, 1000, 20, 75
Jonathan Hearn, 60, 12, 600, 5, 20
Joshua Elliott, 70, 50, 800, 20, 100
John Wootten, 78, 80, 2500, 30, 250
Thos. LeKate, 70, 50, 1000, 20, 50
Wm. King, 100, 100, 2500, 80, 2000
Wm. Gordy, 100, 40, 1500, 80, 500
James Chapman, 28, 72, 3000, 50, 152
John Conner, 16, 5, 175, -, 12
James Graham, 50, 18, 200, -, 12
Mesha M. Elliott, 100, 125, 5000, 100, 250
Norris Palby (Pully), 10, 110, 700, -, 25
Wilson Knowles, 27, 66, 5000, 24, 156
Gibson Windsor, 20, 15, 300, -, 15
Wilson Sligo, 55, 64, 600, 50, 175
Henry Williams, 18, 20, 200, -, 16
Joseph Boyce, 50, 200, 500, 500, 500

Names Below appear to be repeats although different page number with some numbers on some names transposed or new numbers in same name sequence. This was very confusing.

Leonard Ward, 140, 100, 1000, 20, 70
Jacob W. Elliott, 70, 45, 700, 25, 160
Hughett Elliott, 30, 25, 500, -, 15
Mary Elliott, 10, 10, 200, -, 26
Stockly Elliott, 66, 66, 1000, 60, 150
Burton Elliott, 50, 50, 600, 20,75
Elias Elliott, 45, 22, 500, 50, 120
Josiah Collins, 75, 54, 500, -, 25
Geo. W. Hearn, 150, 170, 5000, 50, 125
Thos. N. Hearn, 125, 160, 3500, 50, 200

Wm. W. Elliott, 60, 50, 700, 10, 50
Benjamin Ward, 150, 150, 1800, 60, 500
Benjamin Carmean, 75, 80, 1000, 20, 60
Josiah Downs, 30, 20, 500, 15, 100
John C. Gordy, 60, 110, 900, -, 20
Burton H. Gordy, 200, 175, 3000, 100, 200
Philip W. M. Carmean, 100, 50, 1000, 25, 170
Chas. B. Green, 50, 110, 400, 60, 500
Wm. W. Hall, 8, -, 110, -, 20
Philip Truitt, 175, 82, 2000, 50, 150
Wm. Colbourn, 8, -, 100, -, 10
Chas. J. Hitchins, 15, 50, 200, -, 12
Isaac Vincent, 50, 40, 700, 50, 125
Sarah Wootten, 100, 50, 800, 50, 150
Joshua S. Elliott, 100, 40, 1500, 50, 200
Isaac Adams, 50, 30, 1000, 20, 75
Wm. Carmean, 10, -, 10, -, 18
Joshua Harting, 75, 65, 750, 30, 150
Jos. G. White, 50, 75, 1200, 5, 50
Henry Easum, 10, -, 150, -, 18
John Hosea, 300, 150, 3500, 200, 600
Cyrus Carmean, 50, 30, 500, 36, 75
Shedrach Hall, 50, 50, 1200, 20, 125
Marshel Smith, 80, 20, 1000, 20, 150
Elijah Hitchens, 100, 30, 800, 70, 120
Elizabeth Smith, 75, 75, 800, 35, 150
William Smith, 90, 44, 900, 15, 150
Jos. Hearn, 75, 60, 800, 20, 75
Jacob Hearn, 160, 70, 1800, 60, 225
Augustus Davis, 30, 30, 500, 15, 75
Leonard Harting, 50, 50, 800, 60, 250
END OF REPEATED NAMES
W. C. Hopkins, 60, 100, 800, -, 4
Elijah Calloway, 100, 75, 3000, 50, 200
Kendal M. Lewis, 150, 158, 5000, 1000, 300
Saml. R. Riggin, 50, 125, 800, -, 8
Thos. Larner (Larmer), 120, 90, 1200, 20, 50
Saml. Calloway, 100, 100, 2000, 15, 80
Wm. Moore, 125, 75, 2200, 50, 10
Benjamin Brown, 40, 15, 300, -, 5
David Shields, 125, 45, 1000, 25, 60
Chas. Parmer, 100, 150, 5000, 30, 100
Mary Massey, 80, 10, 500, 130, -
Wm. Wheatly, 170, 70, 2500, 15, 460
Thos. Owens, 50, 500, 500, 70, 800
Jonathan Moore, 60, 40, 1000, -, 550
Alen Hopkins, 60, 30, 600, -, 28
Lovy Drain, 70, 5, 400, 150, 3
Harriett Connor, 200, 200, 4000, -, 500
Shepperd Drain, 35, 60, 500, -, 30
Aight Thompson, 10, -, 110, -, 13
Elisha Cary, 15, 10, 175, -, 60
Elijah Colbourn, 7, -, 200, -, 50
Joseph Moore, 100, 50, 1200, -, 125
James Hopkins, 80, 20, 500, -, 35
Bayard Moore, 60, 75, 2000, 25, 40
Burton Morris, 40, 100, 500, -, -
Wm. Bell, 100, 50, 1000, 20, 140
Booz Bell 20, 30, 500, 25, 100
Mary Bell, 50, 200, 1000, 3, 25
Nathan York, 70, 30, 1500, -, 15
N. H. Bell, 50, 70, 600, 20, 100
Elisha Bustin, 200, 10, 3000, -, 170
Wm. W. Cannen, 90, 30, 400, -, 30
John Lord, 10, 60, 100, -, 12
Levin Lauk, 100, 75, 1000, -, 90
Saml. Holt, 30, 19, 300, 5, 70
Danl. Holt, 150, 200, 1200, -, 35
Elizabeth Taylor, 10, 20, 300, -, 17
Elijah Morgan, 30, 50, 600, -, 15
David Williams, 150, 75, 2000, 40, 175
E. A. Morgan, 4, 10, 400, -, 35
Wesley Morgan, 40, 60, 1500, 20, 80
Geo. W. Boyer, 75, 111, 600, 15, 50
Smith Williams, 12, -, 200, -, 15

Wm. Massey, 125, 65, 1500, 12, 12
Eliza Knowles, 150, 150, 1000, 20, 50
James Burton, 16, -, 500, 12, 50
Dennis Phillips, 150, 150, 2500, 25, 175
Benjamin Hearn, 40, 20, 00, 12, 75
James Martin, 140, 75, 2000, -, 5
Kendal Callaway, 10, -, 100, -, 90
Danl. Baker, 35, 10, 300, -, 16
Wm. G. Moore, 75, 75, 1500, 50, 200
Jeremiah Eskridge, 90, 60, 1800, 8, 200
Elizabeth Quillen, 30, 30, 100, -, 35
Zach Clifton, 100, 25, 1700, -, 130
James Riggin, 100, 96, 1800, 8, 102
Jerry Wright, 125, 100, 1200, 20, 80
James Knowles, 50, 75, 1500, 20, 50
Jacob B. Baker, 60, 50,1200, 25, 200
Eccleston Moore, 150, 150, 6000, 100, 700
Danl. Phillips, 100, 150, 1500, 40, 50
Mager Riggen, 80, 6, 2000, 20, 250
Thos. Benson, 25, 25, 400, 20, 100
Jas. P. Owens, 50, 50, 600, 10, 60
Margaret Hearn, 25, 5, 500, -, 60
Tilgm. Spicer, 90, 95, 1850, 40, 250
Jacob Morgan, 100, 50, 1500, 8, 150
Chas. Pack, 50, 50, 700, 25, 75
Isaac H. Kinder, 70, 225, 2700, 30, 150
Wm. Taylor, 150, 250, 1600, 50, 250
Spencer Brackson, 5, -, 100, -, 30
John Wright, 150, 150, 2000, 20, 150
James Scott, 50, 100, 1000, 100, 200
Mitchel Scott, 10, 110, 1000, -, 10
Obed Clifton, 50, 120, 800, 15, 100
John Pusey, 88, 7, 600, 70, 400
Aaron N. Gordy, 50, 75, 800, 75, 175
James D. Cannen, 60, 110, 1800, 75, 140
Jacob Grimly, 30, 115, 1000, 35, 175
James Bowden, 6, 140, 1200, -, 12
M___ Love, 30, 100, 800, 5, 40
Isaac Parmer, 30, 100, 1000, 10, 140
Ephraim Wells, -, -, -, -, 40
Isaac J. Mitchel, 50, 90, 1200, 20, 100
Jeremiah Jones, 30, 10, 300, -, 45
Danl. Mitchel, 20, 30, 300, 10, 40
Elzey Collins, 30, 20, 500, 10, 50
Ann Collins, 30, 23, 300, 10, 10
John Collins, 60, 30, 600, -, 50
Thos. Mitchell, 20, 70, 1000, 25, 72
David Guseby (Ginby), 40, 60, 800, 30, 250
John P. Short, 50, 40, 400, 28, 150
Southy King, 50, 50, 1500, 20, 200
Thos. Parker, 12, -, 120, -, 20
Wm. Chandler, 12, -, 120, -, 10
Isaac _. Mitchel, 18, -, 130, -, 12
Wm. Downs, 15, -, 100, -, 20
Isaac Mitchel, 15, 20, 200, 5, 30
Wm. A. Truitt, 5, 2, 225, 5, 150
L. N. Hearn, 20, 125, 1050, 50, 500
Joshua Gray, 35, 50, 1500, 50, 200
Nathaniel Betts, 20, 50, 300, -, 20
Elijah Baker, 50, 50, 1000, 100, 60
Thos. Lowe, 50, 50, 800, 5, 35
Joshua Workman, 100, 75, 1000, 30, 500
Burton Mitchel, 18, -, 120, -, 25
Clement Harting, 18, 60, 325, -, 20
John Mitchel, 30, 40, 400, -, 50
Elijah Mitchel, 25, 10, 200, -, 50
Rufus Mitchel, 40, 50, 500, 30, 200
Archibald Baker, 25, 17, 500, 45, 200
Matthew Truitt, 40, 60, 600, 15, 100
Robt. Mitchel, 20, 20, 500, 30, 100
George Mitchel, 25, 50, 500, 3, 200
Noah Downs, 18, 10, 200, 10, 200
Isaac T. Jacman, 25, 65, 500, 20, 160
Thos. Drake, 60, 50, 500, 5, 150
Wm. Parens, 50, 100, 1000, 20, 60
Elizabeth Truitt, 40, 160, 1200, -, 120
Nancy Baker, 100, 100, 1500, -, 30
John B. Sheet, 20, 130, 1500, 2, 400

Wm. P. Parsens, 60, 60, 1000, 75, 300
Ebenezer Carvean (Cacuen, Carmean), 50, 32, 1000, 50, 200
Isaac Cannen, 80, 185, 1000, 25, 300
E. H. Pasey (Posey, Pusey), 25, 50, 500, 500, 20
Gillis Jones, 71, 145, 2000, -, 50
Wm. Cole, 40, 40, 600, -, 50
Noah Parsons, 10, -, 100, 10,50
Manuel B. Hadden, 40, 40, 1000, 20, 60
Jest Jones, 75, 200, 1000, 75, 150
Benj. Hearn, 100, 100, 2000, 100, 300
Clement Hearn, 100, 100, 2000, 75, 500
Philip West, 60, 50, 1000, 10, 100
Rhoda Pusey, 60, 40, 800, -, 14
Thos. Hearn, 75, 100, 2000, 5,125
John Parsons, 40, 20, 300, 75, 125
James Parsons, 30, 40, 400, -, 60
Aaron Cordrey, 75, 125, 1500, 50, 100
John N. Phillips, 65, 96, 1200, 30, 150
Wm. E. Cannon, 30, 122, 930, 50, 100
Jos. H. B. Cannon, 6 56, 810, 50, 150
Jos. F. Cannon, 75, 75, 1000, 60, 150
Wm. Tayfal, 100, 100, 800, -, 75
John Ony (Orey), 30, 70, 300, -, 20
Saml. Baker, 50, 90, 1000, 50, 150
John Short, 50, 20, 400, -, 10
Johnson Cannen, 40, 85, 500, 30, 125
Noble King, 50, 20, 500, 5, 73
Nathl. King, 25, 20, 400, 6, 50
Wingate West, 50, 75, 1000, 20, 100
Joshua Downs, 30, 20, 400, -, 25
Mary Taylor, 10, -, 100, -, 10
Maney Pennuick, 40, 10, 300, -, 100
Eben Gray, 75, 50, 1000, 100, 400
Thos. Neal, 125, 200, 3000, 200,2 50
Wm. Bryan, 50, 25, 800, 50, 250
Geo. H. Vincent, 250, 100, 4000, 150, 500
Danl. Phillips, 12, 3, 100, -, 30
Jos. W. Vincent, 58, 15, 500, 30, 90
Ann Elleford, 150, 200, 5000, 75, 300
Wm. W. Morgan, 110, 180, 3000, 100, 300
Thos. Hooks, 300, 110, 3000, 100, 500
Nathl. Messick, 44, 4, 250, 10, 200
Whitfield Danpates (Daupates, Daupater), 105, 150, 1500, 56, 250
John Messick, 125, 25, 700, 5, 75
Isaac Dalby, 80, 20, 400, 25, 125
Joel Messick, 70, 30, 1000, 25, 175
William Messick, 75, 35, 1200, 25, 250
Danl. Hudson, 100, 100, 2000, 50, 420
Gideon B. Hitchens, 30, 8, 500, 5, 100
Robt. L. James, 100, 100, 1000, 50, 150
Henry Messick, 12, -, 110, -, 10
Joshua James, 300, 200, 2500, 50, 300
Sebley Conway, 100, 100, 2000, 75, 300
Colwell James, 140, 100, 2500, 75, 500
Geo. M. Collins, 100, 100, 1000, 30, 200
Branson James, 100, 100, 1200, 15, 100
Saml. J. Jakes, 100, 100, 1500, 5, 50
Nathl. Melson, 75, 88, 1400, 15, 100
Benjamin Melson, 100, 250, 2500, 50, 200
Walter Melson, 6, 94, 500, 10, 40
William Melson, 50, 50, 500, 15, 40
Rachel Cannen (Carmen), 40, 110, 1200, -, -
Ann Wright, 70, 50, 1000, -, 15
Lewis Workman, 125, 10, 2000, 20, 200

Nathl. Hitchens, 20, 70, 300, 5, 100
Smith Hitchens, 80, 15, 1200, 30, 300
John Melson, 100, 50, 1200, 25, 100
Sarah Johnson, 18, -, 120, -, 10
Hannibal Torbert, 70, -, 500, 15, 80
Mary (Many) A. Torbert, 25, 25, 500, -, 30
Philip Neal, 100, 100, 1200, -, 200
Mary R. Truitt, 80, 50, 800, -, -
W. T. Warrington, 100, 50, 800, 50, 200
Elizabeth Walter, 100, 100, 1000, 10, 150
Joseph B. Camean(Cameron), 350, 250, 3000, 100, 150
Jacob W. Cameron, 75, 42, 500, 25, 100
John S. Pasey, 50, 50, 800, 6, 60
Peter Hitchens, 5, 61, 800, 15, 25
Wingate Matthews, 25, 20, 400, 10, 50
Benton Camron, 200, 125, 2500, 100, 275
Robt. Frada, 100, 50, 800, 10, 100
Isaac West, 80, 50, 80, 10, 100
Giles Hitchens, 70, 40, 800, 30, 150
Robt. M. Rhody, 100, 100, 1200, 50, 125
Danl. Rodhey, 60, 70, 500, 20, 40
Geo. W. Pasey, 12, -, 100, -, 6
Zedekiah P. Harting, 130, 72, 1000,30, 300
Robt. West, 30, 25, 400, 20, -
Henry B. Truitt, 100, 200, 1000, 30, 200
James Danahr, 35, 35, 1000, 30, 100
Wm. D. Reads, 50, 50, 1000, -, 100
James Truitt, 200, 300, 1000, 30, 300
Wm. Otwell, 75, 75, 1000, 20, 100
John Truitt, 75, 100, 1200, 50, 150
Greenbery M. Truitt, 50, 60, 500, 50, 100
Burton Truitt, 50, 50, 800, 25, 100
Elizabeth Carmeon, 40, 10, 500, -, 30
Mary Collins, 50, 40, 500, -, 50
Davis Lowe, 100, 100, 1500, 60, 225
Robt. K. Smith, 60, 69, 500, 30, 160
Jos. A. Smith, 70, -, 160, 4, 20
Jackson Smith, 20, 200, 300, 50, 125
Ephraim Collins, 10, -, 100, -, 15
Joseph Torbert, 30, 30, 400, 12, 75
John Gordy, 25, 69, 500, 2, 75
Joseph Wells, 20, 10, 250, -, 15
Wingate Drans, 50, 50, 1000, 5, 120
Levin Duncan, 12, 12, 200, 8, 40
John P. Baily, 72, 12, 200, -, 40
Elisha English, 100, 200, 2500, 90, 400
Isaac Collins, 35, 50, 1000, 50, 100
James Mitchel, 80, 40, 500, 20, 75
Abel West, 50, 50, 900, 25, 150
_dward Dill, 25, 25, 300, 3, 20
Anderson West, 12, 30, 600, 8, 16
Cornelius West, 15, 153, 600, 50, 100
Benjamin West, 50, 100, 600, 30, 300
Elizabeth Oliphant, 70, 70, 500, -, 20
Burton West, 70, 80, 1000, 50, 125
Wm. Wainwright, 15, -, 120, -, 20
Levi Timmons, 10, 15, 120, -, 40
Ephr. Short, 50, 200, 2400, 30, 100
Nancy Short, 40, 30, 500, 40, 125
Isaiah Short, 40, 60, 800, 75, 300
James Owens, 20, 15, 100, -, 10
Southy King, 100, 50, 1800, 50, 150
Ezekiel Timmons, 60, 60, 700, 20, 100
John Lokey, -, -, -, 10, 50
Jacob Jones, 100, 200, 1000, 30, 200
Isaac Jones, 10, -, 100, 5, 10
Sarah Jones, 12, -, 100, -, 25
Danl. LeKate, 22, 4, 150, -, 20
Jeremiah LeKate, 40, 43, 800, 15, 75
Sarah Magee, 50, 50, 600, -, 25
Nancy Timmons, 60, 40, 600, 20, 50
Wm. Foskey, 7, -, 50, -, 18
Min__ Collins, 70, 70, 1000, 20, 100
Ephraim Collins, 70, 50, 800, 20, 125

John Jones, 50, 70, 1000, 30, 125
John S. Matthews, 22, 80, 900, 85, 150
Peter Gordy, 50, 50, 400, -, 75
Nathan S. Matthews, 12, 9, 250, - 100
Ca__ Wingate, 100, 150, 2500, -, -,
Cyrus Cannon, 100, 106, 2000, 30, 250
John West, 125, 75, 1500, 70, 500
Eliza Connerly, 300, 200, 300, 20, 50
Amelia Phillips, 150, 150, 2400, 60, 500
Greenbery Hall, 100, 100, 1500, 30, 35
Benjamin Reddin, 80, 120, 600, -, 25
Phillip Graham, 80, 100, 820, 35, 500
Wm. Oneal, 80, 100, 1000, 25, 125
Thos. Oneal, 80, 100, 2000, 75, 200
Jesse Clifton, 20, 20, 250, -, 18
Jas. L. Wainwright, 18, 9, 500, 20, 225
Curtis Spicer, 50, 112, 1000, 60, 250
David H. Boyd (Boyce), 125, 125, 2000, 60, 200
Nancy Benson, 30, 30, 200, 15, 40
Peter LeKate, 100, 167, 1500, 20, 125
John W. Walker, 80, 100, 1200, 55, 100
Rebecca Willy, 80, 12, 500, 4, 50
Lazarus Turner, 100, 45, 1000, 50, 250
John Windsor, 100, 60, 500, 8, 75
John Chipman, 50, 60, 1400, 100, 300
Samuel Oneal, 40, 60, 2000, -, -
John Taylor, 200, 200, 5000, 80, 200
Jno. Harting, 40, 50, 500, -, 20
Jno. W. Jarman, 80, 100, 1800, -, 12
Isaac Hall, 10, -, 100, -, 100
Elias Taylor, 100, 120, 1500, -, 600
Wm. Covington, 10, -, 100, -, 25
Geo. Pance, 80, 120, 1500, 75, 250
Samuel S. Vaughn, 120, 150, 2000, 100, 400
Jno. C. Vaughn, 60, 110, 1000, -, -
Thos. Buckman, 50, 60, 1000, 30, 150
Geo. W.James, 50, 40, 700, -, 100
Wm. Hitch, 80, 300, 3000, 125, 500
Geo. Magee, 80, 75, 1200, 20, 100
Asberry Harting, 10, 10, 250, -, 100
James Riggin, 100, 100, 2000, 20, 80
Henry Pepper, 60, 60, 1000, 50, 250
Sarah Downs, 55, 50, 500, -, 70
Benjamin Swain, 50, 60, 500, -, 75
William W. Smith, 100, 40, 1500, 50, 250
Wingate Matthews, 100, 100, 1400, 30, 180
James Pusey, 70, 29, 375, 5, 30
Philip Matthews, 100, 100, 1000, 50, 250
Eliza Pusey, 12, 16, 400, 5, 80
Eleanor Pusey, 50, 50, 1000, -, 35
Hezekiah Matthews, 100, 80, 800, 50, 250
Sovereign Hitchens, 75, 75, 600, 25, 150
Stephen H. Gordy, 80, 20, 1800, 125, 500
Alfred Calloway, 80, 50, 1200, 20, 125
Matilda Tr_e, 50, 50, 700, -, 75
Peter Truitt, 15, 10, 200, -, 55
Josiah M. Melson, 30, 30, 500, 25, 150
John P. Torbert, 40, 20, 500, 20, 60
Watson Torbert, 60, 60, 800, 25, 200
Edward Hitchens, 30, 50, 300, 71, 250
Elison Boyce, 50, 60, 700, 50, 150
James H. Boyce, 100, 102, 2000, 50, 250
Sovereign Hall, 15, 5, 160, -, 15
Jno. W. Calloway, 125, 125, 3000, 100, 500
Edmond Hitchins, 54, 50, 400, -, 150

John Matthews, 250, 350, 3000, 20, 550
James Oneal, 50, 50, 700, 25, 100
Isaac Giles, 75, 125, 1200, 20, 150
John Short, 50, 50, 400, -, 70
Wilson Messick, 90, 10, 600, 20, 80
Sol. Short, 100, 40, 1400, 100, 520
Ann Moon, 150, 100, 2000, 50, 60
Nehemiah Messick, 100, 60, 1200, -, 150
Philip Short of J, 150, 150, 2000, 20, 400
Philip Short of N, 50, 10, 300, 50, 125
James Messick, 100, 100, 1500, -, 850
Kenzie J. Hill, 40, 40, 400, 50, 100
Eleanor Moore, 10, 10, 150, 25, 15
Henry Hudson, 125, 360, 2000, -, 200
Mary Boyer, 160, 80, 800, 28, 100
Wm. Kennikin, 50, 40, 500, 10, 100
Thos. Short, 100, 25, 1000, 25, 200
Edward Taylor, 40, 40, 400, 50, 100
Greenberry Cam (Carnor, Cannor), 2, -, 300, 30, 20
Augustus Pavis (Paris), 2, -, 150, -, 30
George W. Green, -, -, -, -, 400
Jno. S. Smith, 2, -, 300, -, 15
David Cal__ow, -, -, -, -, 200
Mary A. James, -, -, -, -, 40
Eliza Adams, 100, 80, 1500, -, 60
Anthony Law, 2, -, 100, -, 30
Moses Tubbs, 80, 190, 1500, 15, 150
William Graham, 100, 15, 1000, -, 100
James Pettyjohn, 30, 10, 150, 5, 100
James Martin, 50, 210, 1500, 25, 300
Allice Martin, 100, 300, 3000, 20, 200
James Prettyman, 50, 100, 1000, 15, -
Barkley Wilson, 100, 200, 1500, 26, 300
Thomas Walls, 65, 65, 1000, 20, 200
Ash Walls, 60, 65, 700, 25, -
William Massey, 125, 225, 1500, 25, 100
Samuel Warick, -, -, -, -, -
Erasmus D. Marsh, 52, -, 500, 20, -
Paynter Turner, 100, 100, 1000, 15, 30
John Dalton, 200, 50, 1000, 15, 150
John Donovan, 50, 10, 500, 20, 50
Moses Heavlor (Heavlon), 80, -, 300, 10, 150
David H. King, 100, 50, 1000, -, 100
Warrin Wright, -, -, -, -, -
Isiah Salk, 60, 17, 1800, -, 150
John Corsey, 18, -, 1000, -, 203
William White, 100, -, 600, 10, 150
Mary Prettyman, 100, 75, 2000, 20, 400
David H. Wampler, 100, 25, 1500, 10, -
George Walker, -, -, -, -, -
Pete Manlove, 100, 44, 2500, 10, 300
Elizabeth Gee, 75, -, 1500, -, 400
William Brian, -, -, -, -, 150
Jessie Lewis, -, -, -, -, 10
Wolsley Madson, 75, 2000, 1000, 10, 100
John Swain, 3, -, 300, -, 75
Asbury Prettyman, -, -, -, -, -
George Chambers, -, -, -, -, -
William Russell, -, -, -, -, 25
Peter Warrington, -, -, -, -, 25
William Rusket, -, -, -, -, 75
Hudson Burr (Barr), -, -, -, -, 100
Thomas Norman, -, -, -, -, 75
George Hickman, -, -, -, -, 450
Purnal Norman, -, -, -, -, -
Robert West, -, -, -, -, -
Thomas Hudson, 95, 15, 1500, 20, 300
Lewis West, 40, -, 1500, -, 300
Benjamin McIlvaine, 30, -, 2000, -, -
Reece Woolfy, 400, 100, 8000, 25, 500
Henry Wolf, -, -, -, -, -

Jesse Pool, -, -, -, -, -
Robert Robertson, -, -, -, -, -
Thomas Carpenter, -, -, -, -, -
Joshua Burton, 32, 30, 1000, 10, 250
James Duffall, 50, 100, 1000, 10, -
Alfred Willbank, 200, 200, 5000, -, 250
John Rodney, -, -, -, -, 100
Joshua Burton, 120, 80, 1200, -, 150
John Parker, -, -, -, -, 50
Daniel Simple, 121, -, 3600, -, 150
James Russel, 100, 20, 200, 10, 100
James Bouland, 100, 10, 2000, 10, 100
Thomas Maule, 25, -, 120, 15, 200
William Maule, 42, -, 1200, -, 150
James Maule, -, -, -, -, 100
Peter Maull (Maul), 10, -, 100, 10, 300
Benjamin Bolins, -, -, -, -, -
William Russel, 250, -, 7000, -, 400
Jacob Buttons, -, -, -, -, 73
Cornelius Moslade (Mastock), -, -, -, -, -
John H. Burton, -, -, -, -, 25
Nathaniel Clark, 150, 50, 2100, 20, 200
Thomas Kellum, 40, -, 600, 10, 150
S. M. Joseph, 150, 100, 2000, 25, 500
Mitchell Lamb, 100, 50, 1500, 25, 250
Samuel Hazzard, 10, 20, 100, 20, 100
James Lauk, 50, 25, 500, 10, 200
Gideon Phinas, 125, 25, 1500, 25, 250
Eli Wilson, 100, 200, 1000, 15, 150
Nancy Black, 100, 50, 1000, 25, 150
Jacob Hookins, 100, 50, 600, 1, -
Asa Manlove, -, -, -, -, 25
Zachariah Price, 150, 50, 1510, -, -
James Dickerson, 200, 50, 2000, 25, 250
Joseph Dodd, 500, 200, 2500, 30, 500
Cornelias Holland, 200, 30, 2000, 50, 500
Elijah Holland, 200, 50, 2000, 30, 400
Peter Mocet (Macet), 175, 75, 1500, 20, 400
Temperance Records, 100, 50,500, 10, 350
James Marsh, 70, 7, 350, 25, 150
Lemuel Marsh, 100, -, 300, 10, 250
Rhoda Thompson, 100,7, 1000, 25, 350
Burton Stockly, 100, 5, 1500, 50, 300
Woodman Stockly, 100, -, 1200, 30, 350
Peter Marsh, 105, 21, 100, 25, 50
James Davidson, 175, 25, 1000, 10, -
Greenbury Brown (Broom), 7, -, 300, 10,200
Laben Hudson, 100, 25, 1000, 15, 150
John Parsons, 100, 50, 1000, 35, 200
Elizabeth Dickson, 200, 8, 1000, 50, 300
Josep Morris, 75, 20, 500, -, 200
Zadock Milby (Milley), 100, 40, 1200, 50, 350
Joseph Cannor, 6, -, 150, 5,100
James Fletcher, 125, 125, 3000, 75, 500
Benjamin Benton, -, -, -, -, -
Daniel White, 6, -, 150, 20, 200
David R. Burton (Benton), 150, 20, 2000, 30, 75
Rice Paynter, 75, -, 700, 15, 200
Lewis Hopkins, 40, -, 200, 15, 150
Samuel Lodge, 200, 70, 5000, 75, 500
Isaac King, 28, -, 1000, 10, 50
Thomas Stockley, 100, 20, 1500, 20, 200
David Hepson (Hefson), 50, 10, 1000, 25, 300
Daniel Paynter, 5, -, 100, -, -,
Miles Corsey, 22, -, 250, 10, 350

Thomas Wilson, 50, 50, 2000, 75, 500
Benjamin O. Wapler, 64, -, 1000, 25, 250
Zachariah Joseph, 100, 20, 100, 50, 300
John Benton, 60, 10, 1000, 30,150
William Robinson, 275, 25, 2000, 30, 200
Samuel Vak__, -, -, -, -, 40
Selby Stockley, -, -, -, -, -
John Roach, 100, 25, 700, 50, 200
Reuben Cannon, -, -, -, -, 200
Samuel Hudson, 100, 25, 1000, -, -
Mary Griffith, 40, 24, 350, -, 150
David Hopkins, 60, 20, 1000, -, 150
Elizabeth Reid, 20, 15, 200, 50, 390
William Hopkins, 75, 25, 1000, 30, 202
Lemuel Draper, 100, 400, 6000, 100, 1200
John Campbell, 250, 200, 6000, 130, 100
David Holland, 150, 100, 450, 50, 1000
John Townsand, 170, 135, 2500, 30, 200
Mary Bennett, 100, 100, 2000, 75, 200
Alexander Draper, 50, 50, 1000, 70, 150
Purnal Bennett, 60, 50, 1000, 60, 200
William B. Watson, 75, 30, 1500, 10, 150
Parnal Townsand, 253, 40, 3000, 20, 150
Cornelias Watson, 100, 30, 1000, 60, 300
John Bennett, 200, 25, 2500, 30, 300
Athonny Ingham, 50, 50, 1000, 70, 300
John Ingram, 60, 70, 800, 60, 300
Daniel Young, 80, -, 360, 25, 250
Nehemiah Prettyman, 60, 40, 800, 30, 253
William Hickman, 40, 50, 800, 60, 100
David Warren, 25, 25, 30, 30, 150
Jonathan Milmer, 200, 200, 100, 30, 250
George Clendance Sr., 150, 150, 1500, 40, 300
George Clendnace Jr., 40, 20, 600, 50, 150
John Lofland, 10, 11, 300, 20, 125
David Lofland, 150, 100, 1300, 30, 300
John Lofland, 131, 30, 700, 60, 153
Samuel Shockly, 150, 50, 1500, 80, 400
William Layton, 75, 30, 600, 60, 300
Muriah Davis, 200, 50, 1000, 30, 100
James Ward, 50, 400, -, 70, 400
William V. Colter, 150, 150, 6000, 150, -
William Hickman, 250, 100, 1000, 50, 350
Zachariah Deputy, 100, 50, 1500, 30, 300
Robert Riley, 100, 50, 1000, 70, 150
Reubin Carpenter, 200, 1000, 25000, 60, 300
Jacob Burtin, 57, 31, 300, 30, 150
Edward Davis, 300, 150, 2000, 90, 200
Mollin R. Daniel, 80, 25, 1200, 80, 200
George Bennett, 150, 50, 1500, 70, 150
William Williams, 275, 25, 2000, 60, 200
Minas Lynch, 100, 50, 1000, 50, 150
James Williams, 50, 60, 800, 50, 150
Booz Poswaters, 100, 60, 600, 50, 175
Pery Prettyman, 100, 50, 1000, 25, 150
Thuen Warren, 210, 30, 1000, 30, 200

Peter Calhoon, 150, 50, 2500, 50, 300
William McDaniel, 100, 200, 1500, 50, 200
Jonathan Draper, -, -, -, 50, 100
Got. W. Davis, 80, 120, 250, 50, 300
Isaac Cook__ (Coowitz), 125, 25, 1500, 30, 250
David R. Truitt, 100, 25, 700, 30, 250
Robenson Shockly, 100, 50, 600, 30, 200
James M. Shockly, 32, -, 400, 20, 10
Myus (Myres) M. Reynolds, 150, 50, 2500, 26, 250
Thomas R. Wilson, 125, 35, 2500, 30, 40
James Kno__, 200, 100, 3000, 50, 400
Thomas Roach, 100, 50, 1000, 20, 300
John Abbott, 52, 150, 1000, 30, 200
James Samis, 100, 50, 1500, 50, 550
Nathaniel Davis, 70, 50, 1200, 60, 300
Robert M. Davis, 70, 50, 1000, 75, 451
John Wilcott, 60, 50, 1000, 80, 500
John S. Holland, -, -, -, 75, 350
Edward Roach, 200, 100, 1000, 40, 150
Phillip Cordithy, 100, 50, 1000, 30, 150
Henry Vans (Vaus), 50, 20, 1000, 90, -
John Wapler (Waples), 100, 50, 1500, 50, -
John _. Shockly, 30, -, 1000, 1000, 100
Elias Russel, 150, 100, 4000, 200, 300
Minus Draper, 200, 150, 1800, 75, 750
Marica Draper, 150, 50, 2500, 100, 450
Henry Draper, 100, 50,1800, 40, 350
David Lindle, 160, 50, 1000, 745, 250
Joseph M. Davis, 500, 8, 1000, 50, 250
Thomas S. Davis, 100, 40, 2000, 70, 300
Joshua Hurley, 75, 50, 1000, 40, 175
Joseph Argo, 80, 50, 1000, 50, 150
Nathaniell Ingram, 65, 35, 1000, 90, 250
Burth Walston, 75, 25, 650, 30, 100
Littleton Clendance, 60, 50, 1500, 20, 300
Elias S.Townsand, 200, 100, 3000, 30, 250
Thomas Deputy, -, -, -, -, 130
George R. Fisher, 250, 280, 3500, 100, 500
Thomas Ross, 100, 50, 800, 25, 200
John Donovan, 75, 75, 1500, 50, 300
Elias Coverdale, 100, 50, 1500, 50, 30
Gibson Donovan, 100, 50, 800, 70, 150
Joshua Purce (Pierce), -, -, -, 50, 400
John Brittinham, 100, 50, 1500, 50, 250
Luther Walsh, 70, 20, 500, 50, 300
Arpy Warren, 100, 100, 1000, 50, 250
Jonathan Milmer, 100, 20, 800, 50, 200
Peter Melby, 200, 100, 1000, 75, 300
Stephen Warren, 10,30, 450, 50, 300
William E. Joseph, 25, 149, 850, 25, 150
Walter Hudson, 75, 25, 1000, 24, -
Eli T. Joseph, 150, 50, 1000, 25, 200
Benjamin F. Wapler, 90, 180, 3000, 100, 500
Jacob Wilson, 10, 50, 1000, 25, 250
Nehemiah Messick, 40, 25, 600, 25, 150
William Burrows, 150, 20, 2000, 25, 150
John J. Davis, 200, 75, 1200, 50,300

Thomas P. Jefferson, 105, 65, 500, 50, 500
Nehemiah Casey, 100, 50, 500, 10, 350
Samuel Brittenham, 60, 140, 1000, 30, 200
Absolum Lynch, 75, 750, 800, 20, 100
William Shepherd, 60, 40, 300, 20, 150
Joshua Lofland, 75, 60, 800, 20, 250
Booze Warren, 75, 25, 1200, 25, 250
William Warren, 100, 50, 600, 30, 200
William Carpenter, 400, 34, 1000, 25, 100
Littleton M. Lofland, 200, 20, 1000, 25, 150
Purnal Walls, 150, 50, 1000, 30, 150
Wilson Shockly, 200, 150, 500, 25, 25
William Shockly, 100, 20, 800, 35, 25
Elias Shockly, 75, 20, 600, 30, 30
Daniel Deputy, 100, 80, 1500, 50, 350
Robert Fowler, 50, 25, 600, 32, 350
William Hickman, 75, 25, 600, 25, 350
Benjamin McIlvain, 40, 20, 400, 100, 200
Riley Bennett, 100, 75, 1000, 51, 350
Elias T. Bennett, 100, 855, 1200, 35, 200
John C. Dawson, 50, 45, 1400, 72, 400
Nathaniel Coverdale, 75, 50, 1500, 80, 300
Robert Campbell, -, -, -, 25, 200
Jesse Web, 40, -, 60, 25, 200
Henry Quinn, 60, 24, 1000, 50, 350
David Spurier (Spencer), 150, 60, 1000, 30, 150
Jacob Web, 100, 50, 1000, 25, 150
Bennah Sharp, 200, 20, 2000, 50, 500
Matthias Web, 75, 20, 2000, 20, 100
Iles Mills, 100, 75, 2000, 50, 300
William H. Faunton (Taunton), 200, -, 1200, 50, 250
Isaac Carpenter, 150, 50, 2000, 50, 400
William Clendance, 45, 100, 800, 25, 250
William Jacobs, 100, 30, 800, 15, 153
Thomas Clark, 100, 53, 1000, 40, -
Michel Melman, 100, 50, 1000, 100, 300
Noah Transom (Tourson), 300, -, 4000, 100, 400
Thomas Davis, 250, -, 300, 40, 500
Mary Watson, 50, 25, 500, 40, 250
Celia Potter, 100, 50, 2000, 50, 300
William Parks, 80, 50, 1000, 30, 100
Robert McCauly, 175, 25, 1500, 40, 150
Elias Daniel, 50, 25, 1000, 25, 250
Peter Smith, 50, 25, 800, 30, 251
Robert Ingram, 75, 25, 1000, 35, 300
Samuel Collins, 30, 25, 600, 25, 400
William Price, 50, 50, 1200, 30, 250
William Watson, 150, 75, 1200, 25, 300
William VanKirk, 100, 40, 600, 20, 300
Stephen Casey, 75, 25, 1000, 40, 150
Nunn Daniel, 75, 50, 1000, 15, 75
William Lofland, 100, 50, 1000, 20, 100
Davis Smith, 50, 50, 1000, 50, 20
Joseph Wason, 100, 89, 1500, 30, 30
Nutter L. Davis, -, -, -, -, 10
William Revel, -, -, -, -, 15
Thomas Watson, 200, 37, 1000, 53, 307
John J. Truitt, -, -, -, -, 300
Sarah Dickerson, 100, -, 500, 40, 300
Mary Jestee, 75, 25, 500, 30, 200

Stephen Warren, 100, 25, 600, 50, 150
Luke Elensworth, 100, 25, 500, 70, 125
Isaac Betts, 100, 100, 3000, 100, 500
Alfred Short, 150, 60, 1100, 150, 300
Solomon Deputy, 75, 30,600, 20, 150
Mason Web, 75, 25, 500, 15, 200
Nancy Web, 125, 75, 1500, 10, 200
Charles Web, 75, 75, 1000, 1, 100
William Walls, 75, 75, 1000, 14, 150
Isaac Briam, 50, -, 350, 14, 150
Job Johnson, 75, 50, 700, 30, 75
Joshua Tenant, 145, 35, 2000, 512, 40
William Voss, 100, 40, 1000, 51, 350
John Web, 200, 100, 2000, 50, 250
John M. Collins, 100, 25, 1000, 71, 200
Daniel Jones, 100, 90, 1000, 41, 175
Curtis Macklin, 100, 50, 1000, 62, 150
James Wheeler, 75, 25, 600, 21, 300
John Dawson, 175, 50, 1000, 32, 100
Charles Collins, 150, 50, 800, 15,10
Elisha Johnson, 100, 50, 1000, 41, 50
John W. Clifton, 200, 130, 4000, 64, 350
Thomas Hays, 125, 150, 2000, 32, 500
Theodore Johnson, 100, 75, 1000, 22, 350
John Hays, 125, 45, 1500, 25, 150
James H. Carlisle, 100, 24, 600, 15, 100
Levin Pettyjohn, 80, -, 200, 50, 350
James O. Day, 50, 150, 1000, 40, 350
David R. Smith, 90, 280, 1500, 50, 350
Charles Dean, 100, 200, 1500, 30, -
Lot (Got) Rollins, 200, 200, 3000, 75, 500
Joseph Isaacs, 100, 200, 1800, 75, 350
Joseph Collins, 120, 75, 1000, 30, 358
Jonathan Collins, 175, 68, 1200, 50, 600
James Welsh, 150, 100, 2500, 60, 400
Henry Smith, 100, 40, 2000, 60, 400
Samuel Williams, 350, 150, 2000, 75, 350
Nathaniel Short, 225, 75, 1000, 30, 350
Samuel J. Willey, 250, 290, 3264, 50, 650
George Reynolds, 100, 60, 1000, 30, 350
Richard Lewis, 200, 100, 1000, 50, -
John Davis, 100, 100, 1000, 60, 300
Jacob Smith, 100, 100, 3000, 60, 200
James Hardle, 200, 100, 500, 30, 300
Nathan Ratliff, 125, 50, 1000, 70, 450
John Morris, -, -, -, -, 1600
Elias Lofland, 50, 100, 300, 50, 300
William Gosley, 30, 100, 300, 20, 300
Zebody Nutter, 10, 20, 150, 10, 150
Joshua Sharp, 260, 200, 2000, 30, 300
Isaac Owens, 200, 200, 2000, 40, -
William Willey, 60, 100, 1000, 30, -
William Ryon, 125, 25, 4000, 20, -
William B. Ryon, 125, 175, 1500, 40, 450
Covinter Messick, 65, 100, 600, 40,200
Waitman Willey, 30, 60, 500, 2, 300
Richard Co___k, 20, 30, 300, 10, -
Minus Willey, 30, 50, 500, 30, -
William H. Griffith, 100, 30, 1000, 20, 500
Taumer Griffith, 100, 50, 1500, 30, 450

William G. Carlisle, 10, 115, 2000, 30, 500
George Polk, 200, 250, 400, 40, 500
John M. Layton, 200, 100, 3000, 30, 500
Mary Fisher, 100, 150, 1350, 40, 600
John Tindle, 175, 135, 2000, 50, 500
Solomon Vincent, 125, 75, 1500, 40, 500
Theopholus Collins, 300, 300, 1000, 30, 1200
Sarah E. Spicer, 75, 35, 1200, 30, -
Cyrus Jefferson, 75, 50, 700, 40,300
Amy Morgan, 130, 200, 2500, 75, 605
James Conaway, 150, 170, 3500, 75, 400
Nickolas N. Adams, 125, 125, 1500, 30, 250
James D. Stuart, 50, 30, 500, 30, 100
George Stuart, 30, 30, 300, 40, 250
James Tindle, 150, 100, 2000, 30, 450
Isaac Conaway, 200, 150, 2000, 150, 900
Levi Hall, 200, 50, 1500, 50, 400
James Hufington (Hafington, Hassington), 100, 50, 1000, 100, 450
John W. Jefferson, 30, 40, 500, 10, 150
Robert L. Conwell, 150, 150, 1500, 40, 250
Jacob Knowles, 100, 50, 1500, 30, 400
Stephen Morgan, 125, 100, 2000, 100, 550
Minus Isaacs, 100, 100, 600, 40, 350
Thomas A. Jones, 200, 150, 200, 50, 500
Thomas York, 200, 200, 400, 40, 550
William Slayton, 175, 30, 2500, 100, 450
Joshua McCauley, 80, 120, 1500, 100, 450
John Day, 100, 50, 1000, 100, 500
John Isaacs, 66, 65, 700, 50, 100
Rachel Owens, 211, 100, 4000, 100, 500
Esekial Jones, 100, 50, 3000, 50, 300
Tilman Layton Jr., 100, 50, 3000, 50, 400
Benjamin Morgan, 75, 50, 1000, 50, 200
Lorenzo D. Morgan, 50, 25, 500, 30, 150
Rachael Owens, 150, 50, 2000, 110, 300
Henry B. Fidemer, 220, 20,10000, 200, 550
James Hudson, 200, 50, 2000, 100, 350
William Walls, 100, 200, 2000, 30, 300
James Houston, 100, 50, 1000, 20, -
Henry Huston, 220, 55, 90, 30, -
Alexanell Truitt, 100, 100, 1000, 20, 200
John Clendance, 100, 100, 1000, 30, 250
Henry Macklin, 55, 50, 500, 20, 150
John Macklin, 100, 50, 1000, 30, 200
David R. Smith, 100, 120, 3000, 30, 150
Jacob Clendance, 75, 40, 1000, 20,200
William Hemmer, 100, 75, 600, 30,300
Purnal Sharp, 75, 100, 500, 30, 250
Clausbury Clark, 645, 75, 1000, 30, 300
Job Sharp, 100, 50, 300, 50, -
Benton Sharp, 50, 75, 1000, 20, 200
James Plumer, 75, 50, 600, 30,175
Joseph Russel, 100, 100, 1500, 50, 150
Peter Donovan, 100, 100, 2500, 20,400
Burton Fowler, 75, 50, 500, 30, 250
Albert Macklin, 75, 200, 1000, 20, 350

James Plummer, 100, 200, 1000, 20, 250
Robert Lynch, 75, 200, -, -, -
Peter McCauley, 75, 25, 500, 50, 250
William Carpenter, 175, 225,700, 30, 175
Josiah Marvel, 100, 60,1500, 40, 300
Daniel Knowles, 1500, 1200, 6000, 200, 800
John Fleetwood, 50, 50,1000, 400, 300
Cyrus Fleetwood, 100, 100, 1200, 50, 300
Noble Conaway, 200, 200, 1500, 50, 400
James Hurley, 100, 200, 2000, 60, 300
Owen Isaacs, 75, 30, 1500, 30, 200
Noah Isaacs, 100, 125, 1000, 30, 300
Joseph Elliot, 50, 10, 700, 40, 300
John Spencer, 75, 75, 500, 10, 150
Dixon Conaway, 35, 60, 1000, 10, 175
Nathan Conaway, 129, 125, 500, 30, 150
Charles Colleck, 150, 150, 3000, 40, 300
Gray Short, 100, 100, 3000, 50, 350
William Swain, 100, 100, 1500, 30,150
Daniel Short, 130, 130, 2000, 60, 450
Walter Swam 25, 50, 800, 30, 300
Parnal Short, 100, 50,1800, 30, 250
Abel Swam, 75, 75, 800, 30, 350
David Short, 100, 50, 1000, 35, 450
Gilly Short, 75, 75, 800, 30, 750
Nancy Torbert, 30, 50, 500, 20, 25
Robert Barr, 100, 200, 2500, 40, 530
Jacob Prettyman, 30, 40, 400, 30, 20
Robert P. Barr, 50, 50,1000, 50, 300
Daniel Turner, 75, 50, 500, 30, 350
John Swam, 50, 50, 1000, 20, 150
John Dolby, 50, 50, 1000, 30, 300
Jese Pastwaters, 40, 50, 100, 20, 200
C. M. G.. Smith, 120, 40, 1000, 30, 350
John Paswaters, 100, 75, 1000, 30, 300
Miles Messick, 150, 210, 3000, 50, 45
Minus T. Conaway, 100, 200, 200, 60, 550
John Smith, 75, 50, 800, 30, 350
Curtus Conaway, 75, 75, 1000, 50, 300
John Tindle, 40, 20, 600, 30, 250
Miles Tindle Jr., 10, 15, 1000, 30, 150
Milfred Danahoe, 100, 200, 2000, 30, 50
Peter Danahoe, 200, 100, 1000, 30, 170
Josiah Marice Sr., 100, 150, 1500, 40, 150
William Fleetwood, 50, 80, 1000, 50, 400
Gunneset Harris, 40, 40, 500, 30, 300
William Morris, 30, 30, 350, 20, 250
John P. Conaway, 50, 20, 500, 30, 300
Benton Tindle, 50, 100, 1000, 35, 400
James Swam, 30, 40, 500, 30, -
Elzkiel Jack, 50, 30, 500, 40, 200
Horris Tindle 50, 50, 1000, 50, 250
Levin Conaway, 75, 75, 1000, 30, 450
Elizabeth Jefferis, 100, 150, 1500, 3, -
Miles Tindle, 125, 100, 2500, 40, 550
Marcus Hill, -, -, -, -, -
Elisha Dickeson, 50,75, 1000, 30, 300
James Spicer, 75, 125, 1000, 40, -
John Short, 100, 75, 2000, 40, 400
Jacob Carfreed, 40, -, 600, 30, 350
James Anderson, 300, 300, 1150, 20,500

Jesse Paswaters, 300, 400, 2000, 50, 500
David Reynolds, 50, 50, 600, 25, 175
Samuel Warren, 50, 50, 200, 40, 400
John Russell, 150, 750,800, 30,300
Thomas Dyar, 50, 60, 1000, 30, 100
Charles Macklin, 100, 100, 2000, 50, 400
Fisher Willis, 225, 175, 2000, 60, 550
William Sharp, 150, 160, 2570, 50, 450
Clement Sharp, 50, 5, 1000, 20, 200
Obediah Macklin, 50, 60, 503, 30, 300
Joseph Lindle (Tindle), 40, 30, 400, 10, 60
Isaac Corlisle, 100, 25, 600, 30,100
James Mosley (Moxley), 60, 40, 600, 25, 150
Elias Lofland, 40, 40, 500, 20, -
Bayard Sharp, 100, 75, 2000, 70, 400
Clement Cines, 50, 50, 600, 25, 200
Robert Botex (Boten), 40, 30, 400, 20,100
Isaac Web, 75, 75, 800, 30, 450
Samuel Clendance, 50, 35, 600, 20, 300
David Mullar, 40, 40, 400, 10, 100
Cyrus Macklin, 50, 50, 800, 30, 350
William H. Rockets, 145, 65, 800, 40, 350
Charles Rockets, 150, 200, 2000, 75, 600
Johsua Web, 75, 75, 1000, 40, 400
James Web, 75, 50, 100, 40, 30
Purnal Tatman, 200, 200, 2000, 60, 450
George Macklin, 50,70, 400, 20, 250
Lewis Slayton, 30, 35, 600, 10, 150
Thomas Borning (Boming), 100, 50, 800, 10, 300
Elizabeth Clifton, 100, 50, 100, 20,175
Joshua Clifton, 100, 125, 300, 100, 450
Leon Vernt (Verut), 75, 75, 1000, 25, 100
Amos Slayton, 60, 15, 300, 50, 300
Thomas Slayton, 200, 150, 3000, 75, 350
James Johnson, 100, 50, 600, 30, 300
John Price (Pride), 50, 10, 1000, 20, 200
Robert Wilson, 40, 30,700, 10, -
Smith Brother, 60, 60, 1000, 30, 200
George W. Drain, 30, 60, 2000, 20, 300
Zachariah Griffith, 20, 30, 500, 10, 100
Henry Conaway, 60, 60, 1000, 30, 175
Zacharian Griffith, 125, 125, 1500, 50, 300
Joseph Hatfield, 75, 75, 600, 30, 170
Charles Johnson, 75, 50, 1000, 25, 125
Albert Currey, 350, 250, 9000, 200, 500
Joel Coilesle, 250, 100, 4000, 50, 650
John W. Corlile, 10, 25, 811, 30, 100
James Satterfield, 15, 30, 1000, 30, 150
Edward Wiley, 200, 50, 3000, -, -
John Conaway, 100, 50, 1500, 25, 175
Mary Hunt, 100, 750, 800, 30, 300
John S. Rocket, 100, 75, 1000, 20, 200
George W. Collins, 200, 200, 4000, 30, 500
William Willey, 200, 100, 1500, 30, 400
Theodore Milley, 700, 100, 2000, 50, 300
Bartholomew Adams, 100, 150, 1500, 50, 250
David S. Lewis, 151, 100, 1000, 30, 100
Charles Smith, 126, 270, 2500, 80, 400

Mary Smith, 75, 75, 600, 10, -
Stanley Messick, 50, 60, 400, 30, 300
John Messick, 30, 30, 200, 20, 150
Robert Smith, 15, 125, 1500, 3, 300
John Lewis, 75, 75, 80, 3, 300
Robert McCauly, 30, 30, 300, 25, 250
Lemuel Smith, 100, 145, 2000, 30, 150
John _. Swain, 12, 150, 120, 10, 100
Isaac Willen, 100, 100, 1500, 40, 400
George Willey, 50, 50, 500, 30, -
Russel Dickeson, 30, 60, 600, 24,
Russel Dickeson, 60, 30, 300, 3, 300
George Conaway, 30, 40, 200, -, 300
Briant Parker, 47, -, 400, 30, 200
Eleazor B. Vaut, 30, 20, 400, 15, -
Thomas Wilkins, 50, 250, 1500, 30, 400
Joshua Sharp, 400, 300, 4700, 20, 500
Elisha Sharp, 200, 100, 600, 50, 350
Solomon Jester, 200, 100, 200, 30, 300
Clement Houston, 75, 150, -, -, -
Joseph B. Morris, 125, 75, 2500, 60, 275
Asa Johnson, 50, 15, 400, 5, 155
Kitty A. Spicer, 100, 100, 1500, 15, 100
Robert Warren, 150, 150, 3000, 50,300
Robert W. Jefferson, 60, 40, 1000,15, 100
Stephen Warrington, 75, 25, 500, 20,100
Lemuel Davidson, 200, 100, 1000, 50, 230
Robert Jefferson, 200, 50, 1000, 50, 153
William Jefferson, 75, 25, 800, 50,170
Joshua M. Pepper, 75, 75, 2000, 100, 321
Barth. Prettyman, 150, 50, 1000, 12, 50
Robert Prettyman, 150, 200, 2000, 30, 225
Sarah Rogers, 70, 10, 800, 12, 150
Philip Rogers, 75, 40, 800, 30, 245
Stephen Rogers, 100, 40, 600, 12, 130
John Rogers, 125, 300, 1500, 50, 395
Curtis Rogers, 50, 50, 700, 25, 100
Daniel Rogers, 25, 30, 700, 25, 80
George Scott, 50, 50, 1060, 150, 358
Cornelius Prettyman, 50, 20, 800, 25, 185
Purnal Short, 75, 120, 1000, 25, 125
Samuel LaKate, 60, 50, 500, 20, 90
Cyrus Q. Foocks, 100, 25, 800, 20, 120
Raimon Hoster, 15, 75, 200, 20,60
Asa Johnson, 21, 50, 500, 25, 112
Edward Short, 75, 125, 1800, 50, 335
Peter Short, 50, 70, 1000, 40, 235
Clayton Conaway, 100, 100, 1600, 40, 138
John H. Ellott, 70, 20, 400, 20, 75
Thomas W. Short, 60, 40, 800, 30, 125
Isaac W. Short, 25, 15, 500, 50, 90
Tolbert Mathews, 25, 50, 500, 20, 95
John Gunby, 75, 50, 500, 20, 160
Minos Johnson, 75, 125, 1000, 50, 200
William Derbath, 40, 25, 500, 20, 75
Eliza Spicer, 100, 75, 1500, 50, 260
Daniel Messick, 40, 25, 500, 20, 100
Jincey Jones, 100, 175, 1500, 40, 179
Sampson Mathews, 50, 25, 500, 20, 128
Winget Ellott, 75, 80, 1000, 20, 90
Curtis Stern, 50, 50, 600, 40, 272
Mancin Short, 75, 70, 1000, 50, 240
Hiram S. Short, 75, 20, 600, 50, 294
John W. Short, 60, 70, 1000, 30, 130

Lemuel Hood (Hovet), 30, 70, 200, 15, 60
Burton Meers, 50, 100, 600, 8, 93
Robert M. Johnson, 50, 30, 400, 12, 40
Anemius D. Johnson, 75, 100, 1500, 50, 225
John Carpenter, 40, 40, 500, 12, 20
Francis Phillips, 40, 30, 400, 12, 80
Lovenia Johnson, 100, 30, 1000, 21, 91
Benjamin Warrington, 40, 21, 500, 21, 75
Levin Pepper, 60, 20, 500, 20, 500
Winget Rogers, 15, 120, 300, 20, 500
Elizabeth Ennis, 70, 120, 1000, 29, 220
Walter Marvel, 45, 50, 600, 20, 210
Sarah Marvel, 41, 60, 500, 12, 90
Thomas Coffan, 125, 170, 1000, 20, 80
Aaron R. Marvel, 125, 75, 1000, 31, 150
John Marvel, 150, 50, 1000, 40, 160
Joseph Marvel, 75, 50, 1200, 40, 143
James Prettyman, 200, 90, 1200, 50, 200
Benjamin Christopher, 40, 48, 500, 25, 107
Samuel B. Davidson, 45, 85, 625, 20, 75
Noah Lockwood, 100, 150, 1000, 20,290
George Chamberlin, 30, 160, 500, 20, 125
__thaniel H. Rhodes, 145, 150, 1400, 50, 112
Samuel Gibbons, 8, 44, 400, 10, 400
Thomas West, 60, 40, 500 12, 70
William F. Toomey, 30, 70, 600, 12, 125
Henry Godfrey, 75, 75, 700, 20, 75
Elizabeth Short, 140, 60, 1200, 20, 80
Littleton H. Dukes, 50, 200, 800, 50, 108
Sarah Watson, 50, 50, 600, 15, 50
James Freeman, 60, 70, 600, 15, 50
John Chamberlin, 35, 35, 500, 60, 111
William Foskey, 75, 75, 1200,75, 383
Levin S. Hopkins, 35, 60, 900, 50, 128
Grace A. Moxley, 20, 70, 500, 12, 60
Samuel H. Layton, 65, 150, 1000, 30, 190
Manden Gum, 9, 20, 3000, 100, 554
John Mumford, 40, 400, 2000, 100, 175
Lemuel Hudson, 40, 50, 500, 15, 71
William Hickman, 25, 25, 500, 20, 100
John Gat, 40, 50, 500, 12, 69
Jonathan Casey, 50, 20, 1000, 30, 170
Edward McCale, 60, 100, 1000, 20, 95
Isaac Watson, 50, 60, 500, 18, 121
Isaac Timmons, 100, 200, 1500, 20, 160
Joshua Williams, 20, 25, 500, 21, 150
William H. Harting, 40, 125, 600, 18, 70
William Moore, 35, 15, 500, 19, 66
James Sampson, 100, 40, 1200, 12, 70
James Moore, 150, 200, 2000, 40, 271
Thomas R. Toomey, 20, 200, 1500, 12, 71
Isaac C. McCale, 40, 300, 1000, 20
Godfrey Brasun, 125, 100, 400, 20, 150
Joseph Davidson, 25, 100, 400, 20, 150
Perry Brasun, 25, 60, 300, 18, 180
Shepherd P. Moore, 25, 100, 400, 19, 91

Elijah Davidson, 21, 140, 300, 18, 84
Purnal Crampfield, 30, 60, 400, 20, 100
Eber Atkins, 28, 200,800, 19, 80
Burton Crampfield, 12, 600, 650, 18, 91
Haislet Thompson, 29, 200, 500, 19, 50
Handy Watson, 45, 300, 400, 21, 111
James Dukes, 60, 200, 1000, 30,185
William Stephens, 100, 100, 1000, 30, 274
Charles Stephens, 100, 100, 1000, 20, 112
Nehemiah, Coffan, 45, 100, 600, 19, 131
Robert B. Houston, 150, 150, 3000, 200, 861
Isaac H. Houston, 125, 150, 2500, 200, 285
Benjamin Joseph, 50, 60, 1000, 20,188
Joseph Kollock, 125, 450, 6000, 150, 442
John P. Hudson, 40, 30, 300, 20, 130
Wollard Thompson, 30, 30, 300, 20, 180
Peter Marvel, 100, 150, 1400, 31, 250
Solomon Willey, 75, 50, 500, 20, 85
Burton J. Winget, 75, 50, 600, 20, 160
Jacob Burton, 30, 40, 1100, 26, 275
James Moore, 20, 30, 300, 21, 121
John Andrew, 50, 40, 600, 20, 101
Henry Stephens, 30, 30, 400, 22, 120
William Jones, 100, 300, 2000, 50, 329
Mary A. Maples, 200, 405, 6000, 100, 429
Nehemiah H. Thomas, 60, 100, 1000, 30, 121
Daniel Timmons, 50, 100, 1000, 21, 119
Nathaniel Davidson, 50, 10, 400, 20, 600
Thomas Wapler, 75, 160, 1000, 20, 175
Levin M. Courmecan(?), 100, 200, 2000, 21, 78
Joseph Vickers, 50, 200, 1000, 25, 80
Daniel Steel, 40, 100, 600, 20, 90
Luther Courmecin (Cornmean?), 30, 60, 700, 21, 101
Elizabeth Steen, 60, 52, 600, 20, 80
Charles Bounds, 30, 60, 300, 25, 97
John Bivins, 50, 100, 600, 22, 700
Peter Engram, 30, 100, 700, 23, 221
Thomas Moss of R, 70, 30, 400, 20, 78
Robert Mumford, 50, 200, 1200, 20, 180
Samuel Mumford, 60, 75, 1000, 21, 156
James Walls, 41, 70, 400, 22, 161
Peter Tingle, 50, 40, 600, 12, 700
Benjamin Newton, 72, 74, 500, 25, 80
Thomas A. Meers, 70, 60, 1000, 20, 170
Robinson P. Meers, 60, 50, 800, 20, 90
Eliza Hopkins, 40, 60, 300, 18, 67
Charleze Phillips, 20, 25, 450, 18, 110
Robert Prettyman, 100, 123, 1700, 150, 39
Thomas Phillips, 100, 110, 1000, 50, 152
Levin Hopkins, 100, 180, 2000, 40, 2223
Joseph Phillips, 150, 135, 1500, 29, 400
Phillip Short, 50, 26, 500, 200, 100
William M. Lingo, 40, 60, 400, 12, 70
Woolsey Casey, 50, 60, 500, 15, 166
Levi Short, 50, 150, 1000, 60, 234
Benjamin Ellingsworth, 60, 60, 400, 21, 171
Jonas Webb, 20, 50, 300, 12, 51

Elizabeth Layton, 25, 25, 200, 12, 89
Michell Scott, 50, 51, 300, 18, 66
Thomas Phillips, 100, 200, 1500, 75, 371
Burton Phillips, 100, 100, 200, 50, 236
Elihu G. Phillips, 40, 100, 800, 20, 100
Samuel Warrington, 100, 20, 500, 20, 157
Biard Otwell, 300, 400, 3000, 310, 254
Ephraim Steen, 50, 74, 500, 15, 135
Renatus Warrington, 40, 60, 400, 12, 111
John C. Otwell, 100, 39, 1000, 65, 275
James Truitt, 80, 70, 1300, 50, 305
Thomas Warrington, -, 20, 400, 18, 66
Eli P. West, 50, 200, 1000, 25, 325
Samuel Betts, 50, 60, 500, 20, 225
William Long, 40, 50, 300, 12, 260
Kendal Bowden, 30, 39, 400, 15, 71
David Long, 20, 41, 300, 12, 70
Jonathan Hickman, 140, 100, 400, 16, 240
William Tingle, 20, 41, 300, 18, 80
Stephen Laws, 50, 70, 400, 21, 181
Samuel Mihell (Mitchell), 70, 160, 1200, 12, 75
William Wells, 60, 80, 500, 12, 81
Perry Lewis, 20, 40, 300, 11, 91
George Lewis, 30, 30, 600, 21, 110
Albert O. Baker, 30, 30, 500, 20, 96
Elisha Parker, 30, 20, 600, 21, 75
Johnathan Baker, 30, 31, 500, 200, 111
George Parker, 60, 40, 600, 15, 175
Thomas Brimer, 25, 75,600, 30, 81
Joshua Hudson, 50, 175, 1000, 31, 96
Payne R. Johnson, 75, 50, 800, 30, 150
Jehu (John), West, 150, 400, 1000, 25, 300
Peter D. Shockly, 200, 150, 2000, 75, 520
Lemuel Littleton, 20, 30, 250, 12, 91
Elijah Carey (Casey), 100, 100, 6000, 26, 450
Tilman S. Johnson, 20, 100, 1000, 20, 200
Samuel Betts, 35, 35, 600, 13, 71
William S. Carny (Caney), 40, 60, 500, 20, 137
Nathaniel Phillips, 50, 150, 1000, 31, 250
Jacob Hopkins, 25, 100, 500, 12, 111
Joshua Phillips, 150, 70, 1000,400, 718
John T. Casey (Carey), 30, 150, 600, 12, 6
Peter P. Marvel, 41, 100, 400, 11, 121
Winget Prettyman, 61, 80, 500, 12, 200
Johnathan Dickenson, 25, 56, 800, 25, 79
William G. Willey, 80, 20, 450, 20, 150
James Walls, 100, 100, 800, 20, 75
Jacob Phillips, 75, 25, 500, 12, 40
John Jones, 100, 56, 800, 50, 204
Eleanor Messick, 50, 60, 300, 12, 81
Charles LeKates(LeKites), 60, 80, 800, 19, 75
Alfred Hitchings, 75, 100, 700, 12, 100
Henry Lawson, 100, 100, 600, 16, 87
Isaac Akins, 60, 70, 500, 18, 100
Samuel Warren, 30, 4, 300, 12, 150
Nathaniel Engram, 30, 60, 800, 26, 460
James Dickerson, 70, 38, 400, 30, 86
John H. Messick, 50, 60, 500, 20, 112
Jesse Green, 150, 150, 2000, 50, 350
John Sh__don, 50, 61, 400, 12, 81
George McGee, 41, 50, 800, 25, 300
John Martin, 50, 30, 800, 25, 300
Charles Engram, 40, 30, 600, 18, 81

James J. Webb, 35, 40, 300, 12, 91
Winget Johnson, 50, 50, 1100, 15, 87
Nelly Massey, 31, 20, 300, 12, 61
George Carpenter, 60, 40, 500, 21, 300
William N. Thoroughgood, 61, 3, 350, 30, 281
Minas Messick, 71, 44, 800, 21, 161
Eliza Harmon, 60, 200, 200, 22, 194
Michell Rickard, 30, 21, 600, 18, 186
Joseph K. Cannon, 40, 50, 1500, 75, 400
Homelite Kirkpatrick, 30, 60, 600, 21, 81
Nathaniel Robinson, 70, 135, 1200, 30, 206
William O. Short, 40, 200, 1000, 21, 311
John Brasure, 95, 50, 1500, 12, 81
John M. Buntin, 100, 150, 2000, 30, 273
Dingle Hudson, 75, 30, 1200, 12, 61
Philip Short, 112, 130, 2000, 175, 359
Parker Duke, 48, 100, 1500, 25, 201
Clayton M. Hudson, 175, 200, 1200, 20, 275
William M. Atkins, 60, 71, 600, 21, 89
Benjamin Phillips, 75, 71, 400, 22, 160
Sampson Gunnell, 31, 60, 350, 8, 61
Woolsey Burton, 60, 30, 350, 75, 350
Philis Short, 150, 190, 1600, 40, 261
John C. Hazzard, 60, 100, 800, 23, 40
Peter Terence, 41, 81, 400, 12, 71
David Marvel, 80, 120, 1000, 20, 163
John Tingle, 50, 60, 500, 12, 86
Richard F. Hastings, 40, 26, 1200, -, 50
William E. Burton, 70, 100, 3000, 125, 575
Elijah McDowell, 50, 60, 500, 12, 70
Thomas Wapler, 40, 60, 250, 10, 78
Benjamin D. Burton, 125, 125, 3000, 120, 350
Gardiner H. Wright, 70, 150, 1500, 120, 415
Jacob Crampfield, 35, 40, 1500, 12, 60
Job Carpenter, 50, 50, 600, 15, 121
Brinkloe Johnson, 80, 100, 1000, 30, 105
Joseph Carpenter, 40, 50, 500, 15, 75
Eli Norwood, 75, 50, 800, 15, 150
Lemuel Hopkins, 20, 30, 500, 12, 80
Woolsey H. Palmer, 60, 50, 700, 20, 60
Robert Morris, 250, 350, 2500, 50, 420
Nancy D. Beachem, 100, 50, 800, 12, 113
Major H. Atkins, 75, 30, 700, 15, 80
Zippera Morris, 150, 150, 1800, 25, 250
William Spice, 65, 50, 800, 25, 187
Philip Marvel, 150, 150, 1000, 100, 315
Robert Plummer, 50, 100, 500, 15, 100
Thomas Johnson, 150, 100, 1000, 20, 150
Job Marvel, 100, 40, 500, 50, 78
William Morris, 50, 95, 800, 20, 225
Stephen Morris, 100, 200, 1500, 20, 270
Henry Hickman, 40, 40, 1400, 100, 415
William Brasure, 60, 80, 1400, 75, 295
Mary S. Dingle, 100, 125, 3000, 50, 65
Benjamin Gray, 30, 40, 1500, 50, 234
Peter Johnson, 35, 65, 1000, 25, 121
Edward Christopher, 45, 10,700, 25, 132
George Howard, 50, 20, 650, 30, 162

Martha Truitt, 40, 160, 2000, 50, 150
Henry B. Johnson, 50, 55, 700, 25, 160
Perry Pool, 100, 100, 2000, 50, 297
Samuel Hill, 60, 100, 700, 30, 108
Mary Aydelott, 40, 5, 400, 30, 80
Eber Jacob, 40, 60, 500, 20, 75
Ann Thomas, 45, 10, 500, 20,75
Sally M. Tracy, 50, 27, 500, 20, 72
Edward Townsend, 40, 10, 200, 20, 89
William B. Toomey, 80, 150,1 200, 50, 62
Lemuel West, 100, 50, 1200, 30, 126
John Bull, 100, 30, 800, 30, 130
John Banks, 50, 40, 400, 35, 50
Philip W. Helm, 80, 50, 1000, 50, 181
Aron C. Townsend, 75, 40, 1000, 30, 181
John Hutson, 41, 60, 600, 31, 150
Joshua C. West, 100, 150, 2000, 60, 196
Nathaniel Clark, 25, 100, 500, 30, 35
Lemuel Vickers, 50, 50, 500, 20, 31
Robert Derickson, 50, 225, 3000, 25, 75
Ruben West, 50, 22, 600, 20, 157
Parker Michols, 40, 50, 800, 50, 165
Azariah Atkins, 50, 50, 500, 25, 102
James Turner, 50, 50, 700, 30, 95
Nathaniel Tunnell, 110, 240, 2500, 100, 300
James M. Tunnell, 100, 100, 2000, 500, 188
James R. Townsend, 30, 70, 700, 50, 118
Charles Long, 50, 20, 800, 20, 95
Major Townsend, 35, 60, 500, 30, 86
Burton H. Johnson, 25, 25, 300, 10, 50
John Hughs, 25, 100, 600, 10, 80
George P. Johnson, 65, 40, 1200, 50, 114
John McGee, 100, 70, 1200, 15, 75
Peter Johnson, 30, 10, 500, 15, 79
Eben Walter, 200, 100, 3000, 200, 450
Stephen Collins, 40, 67, 1000, 40, 241
Isaac Johnson, 30, 100, 500, 20, 40
Jame Townzend (Townsend), 30, 120, 500, 12, 84
John Calhoun, 125, 215, 2500, 50, 417
Ephraim Calhoun, 50, 75, 500, 30, 100
Peter Banks, 50, 50, 400, 20, 82
Jacob Banks, 40, 90, 950, 25, 90
Jedediah D. Evans, 25, 50, 2000, 100, 240
John Townzend, 40, 60, 800, 40, 50
Elijah Howard, 100, 200, 2000, 50, 65
Peter Townzend, 18, 80, 800, 75, 159
William S. Hall, 30, 150, 1500, 125, 330
Zadock Hill, 50, 50, 400, 20, 76
William H. Wharton, 30, 20, 400, 20, 60
William Holland, 150, 50, 2000, 30, 259
James B. Derickson, 100, 50, 1500, 50, 330
Joseph D. Allen, 25, 60, 700, 50, 230
Michell Hudson, 60, -, 700, 30, 160
Sylvester Carey (Casey), 70, -, 400, 25, 80
William Wharton, 35, 70, 800, 30, 210
Nancy Williams, 75, 80, 1500, 25, 181
Isaac Evans, 50, 45, 500, 25, 197
James Law, 20, 60, 1000, 20, 85
Robert Quillen, 80, 20, 500, 30, 188
John M. Rickard, 50, 50, 1000, 20, 138
Lemuel W. Rickard, 50, 60, 1500, 30, 115
William E. Hall, 35, -, 350, 10, 50
Smiley Calhoun, 50, 30, 500, 20, 80

Isabell Jones, 75, 50, 2500, 50, 218
Elizabeth Hall, 75, 125, 2000, 40, 85
John Burton, 100, 50, 2000, 50, 170
Elisha Evans, 50, 50, 1500, 75, 298
William R. Hall, 20, 100, 200, 10, 40
David Hall, 150, 200, 2000, 50, 178
Stephen W. Evans, 6, 8, 200, 20, 106
David W. Evans, 8, 8, 200, 20, 74
William S. Evans, 100, 100, 1000, 20, 128
Thomas Derickson, 40, 30, 600, 20, 105
Charles Evans, 175, 150, 2000, 20, 100
Lemuel B. Hudson, 50, 75, 600, 20, 60
Ezekiel Evans, 25, 125, 1000, 50, 128
William H. Taylor, 50, 65, 600, 50, 76
John Tunnell, 80, 125, 2500, 100, 425
Peter Quillen, 75, 225, 2500, 20, 200
Thomas Dazey, 30, 75, 400, 20, 151
Henry Tunnell, 115, 75, 1000, 50, 146
Alfred S. Evans, 15, 5, 200, 20, 95
Henry Lathburg (Lathbery), 20, 100, 500, 20, 70
Henry Burbage, 50, 7, 500, 20, 50
Siles A. Rickards, 5, 15, 600, 30, 100
Comfort Hudson, 20, 80, 500, 12, 40
Thomas Dazey, 60, 60, 2000, 300, 224
Henry Hudson, 40, 40, 1500, 100, 300
Isaac R. Wharton, 10, 60, 700, 50, 107
Peter Carey, 30, 30, 600, 75, 230
Upshire Wootten, 50, 50, 1200, 25, 175
Prettyman M. Dazey, 50, 50, 1000, 60, 133
Isaac Hall, 30, 40, 500, 15, 60
John M. Betts, 75, 225, 3000, 125, 306
John Steel, 100, 200, 2000, 150, 375
Philip H. Wharton, 15, 100, 1000, 20, 54
James Purman, 20, 60, 500, 20, 175
James F. Hoalt, 18, 20, 400, 30, 122
Thomas H. Steel, 20, 25, 200, 12, 50
James A. Hickman, 12, 38, 400, 21, 60
Henry E. Hill, 40, 40, 150, 50, 228
William W. Evans, 20, 12, 300, 12, 95
William M. Hudson, 10, 10, 200, 12, 35
Lemuel A. Hall Jr., 35, 135, 2000, 35, 172
James M. Knox, 8, 8, 300, 20, 118
Caleb W. Murray, 40, 60, 500, 20, 90
Thomas S. Simple, 30, 60, 500, 20, 70
Cornelius Carey, 30, 70, 500, 20, 140
David Godwin, 50, 50, 2000, 50, 324
Lemuel A. Hall, 100, 200, 3000, 100, 525
Isaac C. West, 30, 30, 1000, 30, 120
John C. Hall, 20, 10, 1000, 50, 265
John M. Taylor, 15, 35, 400, 24, 36
John P. Furman, 15, 15, 300, 10, 25
Job W. H. Derickson, 50, 140, 2000, 100, 179
Joshua Fossett, 60, 60, 400, 25, 60
Jenkins Miller, 60, 50, 800, 20, 160
James W. Rickard, 120, 30, 1500, 200, 323
William Long, 75, 90, 1000, 30, 160
Ephraim J. Rust, 70, 100, 500, 20, 75
Thomas Evans, 100, 300, 2000, 150, 311
Levin Andrew, 50, 100, 500, 20, 40
George Welbenson, 30, 150, 500, 20, 140
Burton Tier, 50, 50, 500, 20, 100
Robert Otey, 50, 250, 550, 20, 75
George Rogers, 60, 160, 600, 20, 95
William Crapper, 40, 40, 600, 20, 95

James Davis, 20, 15, 220, 20, 75
Mary Williams, 50, 50, 800, 25, 65
John Dazey, 40, 130, 1500, 30, 210
Francis R. Tingle, 50, 70, 700, 50, 179
Mills H. Layton, 20, 30, 450, 25, 123
William H. Godwin, 20, 60, 400, 20, 60
George T. Taylor, 60, 40, 400, 20, 93
Ephraim H. Wilgus, 50, 40, 500, 20, 45
James Halaway, 50, 30, 300, 12, 40
Isaac Rickard of E, 15, 60, 600, 20, 63
George S. Williams, 50, 100, 1500, 50, 172
Mary Davis, 50, 60, 500, 6, 46
Joshua Riley, 100, 100, 1000, 30, 100
John Taylor, 60, 40, 700, 30, 75
Sampson Smallwood, 30, 130, 1000, 30, 190
John Melvin, 75, 100, 1000, 20, 60
James Brasure, 50, 50, 100, 20, 87
Jack W. Williams, 150, 125, 1500, 40, 296
Ebenezer F. Williams, 50, 125, 1000, 20, 107
Ananias D. Derickson, 37, 43, 700, 50, 165
Laben Rogers, 150, 100, 900, 40, 212
Jacob H. Brasure, 50, 20, 500, 30, 222
John Macullum, 100, 75, 200, 20, 150
Collen Truitt, 15, 35, 200, 12, 60
Jacob Rogers, 50, 30, 400, 12, 30
Charles Hall, 50, 150, 1000, 20, 100
Peter West, 40, 60, 500, 15, 120
Thomas Atkins, 30, 30, 400, 20, 55
Eber Calloway, 30, 60, 400, 20, 50
Ezekiel C. Williams, 25, 25, 500, 50, 145
Young Brasure, 30, 30, 500, 20, 200
Burton H. Brasure, 50, 70, 800, 50, 140
James Bishop, 200, 50, 2000, 30, 378
Aaron Callaway, 90, 20, 700, 20, 125
Samuel M. Hudson, 40, 70, 1500, 20, 121
Della Hudson, 40, 40, 400, 20, 78
Levin Holdston, 50, 60, 400, 12, 40
Joshua Rogers, 15, 25, 200, 12, 65
Caleb Lockwood, 30, 31, 300, 12, 40
Anda E. Rickards, 50, 61, 800, 20, 42
Jeremiah Hudson, 40,70, 300, 20, 50
James A. Carey, 30, 20, 200, 20, 85
Elizabeth Evans, 20, 30, 400, 21, 60
Zadock Townsend, 100, 30, 1000, 25, 280
Benjamin Derickson, 25, 65, 1000, 20, 92
Daniel _. Walker, 30, 40, 800, 20, 92
James Chincks, 30, 230, 1000, 40, 90
John Watson, 30, 40, 400, 40, 90
Edward Dorey, 30, 125, 1000, 25, 85
Kendal Rickard, 50, 6, 1000, 50, 140
Stephen R. Evans, 30, 30, 800, 50, 332
James Derickson, 35, 40, 1000, 25, 197
Elijah Lynch, 40, 10, 800, 50, 232
Stephen Rickards, 20, 20, 800, 20, 169
Elisha R. Rion, 25, 50, 800, 20, 100
Thomas D. Carey, 30, 20, 500, 20, 50
Eber A. Dazey, 35, 21, 500, 20, 60
Jeremiah Hudson, 30, 20, 1000, 20, 152
Benjamin Hudson, 40, 30, 1500, 50, 261
James Rickards, 40, 50, 1000, 5, 75
George T. Derickson, 15, 25, 600, 20, 60
William Layton, 30, 15, 500, 20, 120

Richard Hickman, 50, 12, 1500, 30, 245
Lemuel Lynch, 40, 80, 1300, 30, 213
Jacob Evans, 25, 25, 800, 20, 139
William T. Evans, 35, 15, 600, 12, 100
Isaiah Ellis, 30, 40, 1000, 30, 142
William Rickards, 20, 40, 800, 20, 98
John Evans, 60, 40, 800, 20, 50
James H. Bishop, 30, 110, 1000, 12, 65
Josiah Mumford, 15, 53, 800, 20, 914
Gilbert Lynch, 15, 100, 800, 12, 38
Stephen Collens, 15, 100, 800, 12, 50
Joshia Lynch, 25, 50, 800, 20, 100
Ezekiel Lynch, 50, 40,700, 20, 117
Clemment Lofland, 30, 30, 500, 12, 38
Ruben Lynch, 40, 10, 600, 12, 153
Nathan Williams, 45, 60, 1000, 40, 140
John B. Furman, 75, 125, 1000, 100, 350
Joshua Halloway, 75, 100, 1000, 12, 75
Isaac Carey, 20, 120, 800, 20, 111
Handy F. Timmons, 30, 70, 1500, 12, 146
Elijah Warrington, 50, 250, 1000, 50,2 03
Michell Morris, 35, 65, 1000, 20, 120
Merril Buntin, 60, 100, 1500, 50, 179
Elijah Buntin, 20, 80, 1000, 12, 55
Eliza McCabe, 30, 70, 500, 20, 111
Sampson Selby, 20, 30, 1000, 60, 184
Garrison McCabe, 100, 150, 2500, 60, 231
John McCabe, 50, 40, 1500, 12, 177
James Campbell, 20, 40, 400, 12, 91
Eber Layton, 20, 25, 500, 12, 81
Robert Layton, 30, 20, 600, 20, 130
Mary Murray, 30, 20, 600, 20, 111
Joseph Morris, 30, 5, 500, 12, 79
Joseph McCabe, 50, 100, 1000, 40, 171
Garrison McCabe, 20, 4, 2000, 20, 60
William McCabe, 50, 60, 500, 12, 60
Stephen Murray, 40, 40, 500, 20, 120
Laben H. Murray, 30, 30, 500, 20, 120
Levin Morris, 11, -, 500, 20, 800
James Hickman, 50, 50, 1500, 20, 80
Selby Hickman, 20, 20, 500, 20, 120
Littleton Hickman, 50, 50, 1500, 20, 80
William P. Hickman, 35, 20, 1000, 20, 150
Aaron Lynch, 30, 15, 700, 20, 75
Isaiah Lynch, 50, 25, 800, 20, 80
Gilbert Lynch, 25, 25, 500, 21, 111
John Godwin, 40, 24, 1500, 60, 50
Nathaniel W. Evans, 50, 24, 1500, 60, 268
James Anderson, 10, 37, 600, 20, 312
Jacob A. Wilgus, 40, 45, 1500, 50, 303
Samuel D. Bennett, 40, 100, 1000, 50, 329
Burton R. Tubbs, 30, 30, 900, 20, 100
Isaac Rickards, 40, 40, 600, 20, 150
Mary Morris, 40, 40, 600, 20, 111
Levin Derickson, 40, 70, 1200, 20, 280
John Bennett, 100, 400, 2000, 50, 772
John B. Derickson, 60, 100, 1500, 40, 212
James Bennett, 60, 50, 2000, 30, 140
Charles Bennett, 40, 30, 800, 20, 75
James M. Dale, 50, 500, 1000, 20, 250
Josiah Bull, 14, 7, 300, 20, 60

William Macormick, 100, 200, 3000, 25, 80
Thomas Mccormick (Macomerick), 50, 60, 1200, 50, 190
Peter W. Halloway, 25, 30, 1000, 35, 106
Thomas R. Moore, 25, 50, 600, 25, 140
David Moore, 50, 50, 1000, 100, 332
James Jacobs, 8, 90, 1000, 20, 95
Henry W. Long, 112, 100, 1500,125, 415
Isaac McCabe, 150, 157, 3000, 150, 429
John Davidson, 75, 125, 2000, 10, 40
Sydenham Long, 50, -, 200, 20, 121
Elijah Brunay, 30, 90, 1000, 12, 158
Baretine T. McCabe, 50, 250, 1200, 12, 75
Joshua Hudson, 40, 50, 200, 12, 60
Milby Buntin, 50, 100, 500, 21, 75
Wilson Campbell, 60, 120, 1000, 21, 96
Robert Hudson, 40, 120, 1500, 18, 81
Seth Long, 75, 200, 1000, 20, 75
George Warden, 25, 120, 700, 21, 71
Simpson Campbell, 21, 150, 860, 18, 111
Benjamin Long, 75, 45, 1000, 60, 250
Lambert Campbell, 75, 200, 2500, 50, 135
Edward McCabe, 40, 80, 1000, 30, 195
Warnel Dukes, 75, 100, 1200, 50, 255
Eshher Timmons, 80, 200, 2000, 20, 170
Isaiah Hudson, 50, 50, 500, 20, 210
Joseph Long, 70, 35, 1000, 20, 50
Lot Murray, 100, 26, 1500, 20, 180
Elijah Carey, 50, 60, 1500, 100, 125
James Brumby, 60, 70, 700, 20, 60
Seth Hudson, 30, 40, 600, 20, 70
Aaron Halloway, 100, 360, 2000, 50, 215
Lacher Murray, 150, 101, 1300, 30, 168
George Murray, 35, 15, 500, 20, 152
Nancy Murray, 20, 40, 300, 20, 50
Milburn Murray, 30, 20, 500, 20, 190
Robert Campbell, 40, 10, 500, 20, 111
Thomas Stephens, 30, 20, 400, 21, 75
Ze___ Long, 25, 25, 400, 21, 78
Isaiah Lockwood, 25, 30, 500, 21, 111
Elisha Long, 50, 40, 500, 12, 81
Lemuel Hudson, 60, 20, 700, 20, 367
Levin Hudson, 60, 40,700, 30, 70
William T. Clark, 30, 48, 1000, 20, 80
John McGee, 25, 25, 500, 21, 95
Thomas T. Rogers, 50, 50, 1000, 21, 125
Isaiah Long, 20, 60, 1500, 20, 190
Joseph Harris, 30, 67, 1800, 20, 146
William Rogers, 12, 21, 200, 12, 55
Jacob Hickman, 20, 110, 1200, 20,150
David Hudson, 75, 25, 400, 21, 161
Eli Davis, 65, 340, 10000, 55, 237
Ephraim M. McNeal, 75, 25, 3000, 50, 231
James W. Timmons, 37, 38, 1000, 21, 80
George R. Halloway, 60, 33, 800, 50, 199
Thomas Halloway, 25, 25, 250, 18, 75
David Murray, 40, 20, 500, 25, 228
Miers Moore, 30, 40, 500, 21, 81
Peter Blizzard, 30, 40, 500, 21, 50
James Williams, 20, 30, 400, 21, 121
Rufus Murray, 25, 50, 400, 21, 150
James Lynch, 50, 55, 1500, 20, 145
Caleb W. Murray, 14, 36, 80, 30, 178

William Roberts, 40, 100, 1800, 50, 262
Lydia F. Banks, 20, 30, 500, 12, 60
Joseph Lynch, 20, 25, 500, 25, 120
John Hickman, 18, 27, 500, 20, 65
William Howard, 40, 4, 700, 29, 100
Eber D. Gray, 33, 92, 1200, 20,195
David Hudson, 50, 150, 1200, 30, 151
John Tingle, 40, 60, 1500, 21, 125
Eber Jones, 10, 60, 1000, 12, 60
Robert Piper, 12, 60, 400, 15, 83
John C. Lynch, 30, 45, 1800, 40, 195
Levi Lynch, 40, 32, 1200, 40,70
Henry Evans, 25, 25, 900, 25, 75
Johnson Gray, 75, 71, 1000, 21, 100
William Truitt, 40, 50, 1000, 20, 121
Clemment Evans, 25, 25, 500, 30, 115
James Andrew, 25, 30, 1000, 25, 70
William Derickson, 40, 40, 1000, 20, 70
John Derickson, 80, 30, 1000, 21, 81
Nathaniel Derickson, 40, 50, 1000, 150, 390
Benjamin Hudson, 50, 150, 1500, 75, 104
William Simple, 50, -, 200, 12, 150
William Barker, 50, 100, 800, 12, 100
William Lynch, 100, 200, 1000, 20, 125
Jesse P. Joseph, 20, 30, 300, 14, 50
Philip Green, 60, 20, 400, 15, 150
Henry C. Hood, 150, 50, 2000, 20, 208
John Lynch, 80, 50, 600, 20, 1000
George W. Green, 100, 30, 700, 20, 155
Priscilla Baker, 100, 120, 2000, 40, 327
Thomas M. Burton, 55, 50, 1000, 15, 95
Jacob Wilson, 175, 30, 1600, 30, 308
Theopholus Street, 150, 100, 1200, 20, 150
Maria Burton, 150, 50, 1600, 21, 115
William T. Burton, 250, -, 1500, 75, 120
John R. Burton, 350, 30, 2000, 125, 844
Nathaniel Commeen(?), 200, -, 1200, 25, 50
Clemment Baylis, 25, 50, 500, 10, 90
Winget Street, 107, 15, 500, 25, 179
Daniel Lingo, 150, 300, 2900, 50, 336
Francis A. Burton, 150, 25, 1500, 20, 125
John P. Salmons, 150, 350, 1500, 20, 214
Lewis P. Reynolds, 40, 10, 400, 15, 70
Samuel C. Collins, 70, 30, 800, 40, 195
Peter E. Gosley, 50,60, 400, 24, 140
Levin Sockum, 60, 40, 1000, 50, 312
Peter K. Burton, 175, 120, 5000, 150, 720
James T. Baylis, 170, 130, 3000, 75, 500
Edward McCray, 25, 115, 500, 25, 70
John M. Wine, 15, -, 750, 40, 154
William W. Gosley, 80, 20, 500, 25, 70
Jacob F. Hurdel, 40, 30, 600, 28, 100
Lemuel B. Lingo, 50, 60, 400, 23, 75
Mary Simple, 100, 30, 1500, 20, 140
John P. Burton, 22, -, 1000, 50, 210
William Johnson, 80, 100, 1500, 18, 170
Garrison Harmon, 40, 40, 1000, 15, 75
William L. Morris, 150, 150, 1800, 25, 226
Plymouth Bounds, 50, 100, 500, 15, 105
Paynter Jefferson, 100, 200, 300, 30, 76
William D. Wapler, 130, 100, 2815, 60, 228

Isaac Wapler, 50, 80, 600, 18, 171
Arcada Burton, 150, 70, 2000, 200, 518
Thomas Robinson, 200, 200, 2027, 50, 342
James A. Atkins, 150, 100, 1500, 50, 295
James Carpenter, 70, 106, 400, 12, 131
William N. Thoroughgood, 100, 100, 1000, 60, 190
Noble Johnson, 60, 70, 700, 12, 70
Elizabeth Burton, 90, 48, 800, 15, 86
Ann Johnson, 46, 100, 400, 12, 75
Mary N. Barker, 60, 140, 1000, 12, 70
John B. Vickers, 70, 100, 1000, 21, 170
Robert Bercher, 40, 70, 500, 12, 86
George M. Carey, 60, 100, 700, 18, 111
John Woolford, 80, 15, 600, 15, 126
William J. Lingo, 70, 40, 700, 30, 123
Joseph R. Barker, 100, 100, 1500, 100, 284
William Hopkins, 150, 50, 1000, 16, 160
Philip R. West, 100, 60, 950, 20, 90
John M. Collins, 150, 180, 2000, 50, 191
Burton Johnson, 100, 115, 800, 21, 120
Shepherd Hanger, 90, 121, 600, 18, 71
Cornelius Wapler, 100, 170, 1500, 50, 324
John B. Thoroughgood, 90, 110, 1000, 46, 125
Elizabeth Weills, 70, 60, 500, 12, 60
Elizabeth Prettyman, 100, 50, 1000, 18, 91
Nathaniel W. Burton, 139, 100, 2000, 100, 309
Thomas Walls, 100, 300, 2000, 30, 150
William C. Joseph, 100, 300, 2000, 50, 130
John L. Crague, 80, 100, 600, 12, 191
Winget Pride, 41, 50, 500, 16, 126
George W. Joseph, 35, 20, 400, 20, 71
John D. Martin, 100, 100, 600, 20, 111
Johnathan Joseph, 25, 26, 300, 12, 81
Johnathan J. Wilson, 90, 100, 1000 20, 226
Silas Smith, 70, 50, 500, 15, 100
Lemuel W. Johnson, 80, 90, 500, 20, 90
Henry R. Johnson, 40, 20, 300, 12, 81
William M. Johnson, 25, 15, 300, 13, 75
Truitt Pettyjohn, 75, 125, 800, 31, 171
Noah Joseph, 40, 31, 400, 18, 81
Asarel Johnson, 31, 21, 200, 16, 100
William M. Johnson, 70, 30, 500, 30, 150
Joseph J. Ennis, 50, 20, 400, 20, 80
Thomas McGee, 70, 30, 600, 21, 160
John S. Solomons, 100, 80,1000, 31, 171
John McGee, 100, 60, 400, 10, 90
Joseph C. Hareld, 100, 41, 600, 21, 100
John S. Burton, 100, 50, 500, 20, 186
James F. Warrington, 20, 10, 300, 20, 82
Elisha Lingo, 125, 80, 1000, 50, 184
Robert Warrington, 100, 54, 1000, 30,100
Jesse Lingo, 100, 140, 1200, 30, 80
Paynter Lingo, 100, 60, 1000, 21, 212
Purnal Johnson, 20, 5, 350, 12, 96
Isaac Miller, 31, 20, 200, 13, 71
William W. Dorman, 132, 100, 1500, 25, 135

Thomas P. Collens, 75, 295, 1500, 20,104
Robert Waples, 200, 50, 2000, 100, 500
Hester W. Pool, 100, 80, 900, 12, 60
William Walls, 60, 30, 400, 13, 112
Robert Lawson, 125, 125, 1060, 25, 262
David H. Joseph, 60, 30, 400, 12, 70
Jemima Hangor, 40, 30, 300, 12, 61
Eli Hangor, 40 29, 400, 13, 81
Henry Lingo, 200, 149, 200, 100, 240
Alfred McIlvain, 200, 400, 4000, 75, 287
William T. Warrington, 130, 130, 3000, 100, 256
John Prettyman, 198, 112, 1400, 80, 207
John _. Burton, 150, 10, 900, 50, 129
James D. Warrington, 150, 70, 100, 50, 170
David H. Simpler, 163, 80, 1500, 75, 224
John Green, 26, -, 400, 20, 155
Thomas Rust, 70, 60, 1000, 40, 277
Moses McGee, 60, 40, 800, 25, 269
Peter R. Semples, 80, 27, 800, 12, 137
James Semples, 60, 27, 800, 12, 137
Elizabeth J. Frame, 200, 180, 3000, 100, 400
Joseph Casey (Carey), 60, 20, 500, 12, 129
Woolsey B. Casey, 80, 29, 500, 20, 155
Arcada Semples, 80, 15, 500, 12, 60
William B. Rust, 110, 15, 500, 12, 93
Nathaniel M. Johnson, 20, 52, 500, 12, 70
Josiah Semples, 80, 16, 500, 10, 196
Gideon Walls, 70, 30, 600, 25, 140
George Lewis, 50, 50, 400, 12, 80
Renatus T. Walls, 40, 20, 500, 20, 100
Gideon Blizzard, 100, 40, 1400, 18k 90
Stephen Blizzard, 70, 70, 600, 20, 20
Lewis Blizzard, 60, 10, 900, 210, 140
John Walls, 45, 20, 1200, 50, 225
Elsee Palmer, 100, 300, 2000, 10, 80
Moses Stockley, 60, 40, 500, 10, 65
Burton C. Prettyman, 75, 47, 1000, 60, 300
Thomas H. Joseph, 100, 140, 1300, 50, 308
Elisha Joseph, 90, 15, 320, 25, 55
William W. Hendel, 100, 60, 1000, 50, 162
William Harmon, 70, 40, 500, 10, 45
David M. Prettyman, 70, 70, 150, 75, 257
John Housman, 75, 25, 400, 6, 45
Handy McIlvain, 100, 50, 500, 12, 33
James Davidson, 100, 180, 2000, 50, 209
Robert Crague, 80, 100, 1000, 50, 221
Moses McGee, 150, 50, 600, 12, 150
Henry D. Joseph, 150, 80, 1000, 20, 205
John D. Johnson, 100, 50, 800, 12, 52
Samuel Davidson, 60, 50, 700, 50, 190
Elizabeth Hill, 45, 10, 300, 12, 80
Isaac A. Hart, 50, 50, 600, 20, 140
Peter A. Rust, 50, 60, 400, 15, 60
John W. McGee, 220, 160, 2000, 100, 317
Mary Coffan, 50, 50, 500, 207, 80
Burton Johnson, 100, 100, 1000, 100, 200
Robert Mariner, 50, 50, 500, 12, 90
Mannen B. Marvel, 130, 100, 1300, 30, 193
James F. Burton, 65, 6, 1000, 40, 217

Robert Clark, 150, 150, 2000, 25, 150
Whettleta Johnson, 100, 50, 1000, 30, 191
William Prettyman, 60, 7, 400, 20, 145
Thomas R. Hudson, 200, 60, 1500, 25, 100
Silas M. Warrington, 60, 30, 1000, 50, 275
Jane Hunter, 150, 150, 2000, 100, 405
Benjamin S. Warrington, 250, 150, 4000, 100, 435
John M. Perry, 150, 41, 1000, 25, 180
Bond Harmon, 45, 40, 600, 20, 150
Lemuel Johnson, 80, 17, 1000, 25, 130
Fletcher Lacy, 150, 150, 1600, 50, 216
Cudgo Buntin, 50, -, 400, 25, 100
James M. Fisher, 150, 170, 1900, 307, 70
William J. Wilson, 115, 30, 3100, 40, 275
William Pullin, 100, 90, 600, 25, 60
Henry Braxor (Brewer), 100, 50, 600, 12, 50
Peter W. Burtin, 50, 10, 400, 20, 100
Eli Walls, 100, 50, 1000, 50, 260
Thomas Hood, 47, -, 500, 50, 212
Lydia E. Burton, 50, 25, 400, 20, 120
Warrington O. Wilson, 100, 100, 70, 25, 125
Mary S. Robinson, 70, 30, 700, 25, 175
William F. Wessels, 184, 100, 1200, 75, 335
Bernice B. Wilson, 80, 20, 400, 25, 50
Lemuel B. Burtin, 100, 100, 1500, 100, 300
Harmum Lingo, 100, 100, 200, 50, 170
Elisha Lynch, 50, 50, 500, 12, 80
Jesse B. Stephenson, 90, 10, 500, 20, 100
Wesley W. Stephenson, 130, -, 500, 30, 150
Robert Prettyman, 80, 20, 500, 25, 175
James Pretyman, 40, 60, 800, 50, 150
Mitchel Johnson, 100, 95, 800, 40, 125
John W. Palmer, 50, 50, 500, 15, 100
John Lingo, 100, 100, 2000, 75, 332
Robert P. Stephenson, 150, 250, 2000, 70, 180
Hesekiah Joseph, 100, 273, 1700, 20, 100
Gideon J. Simples, 100, 260, 1200, 30, 80
John Massey, 30, -, 200, 20, 120
James H. B__reta, 60, 70, 800, 50, 300
Henry Robinson, 60, 30, 600, 20, 55
Henry R. Palmer, 100, 20, 500, 25, 60
William E. Haygood, 100, 25, 1000, 25, 225
Nathan H. Joseph, 100, 150, 1000, 50, 100
Jeremiah Harris, 50, 20, 400, 15, 90
Peter Dickerson, 50, 25, 300, 25, 80
John E. Haygood, 100, 17, 500, 25, 134
James Burton, 80, 100, 800, 40, 125
James P. W. Marsh, 150, 150, 1500, 50, 200
Solomon Baylis, 200, 140, 1500, 50, 191
John M. Webb, 100, 30, 1000, 50, 235
Mary Lawson, 100, 20, 800, 25, 112
George Robinson, 130, 100, 1500, 26, 304
John A. Marsh, 200, 200, 1600, 30,158
Abraham Drain, 200, 100, 1000, 20, 140

Brettenham Reynolds, 30, 10, 350, 30, 50
Robert Johnson, 100, 40, 800, 20, 165
Polly Burton, 95, 5, 800, 40, 364
John Hazzard (Haygood), 75, 25, 400, 30, 160
Amos Semples, 30, 10, 200, 30, 252
Thomas Haygood, 40, 60, 400, 30, 75
Parker Robinson, 40, 60, 500, 40, 120
Robert Long, 130, 130, 1100, 50, 240
Dagworthy Derickson, 300, 200, 4200, 150, 824
Caleb S. Layton, 3, -, 223, 40, 250
John B. Wapler, -, -, -, -, 130
Philip E. Jones, 150, 150, 600, 100, 400
Elisha D. Cullen, 8, -, 320, -, 350
John W. Davis, 40, 68, 2000, 75, 350
James Maull, -, -, -, -, 30
Greensbery Rogers, -, -, -, -, 14
Ann Harris, 40, 75, 1500, -, -
Hetty A. Harris, 325, 75, 2000, -, -
William Dunning, 250, 60, 4000, -, -
Caleb B. Sipple, 37, -, 2750, 100, 400
Mathew Roach, 12, -, 600, 20, 150
Thomas B. Sipple, -, -, -, 20, 150
John Richards (Rickards), 30, -, 1200, 15, 200
Adolphus P. Ewing, -, 117, 1000, 75, 450
Chloe Ewing, 15, -, 1000, -, 150
Edward Wootten, -, -, -, -, -
Mary E. Smith, -, -, -, -, 50
Elias S. Rickards, -, -, -, -, 125
Elias Jones, -, -, -, -, 20
Jacob M. Kollock, -, -, -, -, -
Rachael Pride, -, -, -, -, 30
Willber Rogers, -, -, -, -, 33
William O. Redden, -, -, -, -, 100
George Harris, 6, -, 1650, -, 60
Abram H. Marvel, -, -, -, -, 10
George W. Maull, 30, 50, 1000, -, 225
Elizabeth Ridgely, 18, -, 1000, -, 110
William Russel, -, -, -, -, 130
Moses Robinson, -, -, -, -, -
Thomas Hatfield, 12, -, 1000, -, 100
Jacob Kimmy, -, -, -, -, 15
Luther Hastings, -, -, -, -, -
Sally Messick, -, -, -, -, -
George Kollock, 25, 5, 350, 10, 75
James P. Burton, -, -, -, -, 135
Kenda B. Wingate, -, -, -, -, 40
Joseph Hickman, -, -, -, -, 15
James Butler, -, -, -, -, 10
James M. Roach, -, -, -, -, 40
Isaac Adams, -, -, -, -, 12
Wesley Wolfe, 75, 75, 4125, 135, 245
Alfred P. Robinson, -, -, -, -, 80
James Pettyjohn, -, -, -, -, 12
Nathan Messick, -, -, -, -, 6
Stephen Carpenter, 50, 150, 2000, -, 130
John Stockley, 5, 10, 400, 40, 185
Arthur Trahearn, -, -, -, -, 137
John West, 7, 206, 1200, -, 175
Charles Adams, -, -, - -, 300
James S. Chase, -, -, -, -, 500
Isaac P. W. Kollock, -, -, -, -, 50
Laban L. Lyons, 100, 170, 5500, 170, 500
Jonathan R. Torbert, 100, 50, 75, 200, 505
James A. Parker, -, -, -, -, 268
Souder Hammond, -, -, -, -, 18
Isaac Russel, -, -, -, -, 66
Zadock B. Lacey, -, -, -, -, 35
William Hazel, 300, 60, 5000, 50, 250
Fretwelll Wright, 200, 100, 2500, 15, 200
James Philips, 100, 50, 1500, 10, 50
Samuel Paynter, 125, 75, 3000, 50, 300
Stephen Abdole, 75, 50, 1000, 20, 75

David Richards, 80, 125, 2400, 25, 150
Carroll Warrington, 150, 42, 2500, 75, 200
Jacob Wilson, 20, 50, 250, 5, 2
Elizabeth Parker, 25, 50, 500, -, 25
Cyrus Holland, 30, 35, 375, 10, 150
Jacob Holland, -, -, -, -, 20
Theodore Richards, 80, 70, 2000, 35, 250
David Gorden, 22, 15, 630, 15, 175
Robert W. Stephenson, 100, 90, 1500, 20, 250
Nathaniel Juilby(Quilby) Sr., 25, 100, 1500, 20, 100
Doct. William Harris, -, -, -, -, 75
Sarah King, 100, 150, 2500, 25, 250
Hester Johnson, 25, -, 150, 5, 100
Ebe Holland, 300, 30, 4000, 50, 500
David Robbins, 150, 150, 3000, 50, 500
Rhodes Wilson, 40, 50, 700, 10, 100
William S. Short, 80, 120, 4500, -, -
Charles Tunnell, 144, 56, 6000, 150, 325
Peter Pepper, -, -, -, -, 30
George Ratcliff, -, -, -, -, 25
Joseph W. Spaulding, -, -, -, -, 10
Edward Pepper, -, -, -, -, -
Thomas Pepper, 75, 100, 2000, 50, 225
James Steel, 125, 125, 2500, 100, 210
William Tatman, -, -, -, -, 50
Rachael Waples (Wapler), 175, 70, 10000, -, 518
James P. Wilson, 50, 11, 1200, -, 134
John Wilson Sr., 35, 78, 11000, -, 270
Greenberry Lynch, 30,70, 500, -, 80
Major Wilson, 80, 90, 200, 25, 320
George Walls, 70, 70, 300, -, 110
John Ennis, 50, 190, 2500, 25, 120
William Sullivan, 30, 30, 1000, 15, 35
Edward Marvel, 80, 40, 2000, 50, 373
Samuel Derickson, -, -, -, -, 6
Eli G. Roach, 40, 40, 300, 20, 100
Gibson W. Joseph, 20, 80, 400, 6, 44
James R. Chase, 40, 100, 100, -, 35
George B. Chase, 60, 200, 1500, 25, 127
Joshia G. Baker, 23, -, 200, 12, 150
Richard Butelier, 50, 120, 150, 20, 150
John Burton, 170, 205, 2500, 25, 100
Nancy Butler, 30, 60, 250, -, -
Thomas R. Marvel, 40, 150, 600, 10, 100
David Pepper, 150, 150, 5000, 150, 652
Purnal Short, 150, 100, 1000, 1, 150
Daken Parker, 35, 75, 1000, 25, 125
Elizabeth Prettyman, 80, 40, 600, -, 150
Curtis Prettyman, 4, 53, 700, 25, 100
John Prettyman, -, -, -, -, -
Asbury W. Pepper, 125, 85, 5000, 150, 150
Theophilus Salmon, -, -, -, -, 50
Jacob Burton, -, -, -, 10, 50
John Collins, 25, 60, 5000, 25, 125
John Wilson, 30, 70, 450, 10, 50
John Pounder, -, -, -, -, 300
Peter R. Jackson, 10, 40, 1200, 10, 100
Thomas J. Atkins, -, -, -, -, 75
Cornelius Coulter, -, -, -, -, 100
Noble Ellensworth, -, -, -, -, 30
Samuel Martin, 10, -, 300, -, 125
Ephraim Darby, -, -, -, -, 8
Houston Hall, 30, 30, 800, -, -
Capt. Joseph Atkins, -, -, -, -, 100
Charles Manship, -, -, -, -, -
Nehemiah Welch, -, -, -, -, 100
Henry Skidmore, -, -, -, -, -
John Fisher, -, -, -, -, -
William C. Prettyman, 25, -, 500, 10, 30

Benjamin Donovan, 50, 125, 1000, 25, 150
George Milby, -, -, -, -, 30
Elizabeth Terry, -, -, -, -, 25
John J. Morris, 150, -, 1000, -, 1000
Nathaniel Bassitt, 140, 40, 1500, -, 200
Jacob M. White, -, -, -, -, 25
William A. Hazzard, 85, 50, 2500, -, 300
Robert H. Carey, 20, -, 300, -, 30
William Warren, -, -, -, -, 75
Joseph Conwell, 100, 50, 3000, 25, 200
Thomas A. Moore, -, -, -, -, 100
Nathan Clifton, 17, -, 1000, -, 50
David Lofland, -, -, -, -, 125
James McCauley, 200, 495, 3000, 10, 100
Thomas Robinson, 80, 106, 4000, 10, 150
David Roach, -, -, -, -, 150
Haveloc Morris, 100, 200, 2000, 22, 300
John Atkins, -, -, -, -,
Alexander Young, 22, -, 200, 10, -
Nancy Baker, 60, 130, 1000, 10, 150
John Abbott, 150, 125, 1000, 10, 300
John Wilkins, 110, 150, 1500, 10, 150
Henry N. Pepper, 100, 150, 2000, 10, 30
William Roach, 100, 40, 1000, -, 100
Moses Brittenham, -, -, -, -, 50
Henry Carey, 100, 40, 225, 10, 100
Elisha Carey, 60, 40, 1000, 20, 100
John Macklin, 100, 75, 1500, 25, 300
Elzey Truitt, 50, 75, 400, -, 30
John Stephens, -, -, -, -, -
Humphries Brown, 50, 100, 1000, 30, 150
Joseph Robbins, 50, 100, 1500, 30, 200
Nehemiah Abbott, 30, 25, 300, 10, 30
Sarah Haveloc, 25, 70, 800, 15, 50
Philip Workman, 75, 200, 1500, 25, 400
Jonathan Doniphan, -, -, -, -, 100
Selathiel Baker, -, -, -, -, -
George Atkins, -, -, -, -, 75
Elias Coverdale, -, -, -, -, -
Sophia Watson, -, -, -, -, 100
Joseph A. McFerran, -, -, -, -, 50
John C. Hazzard, -, -, -, -, 125
Mary A. Parker, -, -, -, -, 30
Sarah Waples, -, -, -, -, 200
John Fassett, -, -, -, -, -
Theodore W. Parker, 25, 150, 1500, -, 150
Peter C. Parker, -, -, -, -, 100
Daniel Collins, -, -, -, -, 50
Elisha Prettyman, -, -, -, -, -
John Beynum, -, -, -, -, -
Mary Donovan, -, -, 200, -, -
John B. Mustard, -, -, -, -, -
James White, -, -, -, -, -
David Mustard, -, -, -, -, -
Woodman Rust, 25, 100, 1000, -, -
Joseph Oliver, -, -, -, -, 25
George Gordy, -, -, -, -, 150
Bivens Morris, 75, 25, 1000, 20, 125
Robert M. Tall, -, -, -, -, 25
David Hazzard, 150, 100, 8000, 40, 500
Elzey Mosly, 15, -, 200, 10, 75
Bennett Haveloc, 75, 130, 100, 10, 100
David Gorden, 50, -, 300, -, 75
Robert P. Davidson, 100, 100, 1000, 10, 175
Verdin Macklin, 50, 100, 1500, 20, 150
Robert Coffin, 100, 250, 2500, 20, 100
Joseph Gray, -, -, -, -, -
Philip Welby, -, -, -, -, 30
John Workman, 100, 50, 1000, 25, 400
John Dodd, 60, 8, 300, 20, 200
James A. Johnson, 25, 22, 500, 10, -
Francis Willey, 70, 350, 4500, -, 50

John Mosley, 30, 100, 500, -, 50
Roderick Reynolds, 40, 110, 500, 20, 50
William J. Dickerson, 225, 75, 1000, 20, 150
William R. Davis, 150, 50, 700, 25, 250
James R. Donovan, 100, 100, 1000, 20, 150
Boaz Downing, 30, 50, 400, 10, 100
Jesse Dutton, 150, 50, 500, 10, 75
Zachariah Pettyjohn, 25, 60, 800, 10, 150
Maria Dutton, -, -, -, -, -
Kendy Sharp, 100, 300, 1500, 50, 250
William Dutton, 50, 80, 3000, 10, 100
Joseph Fisher, -, -, -, -, 50
Albert Carey, 60, 140, 1000, 25, 150
James Dutton, 60, 100, 1000, 25, 50
Zachariah P. Wilson, 75, 75, 1000, 25, 175
George H. Dutton, 40, -, 250, 10, 25
James Roach, -, -, -, -, 75
Robert Donovan, 60, 50, 350, 10, 750
Byard Donovan, 12, 50, 500, 15, 100
Peter Donovan Jr., 5, 60, 500, 20, 100
Job Donovan, 100, 160, 1000, 40, 150
Peter Donovan, 75, 150, 200, 20, 300
Burton Donovan, -, -, -, -, 150
Nehemiah Donovan, 75, 100, 1500, 25, 100
Kendal Donovan, 75, 100, 1000, 20, 75
Riley Donovan, 30, 45, 400, 10, 125
Asa Wilson, 20, 70, 450, -, 40
William Donovan, 40, 100, 1500, 25, 400
Reuben Donovan, 70, 130, 1500, 30, 1100
James Redden, 90, 100, 2500, 20, 200
Job Lecatt, 30, 20, 100, 10, -
Brinkly Davis, 50, 100, 1500, 25, 300
Levin Blizzard, 50, 46, 800, 20, 100
Tilly W. Burrows, 50, 75, 500, -, 100
John Greenly, 14, 139, 800, 50, 175
Covington Reynolds, 25, 175, 2000, 30, 200
William C. Joseph, 25, 149, 1400, 25, 150
John Messick, 50, 40, 500, 20, 100
George Dickerson, 60, 150, 2500, 50, 300
Benjamin Fowler, 60, 25, 400, 20, 25
Joseph Wilson of E., 50, 150, 200, 100, 150
William D. Adkins, 40, 150, 1000, 25, 75
Paynter Joseph, 40, 100, 500, 10, 50
William B. Blizzard, 8, -, 100, 10, 50
Purnal P. Kimmy, -, -, -, -, 18
Wingate Morris, 80, 40, 700, 25, 200
Paynter Johnson, 30, 40, 400, 20, 30
Mathew Wilson, 15, 95, 1000, 20, 150
Richard Wilson, 30, 60, 300, 10, 35
William Pettyjohn, 30, 310, 500, 20, 100
Levin Dickerson, 90, 15, 300, 10, -
Edward Dickerson, 75, 175, 500, 10, 100
William Martin, 30, 40, 250, 10, 75
David Stuart, 30, 30, 500, 15, 150
Arcada Adkins, 100, 220, 1100, 25, 300
James Cary, 30, 10, 400, 10, 100
Ludwick Atkins, 50, 100, 1000, 20, 150
Peter Rust, 100, 40, 1500, 200, 350
Thomas T. Joseph, 150, 50, 500, 25, 100
Joseph Burrows, 50, 100, 800, 50, 150
Hessie Dodd, 50, 50, 500, 10, 50
Peter Dodd, 150, 50, 500, 10, 150
Thomas Jester, 100, 50, 400, 10, 75

John Simpler, 50, 40, 200, -, -
Robinson Barker, 75, 89, 400, 15, 250
Josiah Veasey, 40, 100, 700, 15, 250
Peter Simpler, 100, 54, 1000, 10, 150
John Dodd, 150, 150, 1000, 15, 100
Sylvester Palmer, 150, -, 1000, 15, 100
Henry O. Beynum, 60, 55, 1500, 25, 250
William Spicer, 75, 125, 1000, 10, 75
John Macklin, 100, 100, 1100, 26, 150
John Robbins, 150, 50, 3000, 25, 200
John P. Robbins, 50, 100, 2000, 20, 200
David Russel, 50, 100, 1000, 10, 5
William Clendaniel, 25, 50, 500, 5, 40
Samuel R. Paynter, 350, 200, 10500, 500, 500
Hugh King, 175, 5, 1000, 25, 200
James King, 50, 50, 1000, 20, 150
William White, 95, 25, 1500, 25, 200
Robert White, 750, 300, 3100, 50, 500
Benjamin White, 60, 20, 1100, 30, 300
Nehemiah Dorman, 68, 50, 1000, 30, 200
William Milby, 34, 10, 800, 5, 11
Absalom Rust, 75, 25, 950, 10, 100
James Foster, 100, 300, 150, 5, 40
Richard Holland, 50, 50, 800, 10, 112
Peter Hopkins, 100, 65, 1000, 50, 200
Reuben Wilson, 100, 140, 600, 20, 100
William H. Simpler, 100, 200, 1500, 20, 200
George Pride, 100, 25, 600, 10, 100
Jesse White, 10, -, 158, 5, 15
James S. Hood, 40, 110, 1500, 10, 50
David Ennis, 40, 150, 1000, -, 300
William Palmer, 75, 150, 1000, 20, 100
John Warrington, 100, 125, 2000, -, 230
Elhanana Reynolds, 20, -, 250, -, 100
Ellen Virden, 100, -, 1000, -, 100
Benjamin Virden, -, -, -, -, 150
Minas Johnson, 120, 10, 800, -, 150
Mary Sherman, 80, 20, 300, 20, 78
George Sherman, 244, 100, 1000, 25, 120
Isaac Prettyman, 100, 30, 1000, 21, 100
William W. Veasey, 130, 23, 675, 30, 100
Andrew Holland, 50, 50, 1000, 20, 130
William P. Palmer, 18, -, 150, 25, 30
William Eddington, 40, 40, 500, 10, 100
William S. Vent, 15, 50, 400, 20, 100
James Coulter, 150, -, 250, 10,50
Shepherd Prettyman, -, -, -, -, 50
Jesse Ennis, 75, 125, 1000, 25, 200
William Wilson, 50, 175, 2000, 30, 300
William Hancock, 75, 25, 500, 25, 150
Jonathan Coleman, 40, 125, 800, 30, 140
George Holhairn (Holhain), 70, 42, 404, 20, 100
Sarah Blizzard, 30, 70, 500, 20, 100
Levin P. Dutton, 75, 10, 1000, 30, 75
William Short, 100, 100, 2000, 30, 150
James Chase, 75, 25, 310, -, 50
James W. Simpler, 50, 5, 1200, 25, 150
David Rollins Jr., 200, 200, 2500, 44, 400
William A. Carnwel, 40, 20, 1200, 30, 150
Donaphin Reed, 50, 100, 4000, 20, 400

James J. Betts, 79, 40, 400, 20, 50
John J. Conwell, 125, 55, 2500, 35, 300
Abraham Reed, 50, 150, 1500, 25, 150
Charles Shockley, 50, 104, 1000, 10, 100
Emaline Simpler, 50, 154, 1000, 24, 150
James Reed, 50, 100, 1000, 10, 150
Henry Jones, 80, 40, 600, 20, 100
David Johnson, 20, -, 500, 10, 75
James Holstein, 8, -, 700, 15, 100
James C. Robbins, 57, 41, 2000, 25, 300
David Wiltbank, 150, 500, 3500, 50, 500
Thomas Williams, 104, 75, 1500, 25, 200
George Collins, -, -, -, -, -
John Sharp, 50, 108, 1500, 20, 200
William Messick, 25, 25, 30, 10, 50
Foster Donovan, 24, 70, 500, 10, 150
Levi Messick, 30, 70, 500, 25, 200
Minos Messick, 25, 75, 500, -, 500
William Walker, 25, 35, 400, 10, 100
John Warren, 25, 50, 300, 15, 11
Seth Haveloc, 200, 400, 5000, 30, 250
Eli Donovan, 200, 400, 5000, 30, 250
Isaac Chase, 50, 100, 1000, 15, 100
Jackson Palmer, 40, 200, 2000, 10, 100
Daniel Burtin, 50, 150, 1500, 20, 100
Silas Reynolds, 40, 125, 1500, 25, 200
Thomas Morris, 50, 100, 1000, 21, 275
Charles Norwood, 50, 100, 800, 21, 200
William Milby, 50, 100, 1000, 10, -
John W. Dean, 50, 200, 1500, 20, 150
Benjamin Reed, 60, 50, 1000, 10, 150
Sylvester Rust, 100, 125, 1000, 10, 200
Robert Russel, 260, 75, 6000, 20, 400
William Wolf, 80, -, 100, -, 200
Samuel Parker, 19, 10, 1000, -, 200
David H. Connel, -, -, -, -, 30
Levi Morris, -, -, -, -, -
Abner Willey, -, -, -, -, -
John King, 100, 100, 1000, -, 30
Anthony Reynolds, 20, 50, 300, -, 25
John Holston, 100, 300, 1000, 10, 55
Edward Short, 250, 50, 5000, 100, 400
Roland Tindal, 35, 65, 350, 10, 55
Solomon Lofland, 20, 40, 250, 10, 100
Benta Carpenter, 100, 200, 2000, 10, 200
Jacob Fossett, -, -, -, -, 75
Lemuel Joseph, 150, 100, 1950, 150
George Waples (Wapler), 75, 50, 1000, 20, 200
John Paynter, 150, -, 1000, 10, 100
Benjamin Kollock (Kollok), 90, 10, 500, 10, 150
Lorenze D. Morton, 75, 2, 300, 15, 150
Cesor Rodes, 6, -, 100, 10, 100
John Rodes, 300, -, 1500, 20, 500
Elijah Burton, -, -, -, -, -
Sydney Marsh, 150, 50, 1000, 10, 300
William A. Dodd, 100, 50, 1500, 20, 150
Peter Hudson, 50, -, 300, 10, 100
Thomas Marsh, 30, 20, 300, 12, 150
Matthew Marsh, 50, -, 350, 10, 100
John Hiaolow (?), 50, -, 350, 10, 150
Alexander Mosley, 63, 5, 800, 20, 150
Robert R. Russel, -, -, 200, 10, 150
John Metcalf, 22, -, 800, 10, 300
Thomas Bell, 2, -, 100, -, 150
Daniel Walker, 20, -, 1500, 30, 250
Thomas Coleman, 8, -, 800, 15, 300

Lemuel Evans, -, -, -, 3, -
William West, -, -, -, -, -
Joseph Walker, -, -, -, 3, -
William A. Conwell, -, -, -, -, -
Robert Pepper, -, 96, 550, 25, 25
Aaron Kimmy, 30, 50, 450, 20, 150
Aaron Marshal, -, -, -, -, 250
Asa Conwell, 80, 250, 2000, 5, 200
Henry Johnson, 50, 64, 1000, 10, 150
Wingate Donovan, 10, 37, 1000, 30, 300
Bennett Johnson, -, -, -, -, 25
Samson Cade, 10, 35, 300, -, 20
Warnus Wiltbank, -, -, -, -, -
Isaac Betts, 6, -, 300, 10, 75
Robert W. Betts, 75, 65, 1000, 20, 200
William Davis, -, -, -, -, 100
James Carey, -, -, -, -, 75
Rouse F. Young, -, -, -, -, 200
Nathaniel Johnson, -, -, -, -, 30
George Hall, 14, 2, 700, -, -
Aaron Marvel, 75, 35, 1500, -, 400
James Martin, 80, 120, 3000, 50, 300
Wingate Salmons, 10, 40, 300, 15, 100
James Reed, 50, 30, 350, 15, 150
Robert J. Lynch, -, -, -, -, 100
Pompy Mitchel, -, -, -, -, -
William Craig, 50, 50, 1000, -, 150
James Coney, 100, 25, 1000, -, 100
John Thomson Jr., -, -, 200, 25, -
John Thomson, 200, 50, -, -, 350
Edward Milby, -, -, 800, -, 150
John Arnold, 17 ½, 9, 1000, -, 200
Henry Hall, -, -, -, -, -
Thomas Walker, 125, 75, 5000, 30, 400
David McIlvain, 100, -, 5000, 20, 300
Edward Watson, -, -, -, -, 200
Davis Millis, 7, -, 500, -, 150
Joel Prettyman, 95, 65, 5000, 30, 400
William E. Marvel, -, -, -, -, -
Prater Wapler, 125, -, 1500, 50, 400
Shepherd P. Hopkins, 300, 150, 10000, 300, 650
Leon Hopkins, 100, 50, 1000, 22, -
Hannah Hood, 100, 50, 900, 20, 250
Richard Paynter, 150, 25, 1000, 10, 400
Charles Mills, 200, 100, 2000, 50, 350
Levin B. Day, 112, 240, 2000, 100, 300
Samuel M. Vaughn, 5, -, -, 40, 72
Timothy Jacobs, 100, 200, 1000, 20, 75
Nathaniel W. Vaughn, 50, 250, 2000, 50, 150
Edward McCauly, -, -, -, -, 84
Joseph B. Vaughn, 50, 180, 2500, 30, 150
Robert West, 30, 210, 1000, 5, 30
James Salmons, -, -, -, -, -
Robert Gumley, 20, 50, 1200, -, 100
Thomas Gumley, -, 60, 600, -, -
William Wilkins, 100, 150, 1500, -,26
James M. Vent, 18, -, 50, -, 75
Robert Warrenton, 60, -, 150, -, 30
Samuel Walls, 34, -, 100, -, -
Lemuel Casey (Carey), 70, 25, 1000, -, 15
Richard B. Cookson, 80, -, 375, 15, 200
Thomas Walker, 200, 1225, 2500, 50, 400
Robert Coffrin, -, -, -, -, 125
James B. Cookson, 30, 20, 500, 15, 100
Jesse B. Cookson, 70, 10, 1000, 20, 200
Charles Vaughn, 70, 130, 3000, 56, 150
Henry Hood, 30, 25, 500, 50, 25
Jobe Reynolds, 20, 60,400, 50, 25
Joseph Wilson, 80, 40, 3000, 10, 40
Purnal Maull, 75, 75, 2000, 20, 150
Minus Messick, 100, 75, 1000, 10, 125

Emily Sharp, 165, -, 500, 20, 125
Mary Lauk (Lank), 50, 25, 500, 10, 100
Thomas Wilson, 40, 25, 250, 10, 30
William Lauk, 45, -, 250, 100, 130
Elisha Holland, 100, 200, 1000, 320, 350
Lemuel H. Stephenson, 60, 70, 600, 25, 95
John Joseph, 150, 50, 1000, 50, 200
James Dickerson, 150, 50, 2000, 20, 300
Hetty A. Prettyman, 100, 25, 500, -, 75
Peter McGee, -, -, -, -, 100
Lyden A. Perry, 150, 350, 3500, 30, 125
Isaac W. Simples, 200, 200, 3500, 30, 200
Elzey Wilson, 70, 20, 800, 25, 200
Thomas Jester, 125, 25, 600, 25, 100

Kent County Delaware
1860 Agricultural Census

This agricultural census was filmed from original records in the Delaware State Archives in Wilmington Delaware by the Delaware State Archives Microfilm office.

There are some forty-eight columns of information on each individual. Only the head of household is addressed. I have chosen to use only six columns of the information because I feel that this information best illustrates the wealth of the individuals. These are shown below:

1. Name of Owner
2. Acres of Improved Land
3. Acres of Unimproved Land
4. Cash Value of the Farm
5. Value of Farm Implements and Machinery
13. Value of Livestock

Thus, the numbers following the names represent columns 2, 3, 4, 5, 13.

The following symbol is used to maintain spacing where information in a column is left blank (-). This symbol is used where letters, names or numbers are not legible (_).

There is no explanation of the use of several abbreviations. However, it appears that the abbreviations: O stands for owner, T stands for tenant, M stands for manager and B cannot determine. Use of these abbreviations in some sections of the county is discontinued, probably due to a different person taking that part of the census. In addition page 23/24 is missing. This may be just a misnumbering between Martin Herrington of Mispellion Hundred and John Costen of Murderkill Hundred, the two marshals conducting the census in these locations.

George P. Fisher O, 160, 121, 8000, 250, 1400
Give Saulsbury O, 5, -, 2000, 100, 400
John B. Pennington, O, 57, -, 1700, 75, 300
Elias S. Reed M, 75, 175, 6000, 300, 200
Thos. Slaughter O, 175, 22, 18000, 300, 2200
Danl. Cowgill O, 8, 5, 1200, 25, 200
John C. Pennington O 10, 3, 1500, 60, 800
Isaac Jump O, 250, 50, 10000, 300, 2000
Geo. B. Dickson T, 100, 14, 10000, 500, 1000
Wm. J. Clark O, 100, -, 3000, 75 300
John Greer O, 100, 5, 5000, 400, 700
Jas. M. Kerbin O, 54, -, 7000, 300, 1000
Thos. B. Bradford O, 450, 300, 54000, 1500, 2500
Thos. B. Bradford O, 90 110, 6000, 300, 450
Wm. M. Jester O, 35, 5, 3000, 100, 300

Saml. M. Harrington O, 130, 20, 2500, 50, 300
Saml. M. Harrington O, 100, 30, 2500, 50, 300
Jos. P. Corney Sr., 11, 8, 7000, 100, 300
Man. W. Bates O, 20, -, 2500, 200, 600
William Walker O, 200, -, 11000, 300, 1500
William Walker O, 75, -, 8000, 400, 1700
William Walker O, 400, 200, 25000, 500, 800
Caleb J. Pennewell O, 150 25, 17000, 200, 1500
Jas. A. Dunning O, 75, 17, 4000, 250, 450
A. S. Lukins O, 30, 60, 3500, 50, 400
Richardson & Robbins O, 120, 48, 6000, 300, 500
Richard M. Jones O, 100, 36, 5000, 200, 1200
Clayton A. Cowgill O, 1, -, 200, 50, 250
Wm. Manlove O, 115, 40, 8000, 200, 500
Elizabeth Milds O, 150, 50, 10000, 100, 800
John Wilcox T, 150, 15, 10000, 200, 500
Francis Register O, 100, -, 8000, 200, 500
Jas. Clendaniel T, 60, 40, 2000, 75, 400
Stephen Catts T, 120, 40, 8000, 300, 500
Danl. Hoffecker O, 300, 40, 25000, 500, 1400
Jos. Smithers T, 125, 50, 9000, 200, 300
Wm. Wilson M, 300, 10, 250000, 500, 1000
Wm. B. Collins T, 160, 10, 15000, 200, 710
Wm. Wharton T, 150, 15, 8000, 200, 500
Joshua Wharton T, 300, 500, 20000, 200, 1000
Bolitha Wharton O, 140, 100, 9000, 200, 700
Chas. M. Wharton O, 150, 10, 15000, 800, 1200
Stephen Levick T, 130, 50, 5000, 50, 600
Henry Wilson T, 140, 100, 6000, 100, 500
Peter Wilson T, 200, 300, 10000, 300, 1000
John K. Morris T, 175, 100, 10000, 100, 1200
Draper Voshell T, 200, 200, 10000, 100, 800
Eli W. Pepper M, 170, 200, 8000, 100, 600
Henry R. Prettyman M, 170, 200, 8000, 700, 1000
George Knight O, 65, 65, 2000, 75, 300
John S. Waller M, 30, 30, 1200, 25, 200
Jackson Lafferty O, 45, 15, 2000, 25, 200
George Mitchell O, 24, 5, 1400, 25, 400
Wm. Parms (Pams, Parris) O, 15, 8, 900, 50, 100
Joseph Millauby O, 100, 35, 6000, 200, 300
Timothy Slaughter O, 200, 95, 12000, 200, 700
James Raymond O, 125, 25, 6000, 200, 500
Thos. Postles O, 200, 30, 10000, 200, 1200
Thos. Postles O, 200, 30, 10000, 200, -
Isaac Herrington O, 150, 50, 9000, 500, 1000
Christopher Ford T, 120, 40, 5000, 150, 225

Jas. D. Kimmy O, 250, 21, 20000, 500, 1000
William Whitesides M, 300, 100, 20000, 1000, 2000
Stephem Mirmer T, 200,150, 9000, 150, 500
Wm. Hutcherson O, 120, 4, 6000, 200, 500
Sarah Cowgill O, 125, 25, 41000, 200, 800
Wm. Slaughter T, 220, 100, 9000, 200, 500
David Argo M, 175, 150, 11000, 200, 1000
Jeremiah Jarold T, 16, -, 500, 50, 100
Edward Pardee T, 107, 75, 8000, 150, 400
Rebecca Slaughter T, 160, 400, 16000, 200, 1200
Nathl. Taylor T, 40, -, 1000, 50, 400
William Adkins O, 112, 12, 3000, 100, 900
Alfred Bell T, 50, 30, 700, 25, 200
William Patten T, 75, 50, 2400, 25, 700
Isaac Fulman T, 50, 100, 2000, 25, 200
Mary Bell O, 80, 140, 1400, 100, 400
James T. Postles T, 130, 12, 16000, 150, 500
James Ratledge T, 180 40, 16000, 150, 600
George Wheatly T, 130, 20, 9000, 200, 400
Robert Welch O, 70, -, 1000, 75, 175
William Williams T, 32, 3, 2000, 100, 400
Alexander Bell T, 275, 20, 4000, 175, 800
Clayton F. Hawkins T, 50, 10, 3000, 100, 500
Outter L. Heverin O, 200, 37, 21000, 800, 1800
Jas. Harrington T, 240, 150, 15000, 500, 1200
Zadock Postles O, 100, 20, 8000, 400, 1000
Zadock Postles O, 100, 20, 8000, -, 100
Jas. L. Heverin O, 300, 200, 25000, 1000, 2200
William Carter O, 165, 40, 12000, 200, 1000
Manlove Calley T, 205, 40, 15000, 500, 100
Nathl. J. Reynolds M, 75, -, 6000, 100, 350
Elijah Bendlen T, 35, -, 2000, 40, 200
Edwd. O. Eccles O, 160, 25, 20000, 400, 1000
Wm. Carrow M, 300, 20, 10000 500, 1200
James Slaughter T, 24, 4, 1000, 50, 250
Carter Dilahay T, 50, 150, 4000, 5, 300
Danl. S. Lingo M, 200, 50, 12000, 100, 700
Moses Merrica T, 150, 60, 5000, 150, 1000
Chas. S. Clements T, 15, -, 1000, 50, 100
George Jewell T, 100, 50, 3000, 100, 500
Manlove Frazier T, 100, 25, 3000, 150, 500
David Lawson M, 80, 17, 6000, 150, 400
Thos. G. Murphy O, 100, 50, 15000, 200, 700
James Adkins T, 100, 5, 4000, 100, 250
John Robinson O, 230, 60, 15000, 600, 1000
Isaac Berry T, 165, 40, 10000, 200, 1000
Robt. Loud O, 75, 40, 2700, 150, 450
Lewis H. Pecky O, 60, 100, 3000, 150, 400

Tilghman Jack T, 13, -, 500, 75, 100
Michael Dempsey T, 50, 160, 4000, 150, 400
Frederick Gregor T, 75, 75, 3000, 75, 200
Elisha Massey T, 45, 30, 1500, 150, 400
Joseph Boges T, 120, 50, 10000, 150, 500
John Snyder T, 150, 50, 10000, 150, 500
John Dennis T, 50, 20, 3500, 100, 600
Saml. Emory O, 35, 11, 1500, 75, 300
William Dean T, 60, 40, 3000, 150, 75
Jordan Argo T, 100, 86, 8000, 100, 300
Thomas L. Simpson O, 80, 100, 5000, 100, 250
J. P. M. Denny O, 100, 55, 6000, 200, 700
Greenbery Becker T, 15, 20, 800, 50, 100
Harvey D. Leonard T, 100, 75, 3000, 150, 400
Wesly Massey T, 40, 20, 1200, 100, 238
Benj. Durham T, 36, -, 1000, 50, 200
Zacanah Johns O, 70, 30, 2000, 50, 200
Annie M. Smith T, 70, 40, 2000, 100, 250
Richard D. Jones T, 130, 20, 5000, 100, 200
Owen Moore M, 150, 20, 4000, 200, 500
William Campbell O, 80, 34, 3500, 75, 750
Greenbery Barcus M, 150, 200, 16000, 200, 1000
Thomas Brunt M, 80, 50, 6000, 200, 400
David Wallace M, 150, 30, 17000, 200, 500
Peter Hinser O, 50, 10, 2000, 100, 400
Levin H. Argo T, 100, 86, 8000, 150, 700
Levin H. Argo M, 50, 25, 2000, -, -
Levin H. Argo M, 75, 34, 3000, -, -
Thos. Saulsbury O, 8, 11, 300, 25, -
Wm. Thomas T, 60, 20, 1600, 50, 200
Wesly Dougherty O, 100, 20, 2000, 50, 200
John Hargadine O, 75, 6, 2000, 125, 600
Chas. Miller O, 65, 25, 2500, 100, 250
Henry Cross T, 42, 8, 2000, 25, 150
Wm. Barcus T, 50, 54, 11500, 25, 100
John Brown M, 75, 45, 3000, 100, 250
William Cassins O, 70, 35, 3000, 150, 400
John W. Ware T, 5, 40, 15, 2000, 50, 200
Henry Goodnight T, 75, 28, 2000, 100, 300
John Kimmey O, 20, 4, 700, 10, 175
Lewis Mason T, 65, 25, 3000, 75, 300
Jacob Tinley T, 35, 334, 2000, 50, 200
Henry B. Senet O, 30, 20, 1200, 75, 300
Philip Cummins T, 70, 50, 2500, 100, 400
Joseph Moore O, 50, 40, 3000, 200, 600
Richard Gibbs T, 6, -, 600, 20, 126
John W. Dunning T, 100, -, 2000, 75, 300
John Knight T, 60, 60, 3000, 75, 300
Brunmet Ford T, 125, 100, 4000, 75, 300
William Morris T, 190, 17, 18000, 125, 600

John W. Smith O, 40, 3, 6000, 50, 200
Nathl. Poutt T, 9, -, 4000, 50, 150
Lewis Weston O, 6, -, 2000, 20, 100
Jno. G. Wapler (Waples) O, 185, 20, 20000, 400, 1500
William P. Coaker(Cooher) O, 60, 50, 6000, 150, 300
Peter F. Smith T, 100, 100, 16000, 250, 700
Thos. Peirce T, 150, 50, 3000, 50, 150
Jos. D. Parker O, 120, 10, 6000, 100, 500
Geo. W. Wells O, 120, 10, 6000, 100, 500
James Donivan T, 50, 50, 5000, 100, 200
Amos Shaw M, 160, 20, 6000, 100, 550
William Webb T, 160, 20, 6000, 150, 1000
William Gorton M, 160, 20, 6000, 100, 300
Jabez Jenkins T, 180, 20, 10000, 150, 600
Robt. Donivan T, 100, 4, 7000, 75, 400
David Mills T, 375, 25, 16000, 200, 150
Elisha Wright T, 200, 36, 6000, 125, 500
Edwd. Mills T, 220, 40, 12000, 75, 600
John J. Niverson O, 90, 14, 5000, 150, 500
Alexander Collins T, 150, 50, 15000, 200, 700
James Crawford O, 100, 40, 8000, 200, 500
Thos. F. Parker O, 120, 60, 5000, 100, 500
Frazer Hewes, 7, 175, 85, 12000, 250, 600
William Lewis T, 100, 50, 6000, 100, 500
Nathl. Slaughter O, 220, 380, 19000, 800, 700
Henry Bratton T, 42, -, 1360, 50, 125
Henry M. Ridgely O, 250, 145, 45000, 1000, 2500
Robert Burton T, 200, 130, 2000, 150, 1500
John C. Wilson M, 75, 200, 5000, 100, 400
Bass Laws O, 40, 25, 1000, 25, 200
Danl. Farrow T, 120, 80, 2000, 125, 200
Peter Miller O, 100, 75, 6000, 50, 300
Stephen A. Miller O, 30, 10, 1800, 25, 300
James R. Powell O, 80, 20, 4000, 150, 350
Moses Keysor O, 30, 38, 1500, 25, 100
Wm. P. Denny T, 60, 40, 3000, 100, 500
Thos. Soward T, 40, 20, 2000, 150, 300
Robt. Wright O, 80, 46, 4500, 150, 600
John Mitchell O, 42, 40, 2000, 25, 200
Thos. Pearson O, 45, 15, 1000, 100, 300
William Hudson M, 121, 80, 6000, 100, 500
Vincent Cubbage T, 50, 51, 2000, 50, 300
John Wyley O, 300, 200, 10000, 100, 400
Henry Bearmard (Bearnard)O, 100, 45, 4000, 150, 600
John M. Frances O, 35, 115, 3000, 125, 700
Isaac Johnson O, 60, 20, 2000, 100, 300
Hewett Mastin O, 100, 51, 3000, 100, 400
Joel Durham T, 20, 5, 500, 25, 150

Abraham Todd T, 200, 200, 12000, 150, 1500
Joseph D. Farrow O, 40, 15, 1000, 25, 200
Nathl. Thompson T, 100, 140, 6000, 150, 300
Henry Moore O, 110, 60, 4000, 100, 600
Isaac Buckingham O, 160, 200, 15000, 200, 1000
John Taylor T, 20,10, 500, 25, 200
Robt. Thompson O, 20, 19, 1100, 50, 400
Jas. Crocker T, 60, 30, 1500, 25, 200
John Thomas O, 50, -, 1000, 25, 200
Moses Rash O, 100, 69, 8000, 150, 600
John Pearson T, 70, 30, 3000, 50,700
Theodore Marvel T, 78, 66, 3500, 75, 200
Johnathan Williams O, 40, 10, 1000, 50, 300
Thomas Chambers T, 200, 360, 10000, 200, 1200
Jruston Mason T, 150, 50, 5000, 100, 500
Wm. S. Short O, 60, 120, 6000, 100, 1000
Christopher Ford O, 22, 3,1000, 25, 200
Elizabeth Rose O, 100, 135, 4000, 50, 200
John Derry T, 18, -, 500, 25, 150
Saml. Derry O, 40, 7, 1000, 150, 600
Wm. Clayton O, 25, 10, 500, 25, 150
Jas. Carrow O, 37, -, 800, 25, 250
Nustor L. Davis, O, 15, -, 800, 20, 300
Mark Rash O, 100, 50, 5000, 150, 600
James Rash T, 100, 35, 4000, 100, 300
Isaac Pearse T, 100, 28, 2000, 100, 500
Saml. Harris, T, 60, 10, 3000, 50, 400
Jacob Clark O, 30, -, 800, 25, 100
Isaac Jacobs T, 73, 20, 2500, 20, 350
Jacob Hallowell O, 50, 40, 1500, 100, 300
Myers Carsons O, 100, 30, 5000, 100, 230
Slighter Harris T, 100, 25, 2000, 50, 150
Jackson Corse O, 60, 60, 2000, 50, 200
Saml. F. Hewes O, 280, 70, 6000, 100, 200
George Hallowell T, 75, 55, 3500, 25, 250
Rachel Craig O, 160, 100, 6000, 75, 600
John Hutcheson O, 20, 30, 2000, 25, 200
Benj. Poore T, 56, 15, 200, 25, 150
William Hirons O, 100, 70, 5000, 100, 500
Obediah Thompson T, 40, 40, 2000, 25, 400
Daniel Tribbett T, 40, 40, 2500, 25, 250
William Thompson O, 90, 24, 3000, 50, 500
Joseph Foreacres O, 80, 150, 8000, 150, 700
Levi Gatt T, 100, 230, 6000, 75, 250
Thos. Hurd T, 100, -, 1000, 25, 300
David Marvel O, 50, 80, 5000, 150, 550
Thos. Wallace T, 70, 60, 4000, 150, 600
Wm. Virden O, 100, 180, 12000, 300, 600
Wm. Virden T, 120, 75, 3000, 100, 1000
Isaac Thomas T, 130, 75, 2000, 200, 800
David Harrington O, 75, 20, 4000, 125, 450
Philip Marvel T, 200, 75, 5000, 100, 1000
Joseph Darnett T, 12, 3, 400, 15, 200

John W. Clark T, 35, 20, 2500, 125, 500
John Clark O, 55, 20, 2500, 100, 250
Thos. Hayes O, 120, 25, 1500, 50, 200
Joshua Lodine T, 80, 50, 3000, 50, 300
Stephen G. Gibbs T, 75, 70, 2000, 100, 200
Isaac Pratt O, 100, 56, 400, 100, 500
Levick Hirons O, 100, 92, 5500, 100,400
John Shorts T, 65, 20, 3000, 50, 300
Mathew P. Green O, 140, -, 5000, 150, 1000
John Townsend O, 100, 25, 2000, 25, 100
Chas. W. Oart T, 15, -, 453, 10, 100
Thos. Faulkner O, 40, 30, 1500, 25, 300
James Simmons T, 18, -, 400, 50, 350
Saml. Craig O, 100, 40, 2500, 100, 500
John Fisher O, 40, -, 800, 25, 400
Eli Kenton O, 30, 10, 1500, 50, 30
Thos. Honey O, 30, 37, 700, 50, 150
Sarah Stant O, 34, 13, 1000, 75, 300
Wm. Atthurs O, 10, -, 1500, 30, 150
Wm. F. Hawkins O, 140, 100, 6000, 150, 1000
Michael Powell T, 50, 50, 2000, 50, 75
Saml. Powell O, 100, 90, 8000, 100, 500
Charles Powell T, 30, 25, 2000, 25, 125
John Greenage O, 75, 55, 2500, 100, 700
James Greenage T, 60, 52, 2000, 30, 200
Wm. B. Duling T, 125, 75, 4500, 100, 500
Wm. Johnson O, 25, -, 600, 20, 200
Levi B. Nickerson O, 120, 70, 3000, 75, 400
Philip Roberts O, 60, 80, 4000, 100, 800
James Foreacres O, 20, 30, 1300, 25, 200
Richard Taylor O, 8, 22, 400, 25, 150
Joseph Rash T, 20, 14, 500, 25, 75
Gre__gel Mooro T, 50, 45, 900, 50, 150
Philomen Scotter (Scotten) O, 60, 50, 2000, 150, 400
James Johnson O, 125, 75, 6000, 150, 1000
William Slaughter O, 130, 80, 6000, 500, 1400
Beauchamp Morris O, 200, 80, 2000, 100, 350
Joseph Barratt T, 125, -, 5000, 100, 800
William Slaughter T, 200, 100, 4000, 150, 700
Jonathan Scotten O, 80, 30, 2500, 75, 400
Wm. Jones O, 75, 25, 2500, 75, 300
Merritt Scotten O, 70, 200, 6000, 200, 500
Charles Scotten T, 60, 110, 6000, 50, 200
Spencer Scotten T, 50, 50, 2000, 50, 300
Nancy Manson T, 75, 28, 4000, 50, 250
Thos. J. Marvel O, 100, 95, 8000, 200, 1200
Thos. Knox T, 75, 75, 2000, 50, 200
Thos. Harrington O, 25, 25, 600, 25, 200
C__ Burkley T, 50, 11, 1000, 40,150
Wm. H. Bokes T, 150, 50, 3000, 100, 650
Philip Hurd T, 25, 8, 600, 25, 175
Nathan Gibbs T, 60, 40, 1500, 50, 40
Jesse M. Jones O, 45, 17, 3000, 50, 50
James Barratt T, 75, 21, 5000, 75, 400

Stephen Gibbs O, 135, 65, 3000, 75, 400
Joseph Duling T, 25, 30, 2000, 25, 200
John Jesten (Jester) O, 20, -, 600, 20, 150
Willam Willis (Nillis) T, 25, 5, 1000, 25, 500
Benjamin Walker O, 40, 12, 1300, 50, 300
James Williams T, 150, 25, 3500, 75, 700
George Slay T, 150, 25, 6000, 100, 700
Alfred Gibbs T, 20, -, 800, 50, 200
Henry Delaney T, 25, -, 600, 25, 100
John Williams O, 65, 25, 2500, 50, 200
Thomas Price T, 65, 25, 2500, 50, 200
Thos. Clements O, 200, 32, 10000, 150, 400
Nathl. Hutchens T, 88, 20, 2000, 25, 200
Amos Hinsley O, 180, 80, 6000, 150, 250
Newton Hubbard O, 60, 20, 2000, 75, 350
John Williams T, 68, -, 1800, 75, 400
James Clark T, 110, -, 3000, 150, 500
Wm. T. Hall T, 30, 20, 1000, 25, 250
Elizabeth Green O, 135, 60, 6000, 100, 700
Jas. J. Voshell O, 160, 28, 4500, 150, 900
Joseph Green T, 60, 56, 2200, 50, 250
Mathias Day O, 107, 45, 5000, 410, 500
John Stevens T, 40, 26, 800, 50, 300
William S. Millburn O, 70, 38, 2000, 95, 400
William Kersey O, 150, 50, 3000, 200, 700
William Boyer T, 150, 50, 3000, 200, 700
William Moore T, 60, -, 1500, 50, 300
William Hall O, 200, 100, 4000, 100, 500
Richard Gott T, 80, 27, 4000, 50, 150
William Griffin T, 120, -, 4000, 100, 350
Wm. P. Nickerson T, 150, 150, 4000, 50, 350
Thos. S. Moore O, 26, -, 600, 25, 100
Wm. Callahan T, 26, -, 600, 25, 100
James Duladaway (Duhadaway) O, 22, -, 1000, 25, 150
Saml. Carter O, 70, 30, 4000, 175, 500
Wm. Wilson O, 150, 50, 6000, 200, 500
John B. Kersey O, 60, 40, 1000, 100, 250
Martin Ford T, 120, 80, 3000, 150, 500
Richard Patten T, 75, 125, 25, 95, 175
Thos. Price T, 75, 65, 3000, 50, 150
Darling Johnson T, 25, 12, 1200, 25, 300
Jacob Rickards O, 90, 60, 4000, 100, 700
William Rickards O, 50, 70, 1000, 50, 300
Isaac Thomas O, 900, 30, 3000, 100, 600
James Cook (Coak) T, 75, -, 1500, 50, 300
Edward Callahan T, 100, 50, 2500, 100, 300
Alexander Calahan O, 12, 6, 500, 50, 200
James Francis O, 15, 15, 2000, 25, 400
James Harwood T, 60, 126, 3000, 50, 250

John D. Voshell O, 75, 50, 3500, 700, 400
Obediah Voshell O, 35, 20, 2000, 50, 125
James Wilkinson O, 18, 3, 1500, 50, 175
Thomas M. Hutchins O, 15, -, 2000, 75, 400
Solomon Price T, 11, -, 400, 25, 150
Saml. Voshell O, 50, 50, 2000, 50, 250
Edward Hubbard O, 50, 50, 2000, 150, 500
Preston Bedwell T, 55, 40, 2000, 75, 400
Jesse Monticue O, 80, 20, 5000, 150, 800
Henry Montigue T, 80, 20, 2500, 75, 400
John Ford T, 50, 8, 2000, 75, 350
William Smith T, 100, 61, 1800, 50, 200
Edward Barber O, 100, 24, 2000, 50, 350
Peter Meredith T, 930, 100, 4000, 900, 300
George Jones O, 110, 50, 5000, 150, 1000
Titus Hobbe T, 75, 20, 3000, 100, 800
Henry Pratt O, 900, 100, 6000, 150, 750
Thomas Slay O, 40, 12, 2000, 100, 350
Jonathan Kersey T, 40, -, 1600, 150, 400
Wm. S. Wallace T, 60, -, 1800, 50, 250
Darling Rash T, 60, 30, 3000, 50, 325
Joseph Smith T, 50, 40, 1800, 25, 175
John S. Smith T, 100, 50, 3000, 100, 700
Joseph Parvis T, 50, 50, 1800, 125, 240
Danl. George O, 100, 95, 4000, 200, 1000
James S. Patton T, 100, -, 3000, 50, 225
James Seager T, 50, 10, 1200, 25, 300
Thomas Purnell O, 50, 10, 1200, 25, 300
John Purnell T, 50, 50, 3000, 50, 350
Moses K. Ford O, 150, 50, 8000, 150, 800
Peter Voshell O, 20, 8, 800, 25, 150
John Taylor O, 80, 70, 3000, 50, 175
John Devon (Deronhan) O, 125, 100, 2000, 150, 600
Emily Morris T, 75, -, 1000, 50, 250
John H. Fields T, 200, 130, 6000, 150, 800
Josiah Steel O, 140, 60, 3000, 100, 600
Joseph Cramer T, 120, 160, 4000, 75, 500
Joble Foreacres T, 70, 20, 2000, 125, 500
John Johnson O, 25, 15, 1500, 50, 400
George Johnson O, 30, 25, 1000, 50, 200
William Pavis O, 15, 22, 1000, 25, 75
William Bedwell O, 60, 34, 2000, 75, 350
Avery Mavel T, 50, 50, 2000, 75, 275
Isaac Lofland O, 160, 40, 8000, 150, 800
John Slay O, 200, 30, 12000, 400, 1000
Edward Wilson T, 40, 20, 1200, 100, 200
Clayton Robinson T, 88, -, 2500, 150, 320
William Bedwell O, 40, 15, 2500, 100, 600
John Goodwin O, 75, 27, 3000, 150, 600

Philip Marvel T, 75, 150, 2500, 75, 400
David Marvel T, 75, 150, 2500, 150, 1200
Hinson C. Goodwin T, 75, 120, 4000, 50, 200
James Powell O, 60, 30, 2500, 125, 500
Nathan Soward T, 50, -, 2000, 75, 275
James Seward O, 25, 12, 1000, 50, 175
Saml. Conner O, 65, 20, 2500, 150, 600
Thomas Semay (Lemay) O, 70, 19, 3500, 150, 550
Powell Aron O, 100, 22, 5000, 150, 900
Gabriel Nickerson O, 80, 20, 3500, 100, 600
Edward Phillips T, 225, 15, 6750, 75, 600
John Seward O, 900, -, 3500, 75, 500
Saml. C. Wallace O, 150, 75, 450, 125, 900
Zadock Patten T, 80, 90, 2700, 100, 200
James H. Green O, 75, 30, 2000, 900, 225
Danl. Hinsley O, 100, 50, 3000, 125, 250
Saml. H. Miflin O, 160, 50, 8000, 75, 500
William Faulkner T, 100, -, 3000, 125, 600
John H. Bateman T, 125, 49, 8000, 80, 4000
Saml. Lockwood T, 200, 25, 8000, 50, 300
Charles Short O, 101, 30, 4500, 100, 450
James Dyer T, 60, 45, 8000, 50, 500
Thomas Pickering O, 590, 10, 12000, 200, 1500
Wm. Dawes O, 125, 75, 6000, 125, 500
John Allen M, 100, -, 4000, 75, 200
William McGonigal O, 125, 37, 9000, 200, 840
Robt. Poore T, 60, 20, 3000, 100, 300
Moses VanBurkalow O, 100, 35, 5000, 150, 300
Thomas Shockley T, 150, 35, 5000, 50, 500
James Jester T, 250, 188, 10000, 150, 800
Caleb Bell O, 10, -, 500, 25, 150
John Foster T, 50, 25, 2000, 40, 100
John Lodge O, 80, 10, 3000, 200, 300
Peter Massey O, 65, 36, 4000, 150, 500
John Voshell T, 27, -, 600, 50, 150
Saml. Wharton T, 150, 17, 1000, 150, 1000
John Plumm O, 160, 20, 12000, 250, 600
Danl. Charles O, 70, 10, 3500, 100, 275
Simeon Blood O, 137, 20, 5500, 200, 450
Edward Jackson T, 80, -, 3200,75, 300
John F. Jackson O, 200, 20, 6400, 20, 300
Eli Ernest O, 30, 13, 2200, 75, 200
Edmund Stout O, 100, 30, 8000, 300, 400
Thomas Draper O, 6, -, 1000, 50, 300
John J. Connor T, 90, 23, 4500, 50, 300
John J. Connor M, 117, 70, 12000, 50, 1500
Charles Emory T, 83, -, 5000, 70, 500
Fisher Lindale O, 85, 22, 5000, 125, 400
Jeremiah Moore T, 160, 20, 5000, 100, 900

John H. Jackson T, 400, 200, 15000, 250, 2500
Samuel Warren O, 200, 163, 20000, 600, 2000
George Warren O, 200, 68, 20000, 200, 1900
Thos. H. McIlwain O, 100, 50, 3500, 100, 710
Robt. Abram M, 100, 100, 4000, 50, 420
Henry Williams O, 150, 50, 6000, 100, 600
Joshua Wharton O, 100, 25, 4000, -, 300
Wm. Collins, 150, 40, 8000, 450, 1100
David J. Murphey, 130, 20, 10000, 400, 2450
Thomas Darps (Davis), 170, 45, 15000, 250, 3165
R. M. C. Lattomus, 138, 18, 8000, 250, 1510
Sarah F. Casperson, 150, 62, 10000, 200, 385
Henry P. Massey, 120, 6, 8000, 400, 840
Thos. Voshell, -, -, -, 50, 206
Anthony Denby, -, -, -, -, 100
John Wright, 200, 80, 15000, 340, 900
John Stevens, 90, 30, 9000, 30, 500
Robt. Mitchell, 275, 50, 12000, -, 1079
James W. Mitchel, 210, 65, 15000, 330, 1182
Daniel Oariel, 100, 100, 25000, 60, 161
Daniel Turner, 100, 50, 8000, 50, 550
C. H. Register, 220, 90, 15000, 220, 1181
Peter Robinson, 250, -, 2000, 60, 215
Isaac W. Connard, 27, -, 700, 75, 105
John Goldsborough, 75, 105, 1500, 100, 260
George Sholtess, 100, 293, 3000, 50, 360
Wm. Morgan, -, -, -, 80, 260
Thos. B. Lockwood, 156, 55, 12000, 200, 950
Tilgham Foxwell, 185, 50, 16000, 200, 780
Saml. M. Jones, 240, 20, 15000, 675, 1160
Wesley Stevens, 150, 100, 15000, 170, 540
James Mannering, 130, 25, 5000, 100, 750
George W. Knotts, 88, 10, 3000, 100, 400
Caleb Jones, 40, 17, 500, 50, 68
John H. Short, 130, 100, 3000, 200, 540
Joseph Sutton, 150, 75, 4000, 110, 429
Wm. Stevens, 100, 40, 4000, 110, 876
Wm. T. Short, 300, 12, 3000, 200, 460
Isaac Short, 200, 800, 10000, -, 240
Edmund Morgan, -, -, -, 100, 440
Wm. T. Campbell, -, -, -, 50, 80
David Lee, 20, -, 1000, 50, 260
Nehemiah Cole, 40, 300, 2000, 150, 290
Wm. Jones, 8, -, 500, -, 40
Jeremiah Robinson, 125, 75, 8000, 200, 590
Job Berrinson, 9, 21, 500, 20, 156
Thos. R. Palmatry, 40, 30, 1500, 50, 162
John W. Reynolds, 60, 30, 1500, 50, 158
John Donovan, 90, 47, 2000, 50, 242
Jeremiah Keen, 128, 4, 5000, 100, 400
Benj. Turner, 73, 7, 1200, 100, 240
James Hoffecker, 60, 15, 5000, 300, 380
Matthew Ford, 100, 10, 5000, 700, 850

Danl Palmatry, 40, 3, 2000, 200, 340
John Wilson, 45, 6, 1000, 100, 125
Nehemiah Slayton, 180, 10, 6000, 200, 625
John Bacon, 100, 5, 3000, 75, 322
Isaac Bulton (Sutton), 100, -, 6000, 200, 452
Alexander Perry, 90, 10, 35000, 100, 472
Enoch Miller, 40, -, 3000, 50, 250
James T. Carrow, 150, 200, 10000, 1500, 580
James C. Robinson, 99, 47, 3000, 150, 550
Thos. L. Sutton, 125, 412, 4000, 150, 1000
Jesse S. Vance, 34, 3, 1600, 75, 290
Jonathan Alles, 164, 30, 4000, 200, 512
Jacob Griffin, 80, 2, 1500, 125, 525
Robt. Miller M, 100, 100, 7000, 100, 390
James Boggs, 150, 150, 7000, 200, 1292
Joseph Hardall, 140, 80, 15000, 400, 1345
Thos. Crosby, 60, 50, 9000, 100, 914
Philip Huffecker, 200, 100, 10000, 300, 1000
Nathan Carrow, 120, 120, 10000, 200, 1039
Joseph Kennedy, 120, 50, 8000, 400, 1057
Job. M. Frazier, 145, 5, 8000, 150, 480
Joseph Faris, 190, 18, 12000, 400, 1235
Thos. D. Jester, 75, 17, 4000, 200, 545
James Voshell, 110, 37, 6000, 200, 525
Joshua Gay B, 6, -, 600, 50, 168
Bayard Brister B, 41, -, 1000, 50, 330
Enoch Brister M, 8, -, 400, 20,100
George B. Cole, 90, 30, 1000, 100, 238
James H. Thompson, 122, 30, 9000, 200, 726
Dennis Hudson, 50, 300, 2500, 50, 265
John R. Cameron, 75, 600, 3000, 150, 1050
James Wallace, 160, 100, -, 50, 233
Abraham Brown, 150, 300, 3000, 75, 850
Saml. Loatman, 14, 130, 1000, 25, 150
Wm.Cameron, 55, 1200, 6000, 69, 252
Nehemiah Jones, 50, 200, 7000, 20, 356
Wm. H. Colescott, 200, 180, 20000, 300, 1210
Presley Ford, 200, 150, 12000, 200, 1140
W. D. Hoffecker, 100, 75, 4000, 100, 450
Robt. Rawley, 150, 100, 4500, 100, 526
Wm. Woodkeeper, 90, 130, 2000, 100, 361
John Rawley, 70, 90, 3000, 100, 596
Wm. Palmatry, 160, -, 7000, 300, 337
John Ferrill, 120, 40, 7000, 300, 1050
Jas. Tumtleson (Tumblinson), 17, -, 900, 50, 202
Thos. Maloney, 80, 20, 3300, 100, 341
John F. Peterborough, 125, 8, 6000, 150, 380
Joseph H. Anderson, 140, 10, 5000, 100, 377
Peter S. Collins, 180, 40, 12000, 500, 1355
John M. Collins, 150, 37, 8000, 100, 345
Owen Ford, 150, 15, 7000, 100, 150
H. Palmatry, 25, -, 1000, 150, 120

Ephraim Jefferson, 150, 50, 7000, 100, 630
Jacob R. Allen, 140, 46, 10000, 250, 753
John H. Hoffecker, 150, 50, 12000, 400, 703
J. V. Hoffecker, 161, 56, 12000, 200, 1125
Elisha B. Green, 262, 63, 20000, 400, 1200
Thos. P. Wright, 75, 16, 5000, 50, 355
Elizabeth Rees, 180, 30, 16000, 300, 1292
Saml. L. Coverdale, 125, 30, 12000, 200, 900
Danl. W. Thompson, 150, 30, 14000, 1500, 100
Stephen Howard, -, -, 1000, 150, 456
Jas. B. Crawford, 250, 6, 10000, 300, 993
Clement Spiltle (Spilettle), 233, 150, 10000, 200, 970
Wm. Burgess, 10, -, 300, 50, -
Thos. L. Poor, 265, 50, 100000, 300, 1300
Thos. Lamb, 125, 75, 12000, -, 440
C. A. Dulin, 125, 75, 8000, 300, 786
Jas. B. Conner, 120, 20, 6000, 100, 497
Jos. B. Boyles, 200, 80, 6000, 100, 521
Chas. Numbers, 80, 16, 3000, 100, 400
Anderson W. Webster, 85, 5, 3000, 100, 333
John Cleaver, 100, 36, 7000, 200, 285
Mrs. Shahan, 300, 150, 17000, 300, 952
John H. Jackson, 50, 6, 1000, 40, 445
Edward Streets (Struts), 119, -, 6000, 20, 555
Isaac Green, 175, 20, 7000, 200, 620
David S. Staton (Slaton), 125, 50, 2500, 100, 452
Benj. D. Burrows, 90, 10, 3000, 136, 300
Saml. Griffin, 126, 25, 9000, 200, 800
Eben Burris, 37, -, 500, 10, 329
F. W. Griffith, 120, 25, 6500, 250, 850
Nathan T. Underwood, 200, 20, 10000, 258, 950
W. D. Burrows, 120, 55, 10000, 300, 1100
George W. Hudson, 170, 30, 1000, 150, 230
John Cloak, 95, 5, 2000, 500, 1122
Clement Reed, 200, 50, 15000, 700, 911
John Kirpatrick, 31, -, 1500, 56, 100
Joseph Reynolds, 130, 10, 10000, 300, 677
Gamial Garrettson, 120, 5, 7000, 150, 797
David Richards, 142, 15, 6500, 200, 471
Wm. R. Long, 125, 22, 7000, 200, 650
A. P. Crockett, 120, 40, 7500, 300, 858
John Baynard, 80, 35, 6000, 200, 448
John H. Rash, 145, 5, 7000, 200, 726
John M. Clarke, 97, 3, 4000, 200,388
Jas. T. Faison, 10, -, 2000, 50, 125
John W. Denny, 170, 25, 8000, 200, 576
Thos. Jackson, 150, 50, 1200, 300, 816
George W. Cummins, 165, 35, 20000, 600, 1600
Francis S. McWhorter, 230, 43, 12000, 400, 1200
Horace Spruance, 250, 30, 20000, 300, 1356
James W. Spruance, 525, 78, 30000, 200, 2000

Obediah Voshell, 150, 50, 6000, 200, 480
John B. Nelson, 150, 50, 8000, 150, 546
Wm. W. Nelson, 250, 50, 12000, 250, 738
Isaac M. Jones, 141, 40, 4000, 100, 520
J. L. Bilderback, 200, 25, 12000, 400, 970
Henry Purse, 100, 20, 5000, 100, 261
Joseph T. Rash, 94, 6, 6000, 250, 461
George View, 60, 5, 1500, 50, 250
John Young, 200, 100, 16000, 400, 155
Wm. C. Jump, 110, 40, 8000, 100, 630
John S. Harris, 760, 25, 8000, 150, 868
Wm. P. Smithers, 200, 40, 10000, 250, 1100
James Jones, 100, 18, 10000, 200, 875
James Harten, 150, -, 15000, 200, 575
Wm. N. Vanson (Ranson,Danson), 150, -, 15000, 300, 590
John P. Rees, 100, 20, 10000, 200, 632
Peter Meredith, 150, 25, 12000, 300, 791
John C. Maybery, 85, 15, 3000, 100, 296
Kees Lewis, 75, 13, 1500, 50, 252
John R. Dickson, 100, 25, 5000, 150, 516
Wm. F. Smith, 90, 20, 3300, 100, 309
Jonathan Brown, 160, 80, 12000, 500, 1256
John Severson, 150, 50, 10000, 200, 740
James S. Boyer, 24, -, 1000, 50, 217
Wm. C. Boyer, 107, 15, 7000, 150, 550
Emory Temple, 10, 33, 2500, 100, 308
Lewis M. Bell, 100, 74, 9000, 200,750
Jane Coverdale, 175, 25, 4000, 150, 392
Nathan T. Sevil, 100, 30, 5000, 150, 700
Thomas English, 60, 30, 2500, 75, 563
Jacob Streets (Struts), 125, 153, 4000, 100, 422
Samuel Johnston, 150, 50, 7000, 75, 315
Samuel R. Powell, 200, 75, 5000, 75, 571
Joseph T. Jones, 110, 40, 3500, 100, 475
James Moore, 100, 35, 3000, 100, 500
John Dulin, 133, 10, 5000, 75,493
Henry Pratt Jr., 225, 207, 10000, 200, 954
Henry S. Pratt, 175, 116, 7000, 200, 500
John Taylor, 350, 150, 18000, 400, 1600
Israel P. Hall, 150, 50, 5000, 150, 913
John P. Shelton, 90, 40, 500, 75, 232
Jane H. Conner, 100, 60, 3000, 100, 500
Edwin Parrey, 52, 8 ½, 3000, 500, 280
Thomas Maybery, 75, 65, 1500, 50, 350
Joseph C. Foreacres, 420, 200, 20000, 300, 1150
John Fenn, 160, 30, 4000, 200, 660
Nanos H. Coverdale, 210, 114, 6000, 200, 1050
Henry Hudson, 11, -, 300, 10, 198
Dellia Hutchison (Hutchinson), 39, 19, 2000, 50, 340
James S. Bilas, 80, 70, 3000, 50, 60

Thomas P. Cops, 180, 100, 4000, 50, 300
Edward Burris, 50, 30, 1200, 50, 170
Benj. F. Blackiston, 235, 150, 12000, 310, 1500
Saml. Hutchison, 120, 65, 6000, 150, 900
Thomas Attex, 70, 30, 1300, 100, 450
Wm. Hutchinson, 110, 45, 5000, 150, 800
James Hall, 60, 38, 2000, 100, 390
James R. Downs, 80, 20, 2000, 400, 240
John W. Hall, 60, 70, 2500, 50, 119
N. K. Leverage, 50, 30, 1500, 50, 325
Henry Stevens, 60, -, 1500, 50, 300
Richard Hutt, 70, 30, 3000, 50, 400
Timothy Carrow, 100, 40, 3500, 100, 420
N. C. Downs, 155, 10, 8000, 100, 70
Thomas Hyliard, 45, 55, 3000, 75, 770
Edward Attex, 200, 100, 10000, 150, 530
Stephen Attex, 60, 55, 3000, 100, 560
Jesse Jones, 130, 100, 3000, 60, 220
J. C. B. Clark, 120, 60, 6000, 75, 110
James D. Clark, 60, 40, 3000, 100, 630
John N. Clark, 75, 8, 3000, 100, 831
James Geussford (Gaussford, Gussford), 170, 30, 6000, 100, 730
Wm. H. Holding, 16, -, 25000, 100, 625
John Geussford, 150, 80, 5000, 125, 450
Mason Bailer, 125, 28, 1000, 75, 750
James D. Wilds, 60, 20, 8000, 200, 750
Thomas Sampson, 200, 117, 16000, 300, 600
Wm. M. Jones, 135, 105, 8000, 150, 700
James P. English, 90, 56, 3000, 100, 216
John Hughes, 61, 9, 1000, 50, 296
Loadman E. Downs, 70, 30, 3000, 100, 292
H. L. Derbron (Derbrow), 96, 6, 3000, 100, 500
Amos H. Streets (Struts), 160, 32, 7000, 125, 300
Robt. Graham, 33, 2, 1500, 100, 200
Joseph Thompson, 50, 50, 2000, 100, 300
John C. Farrow, 150, 20, 1000, 100, 620
Saml. S. Griffin, 60, 40, 2500, 100, 230
John Truax, 175, 50, 9000, 300, 1200
W. J. Blackiston, 150, 156, 7000, 200, 200
Jas. S. Nowell, 200, 100, 16000, 200, 1450
John Carrow, 175, 25, 5000, 150, 480
Robt. George, 60, 15, 3000, 75, 230
Wm. C. Smith, 200, 35, 16000, 300, 450
Jas. Sterling, 150, 40, 12000, 100, 910
James Truax, 170, 80, 13000, 300, 1275
John Williams, 140, 20, 8000, 150, 635
John Reasington, 90, 5, 6000, 100, 360
Alexander Reasington, 127, 10, 5000, 100, 140
Henry F. Hill, 190, 40, 12000, 300, 1400
Saml. Hargedine, 160, 140, 7000, 300, 840
John R. Griffin, 67, -, 6000, 100, 370
James P. Snow, 165, 18, 8000, 200, 1070
John A. Serin (Savin), 115, 19, 4000, 50, 490

Benj. F. Lattimus, 200, 50, 1000, 200, 1300
Jos. Snow, 130, 100, 6000, 200, 850
Robt. P. Collins, 136, 56, 9000, 200, 1160
John Slaughter, 250, 30, 10000, 300, 1400
Henry Slaughter, 220, 25, 25000, 200, 1200
Wm. H. Hoffecker, 175, 150, 12000, 200, 1000
Joseph Voshell, 160, 80, 10000, 75, 800
John Reynolds, 160, 40, 10000, 200, 790
John Jones, 200, 10, 12000, 150, 950
Presly Hoffecker, 140, 40, 12000, 200, 1150
Wm. Jones, 100, 250, 5000, 175, 140
Henry T. Hoffecker, 180, 1000, 10000, 300, 1430
Robt. Davis, 112, -, 9000, 100, 430
Elizabeth Mamson, 150, 5, 4000, 100, 290
Wm. Thompson, 170, -, 15000, 150, 760
Elbert Downs, 150, 150, 4000, 50, 400
David O. Downs, 50, 90, 4000, 150, 400
Jos. C. Downs, 40, 60, 2000, 75, 400
Jas. H. Geussford, 140, 70, 5000, 80, 430
Wm. Taylor, 200, 50, 9000, 150, 1152
Joshua A. Short, 40, 76, 1500, 100, 350
Samuel Kelly, 20, 20, 1750, 75, 230
John W. Boyles, 20, 20, 250, 50, 126
George Ford, 60, 47, 1500, 100, 500
Thos. Scuse, 100, 96, 3000, 100, 490
Oliver B. Hustand, 140, 35, 8000, 150, 640
Wm. L. Hazel, 70, 30, 2000, 200, 780
James P. Hazel, 100, 20, 6000, 50, 350
Jos. Wallace, 50, 58, 3000, 50, 350
Wm. W. Morris, 100, 46, 3000, 100, 245
John S. Rash, 30, 45, 1500, 30, 230
John Numbers, 90, 43, 3000, 100, 300
Jas. H. Shahan, 150, 150, 4000, 100, 300
Randle Brayman, 25, 5, 1500, 50, 150
Enoch David, 85, 83, 4000, 125, 460
Gustavus Carney, 60, 32, 3000, 20, 130
Wm. H. Carney, 30, 30, 3000, 20, 120
John B. Gooding, 150, 52, 8000, 150, 500
Mathew Hazel Jr., 100, 55, 4000, 200, 600
James Ford, 50, 10, 2000, 100, 800
Greensbury H. Short, 45, 20, 1000, 75, 260
Isaac J. Short, 150, 150, 4000, 100, 600
James M. Short, 60, 15, 2000, -, 319
Sarah H. Purge, 200, 42, 4000, 200, 850
Wm. Hyliard, 120, 20, 3000, 50, 263
Peter Biddle, 70, 12, 1500, 10, 210
Wm. H. Walcott, 90, 15, 3000, 50, 330
Wm. Boyd, 145, 55, 4000, 100, 320
Jacob Bryon, 50, 23, 1000, 25, 175
Martin McGralets, 130, 50, 5000, 120, 400
A. M. Marshal, 160, 40, 10000, 100, 1030
T. F. Thurlow, 150, 25, 8000, 200, 581
James Williams, 160, 13, 10000, 400, 1050
Jas. P. Heaton, 300, 100, 16000, 1000, 1600

John T. Everitt, 200, 60, 10000, 200, 490
John Petry, 850, 100, 30000, 500, 2600
Wm. S. Jones, 60, -, 2000, 100, 351
John T. David, 150, -, 5000, 75, 400
Thomas Moore, 65, 80, 3500, 100, 580
Isaac Salmons, 80, 40, 3000, 150, 580
Martin Ford, 60, 45, 2500, 120, 290
D. H. Dunning, 50, 230, 2000, 125, 212
James M. Dodd, 50, 90, 3000, 25, 340
John Moore, 100, 65, 5500, 100, 680
John Hammond, 40, 6, 1000, 50, 152
Robt. Dilmans, 200, 50, 2500, 100, 480
Elisha Durham, 72, 8, 1500, 50, 160
James Barcue, 10, 62, 6000, 300, 420
Wm. Pruett, 275, 110, 8000, 125, 800
Jas. S. Moore, 40, 25, 2500, 50, 415
Wm. S. Bishop, 100, 30, 4000, 15, 680
David N. _. Cloud, 40, -, 3000, 50, 520
Benj. T. Knobbs, 100, 50, 3000, 25, 551
David Boggs, 100, 96, 6000, 125, 520
Nathaniel Barrett, 75, 225, 4000, 75, 326
Ephraim Garrison, 140, 82, 8000, 200, 820
Elihu Jefferson, 100, 60, 5000, 100, 550
Joseph Hoffecker, 62, 10, 8000, 100, 530
Benj. F. Hurlick, 150, 51, 8000, 150, 1200
Wm. Surgent, 150, 50, 8000, 150, 320
Moses Price, 140, 20, 5000, 150, 720
John R. Dean, 150, 25, 5000, 100, 600
Stephen T. Williams, 120, 80, 5000, 100, 600
George C. Simpson, 600, 210, 20000, 370, 1320
Jacob Williams, 135, 18, 10000, 500, 1490
Wm. J. Blackiston, 150, 150, 2000, 200, 00
John C. Slaughter, 80, 15, 6000, 150, 500
John C. Wilson, 143, 30, 10000, 400, 1360
John Barcus (Barcue), 120, -, 4000, 100, 415
Thos. K. Taylor, 190, 40, 8000, 200, 760
Andrew N. Harper, 165, 80, 4000, 100, 500
Thos. S. Harper, 210, 10, 3000, 100, 483
John Clow, 30, -, 2000, 50, 380
James Cummins, 258, 90, 3000, 75, 600
Joseph C. Disch (Dixon), 160, 40, 4000, 100, 500
Edwd. E. Parmer, 60, 40, 5000, 100, 630
George Boyer, 150, 50, 6000, 200, 145
Saml. Daniels, 90, 10, 3000, 100, 280
Thos. R. Boyer, 170, 30, 6000, 100, 270
Wm. Smith, 160, -, 7000, 100, 840
Thos. Keith, 100, 25, 3000, 75, 500
Abraham M. Fox, 100, 200, 5000, 100, 490
George W. Moore, 24, 3, 1400, 60, 250
Benj. Husbands, 70, 12, 4000, 50, 601
Benj. Wallace, 70, 15, 3000, 150,470
Benj. Sherand, 40, 20, 3000, 50, 280

Robt. B. Jump, 200, 25, 12000, 200, 790
Danl. Ford, 6, 7, 2000, 100, 440
Wm. Williams, 160, 50, 400, 50, 700
Wm. Ennis, 160, 50, 4000, 50, 700
Truett Melvin, 150, 40, 3000, 100, 480
John J. Simms, 200, 140, 10000, 100, 690
Mathew Hutchinson, 300, 800, 15000, 150, 1200
John Stafford, 80, -, 3500, 150, 540
Jacob Farrow (Farron), 60, -, 2500, 50, 260
Wm. G. Hazel, 112, 10, 7000, 100, 520
Saml. Hutchinson, 160, 40, 8000, 150, 650
Isaac M. Denny, 140, 80, 5000, 120, 700
James Hammell, 100, -, 5000, 50, 720
Henry K. Hazel, 80, 20, 5000, 50, 5130
Walker Mifflin, 200, 30, 8000, 125, 800
Ann Cubbage, 150, 20, 8000, 300, 685
Lemuel Laws, 200, 70, 6000, 100, 340
Wm. Fowler, 150, 50, 9000, 10, 750
___. Hasington, 156, 1, 8000, 150, 1000
Edward Hasington, 127, 12, 7000, 300, 90
Wm. J. Lank, 250, 50, 8000, 75, 590
James H. George, 120, 40, 4000, 120, 390
John Clark, 70, -, 2000, 175, 346
John Parker, 200, 80, 10000, 200, 1220
Jacob M. Hill, 230, 13, 16000, 125, 1500
A. Tomlinson, 175, 75, 10000, 100, 780
John Parker, 40, 100, 2100, 15, 356
Abraham Moore, 100, 135, 3000, 5, 224
Henry H. Moore, 100, 40, 3000, 100, 720
John Robinson, 100, 20, 2000, 100, 402
Nathaniel Newman, 150, 150, 4000, 75, 647
Wm. Husbands, 200, 100, 6000, 100, 761
Wm. Williams, 100, 15, 6000, 25, 431
Nathan Moore, 120, 30, 6000, 100, 467
George Graham, 97, 30, 8000, 200, 610
Harris Potter, 180, 4, 12000, 500, 850
Wm. Dill, 80, 80, 5000, 200, 450
Houstin McCauley, 80, 20, 2000, 50, 190
Wm. Cowgill, 325, 75, 34000, 200, 2110
Wm. Webb, 160, 15, 5000, 50, 450
Sam. Nowell, 140, 50, 10000, 100, 1070
Mary Courdright (Coodright, Goodnight), 100, 20, 4000, 100, 560
Havet Knight, 22, 35, 2500, 200, 1240
Joseph Seward, 155, 15, 8000, 150, 800
W. Hayes, 250, 30, 18000, 300, 1900
Robt. Collins, 300, 400, 10000, 100, 376
Eben Lewis, 250, 300, 10000, 300, 146
Peter Loper (Losser), 100, 87, 4000, 100, 800
Thos. W. Wilson, 325, 20, 18000, 1000, 1900
James Robinson, 100, 20, 12000, 50, 661
Edward Burris, 90, 2, 3000, 50, 440
John Wordale, 200, -, 12000, 20, 1460

James Veils, 60, 5, 4000, 100, 700
Ezekiel Clark, 40, 15, 2000, 50, 380
Wm. Losser, 50, 80, 2500, 100, 400
Saml. C. York, 80, 75, 3000, 50, 670
George Shelton, 120, 100, 3000, 100, 500
Wm. Daylis, 90, 20, 6000, 100, 500
Danl C. Cowgill, 120, 25, 7000, 230, 1590
Timothy Brown, 100, 50, 4000, 150, 540
Wm. Fox, 70, 25, 5000, 150, 450
Jos. B. Moore, 10, 75, 5000, 100, 600
Parker Handly, 100, 100, 3000, 50, 630
Garret Casperson, 30, 8, 1000, 100, 180
John Anderson, 70, 100, 3000, 151, 470
James P. Cullen, 90, 42, 4000, 50, 375
William H. Bethands (Bethards), 125, 74, 5000, 25, 220
James Sullivan, 125, 74, 5000, 50, 364
James Warrington, 100, 40, 300, 15, 160
Purnell Postles, 208, 115, 8000, 50, 600
Dickerson H. Meredith, 50, 50, 1300, 50, 145
Sylvester Welk (Welb, Webb), 15, 15, 180, 35, 125
Daniel Mitten, 15, 2, 600, 30, 175
Joshua Torbit, 80, 10, 1500, 50, 205
Benjamin P. Needles, 18, 40, 1000, 50, 296
John P. Needles, 40, 12, 1000, 50, 200
Nehemiah Scott, 10, 24, 500, 20, 87
John Erexson, 150, 150, 5000, 200, 825
Mathew Mitten, 150, 100, 5000, 100, 611
Warener Townsend, 100, 300, 5000, 35, 140
William B. Mitten, 125, 50, 3500, 50, 325
Winlock Tomlinson, 100, 76, 4000, 100, 305
John W. Kirby, 84, 20, 4000, 100, 390
Beniah Truitt, 60, 10, 3500, 30, 320
Thomas E. Virden, 30, 77, 2500, 10, 104
William T. Masters (Masten), 84, 30, 6000, 50, 456
Clement Masten, 40, -, 2000, 25, 208
Joseph French, 150, 30, 5000, 50, 482
David K. Watson, 95, 20, 3500, 100, 518
Jacob Quillen, 80, 40, 3500, 45, 486
Dickerson N. Meredith, 150, 50, 4000, 100, 771
James D. Sipple, 60, 40, 3000, 50, 656
John W. Wells, 300, 50, 10000, 130, 495
James Thomson, 60, 14, 1500, 100, 756
David Hall, 30, 10, 1000, 25, 405
Revl Horsey, 100, 75, 8000, 200, 1210
Joshua Bennett, 50, 28, 3500, 125, 600
Thomas W. Welb, 45, 33, 3500, 100, 555
William A. Polk, 200, 500, 10000, 200, 1223
Riley Linch, 80, 20, 11000, 150, 592
John Bennett, 36, -, 1000, 50, 210
William B. Greg, 8, -, 250, 10, 140
Levin Thomson, 70, 130, 2000, 30, 95
Nehemiah Cole, 60, 60, 1500, 15, 223
John M. Bennett, 150, 50, 2500, 35, 250

Gy___ Duniphe, 160, 40, 2000, 20, 242
John Thompson, 200, 120, 7000, 60, 333
Ruben Harington, 120, 22, 3500, 40, 436
Thomas Linch, 110, 70, 4000, 75, 735
William Tomlinson, 191, 25, 4500, 175, 765
Daniel R. Tomlinson, 120, 40, 5000, 400, 350
Clayton C. Moore, 30, -, 1000, 15, 325
Robert Riley, 130, 20, 4500, 60, 240
Mary A. McGinnis, 125, 145, 6000, 40, 310
Samuel Moore, 150, 50, 6500, 40, 483
John H. McGinnis, 85, 20, 4000, 50, 207
Catherine Mason, 60, 10, 1800, 20, 110
Selby Thompson, 100, 70, 2000, 60, 315
Charles Mills, 100, 35, 3000, 100, 200
John Short, 100, 45, 3000, 120, 1130
Lister H. Howston, 80, 18, 1200, 100, 150
Zadock Postles, 70, 39, 4000, 125, 495
Mark Cole, 60, 40, 3000, 50, 622
Andrew Thomas, 15, 4, 300, 15, 160
Alexander Masten, 40, 80,1800, 28, 148
Emanuel Harmon, 20, -, 300, 15, 90
Frederick Wright, 30, 10, 800, 25, 100
John T. Jester, 13, -, 500, 25, 65
John A. Hall, 40, 40, 1000, 25, 232
Manlove Cole, 12, 12, 800, 25, 130
Alfred Driggins, 12, 18, 500, 15, 60
Obediah Macklin, 30, 30, 1000, 20, 212
Charles Jester, 70, 40, 3300, 100, 430
Richard Mills, 55, 78, 1500, 100, 421
A__ Jester, 11, 5, 500, 30, 112
Andrew J. Maloney, 80, 91, 2800, 25, 279
John Coverdill, 80, 100, 3500, 75, 380
James C. Mitten, 70, 40, 3300, 30, 400
James J. Wyatt, 80, 120, 3000, 100, 436
Robert Scott, 14, 3, 320, 12, 240
George Truitt, 30, 35, 1300, 25, 320
Zachariah Duniphe, 36, -, 700, 12, 236
James Hudson, 250, 60, 5000, 100, 1000
John Maloney, 80, 90, 3500, 20, 354
Luther M. Ander (Andes), 200, 60, 300, 75, 550
Isaac Johnson, 40, -, 1500, 20, 300
David Duglass, 60, 36, 2000, 25, 36
John Richards, 50, -, 700, 20, 65
John Davis, 80, 60, 2500, 100, 686
William Walker, 30, 2, 200, 5,140
Eli R. Wadkins, 60, 5, 2000, 55, 481
Charles Webb, 80, 40, 1800, 15, 78
John Maloney, 50, 40, 2000, 20, 283
Peter Torbert (Torbit), 60, 100, 2000, 25, 215
Philip Torbert, 25, 72, 600, 25, 216
John Moseley, 75, 80, 1500, 30, 257
Elisha Moseley, 50, -, 500, 20, 170
Caleb Bell, 70, 80, 3000, 100, 455
William Banning, 40, -, 5600, 35, 35
John W. Hall of Wm., 28, -, 700, 10, 212
William Fowler, 30, 28, 1200, 15, 100
Daniel Clifton, 24, 20, 1000, 20, 55
Samuel Herring, 119, 310, 4000,80, 330
Absalom Hill, 40, 56, 1500, 100, 225
Abner Herring, 50, 20, 1400, 10, 137

William H. Richard, 40,47, 1000, 50,470
Nahaniel B. Thomas, 18, 52, 800, 50, 255
John A. Bickel, 120, 60, 2000, 150, 230
Ezekiel FitesGerald, 45, 42, 1600, 50, 354
William Hall, 120, 300, 8000, 200, 1053
George Thomas, 35, 16, 1000, 104, 335
Henry C. Richards, 60, 30, 1200, 20, 85
John M. Masten, 20, 36, 600, 20, 154
Isaac Jester, 25, 50, 600, 25, 107
Mary Merrideth, 100, 40, 3500, 100, 450
Thomas Evans, 10, 64, 1000, 20, 110
Nathaniel Horsey, 87, 30, 800, 10, 240
Nehemiah Fountain, 40, -, 400, 20, 210
Curtis B. Beswick, 80, 70, 5600, 150, 550
Charles Hudson, 40, 40, 1000, 20, 83
Solomon Hevelow, 33, 22, 600, 15, 94
George Taylor, 50, 5, 800, 15, 120
John W. Hammon, 125, 30, 4500, 200, 654
Nathan Davis, 75, 85, 3000, 20, 112
Peter Walton, 30, -, 450, 15, 70
George Davis, 160, 70, 6000, 50, 600
William B. Daniel, 75, 50, 3500, 50, 500
William Richards, 100, 50, 4500, 30, 260
Elias Hammon (Hammons), 100, 160, 4000, 75, 375
John Redden, 104, 375, 8000, 65, 455
Littleton Daniel, 100, 80, 5000, 75, 208
James Davis, 60, 20, 3200, 150, 458
Thomas J. Sammons, 40, 25, 2500, 100, 550
Eli F. Hammons, 100, 25, 3000, 75, 581
Nicholas D. Hammons, 100, 25, 3000, 20, 153
Isaac Turner, 35, -, 400, 15, 55
Joseph Houston, 75, 25, 2300, 60, 340
Hezekiah Masten, 75, 90, 3300, 90, 557
Davis H. Mason, 100, 60, 4000, 85, 300
James H. Postles, 100, 35, 3500, 75, 545
Joseph Quiller, 130, 35, 6000, 30, 375
Samuel Draper, 133, 75, 6000, 100, 690
James Postles, 92, -, 6000, 170, 600
William Morgan, 70, 80, 2200, 25, 140
Joshua H. Hill, 105, 72, 5000, 100, 270
John Wilkerson, 50, 15, 1800, 100, 431
Evan Morgan, 75, 75, 2500, 30, 150
Rachel C. Stevens, 67, 33, 2500, 60, 335
By__ Henderson, 50,60, 000, 50, 290
Stephen M. Collins, 67, -, 6000, 70, 385
Robert Carpenter, 60, -, 4000, 75, 325
Robert Griffith, 20, 2, 800, 22, 527
William Pasley, 75, 25, 1500, 100, 260
John Redden Jr., 80, 120, 2000, 10, 125
Hillard Griffith, 80, 25, 3500, 40, 147
Joseph Yardley, 150, 61, 10000, 400, 674
James Smith, 75, 30, 3300, 40, 442
Joseph D. Covington, 90, 23, 200, 30, 325

Zachariah Reynolds, 150, 50, 10000, 150, 505
James R. Duniphe, 60, -, 2400, 65, 240
James H. Webb, 100, 25, 5000, 35, 291
Charles Horsey, 110, 57, 5000, 35, 330
George S. May, 160, 115, 10000, 50, 435
Joseph H. Owens, 100, 67, 3500, 50, 320
Charles Schock, 100, 29, 4500, 175, 290
James Vinyard, 87, 52, 3500, 30, 190
Spencer Hitch, 75, 39, 2500, 50, 193
John Quillen, 140, 157, 9000, 160, 760
Robert Lynch, 150, 50, 6000, 65, 300
William E. Torbet, 220, 145, 6600, 100, 427
Joshua C. Craner, 105, -, 2600, 80, 675
Elijah Satterfield, 135, 25, 5500, 100, 765
George McColley, 75, 47, 2000, 75, 420
John Rawley, 160, 101, 4000, 150, 1040
Thomas B. Henderdine, 120, 80, 6000, 250, 430
Mary Masten, 26, 10, 1000, 20, 87
Asa Benton, 53, 16, 2000, 80, 185
Rachel Townsend, 75, 68, 2500, 35, 125
James L. Melvin, 150, 50, 6000, 125, 420
John M. Lofland, 175, 45, 4500, 175, 388
Frances N. Lattimus (Lattimore), 150, 50, 6000, 25, 430
William Tucker, 140, 60, 5000, 60, 500
Aaron Owens, 38, 12, 650, 20, 100
William Clark, 75, 75, 1500, 40, 244
Robert Adkins, 200, 60, 8000, 75, 575
Joseph O. McColley (McCalley), 55, 25, 8000, 150, 408
John H. Hitch, 85, 5, 1000, 15, 165
James Sharp, 140, 135, 9000, 150, 600
William J. Tapp, 75, 325, 8000, 75, 500
Benjamin Duniphen, 65, -, 1200, 35, 245
Broffet Satterfield, 70, 60, 1300, 5, 110
Alfred Newsom, 210, 110, 3000, 100, 136
James B. Ross, 58, 49, 1500, 40, 222
Mager H. Harrington, 120, 60, 3000, 75, 420
Thomas W. Tomlinson, 60, 20, 2500, 20, 185
James D. Tomlinson, 85, 15, 1500, 75, 331
Isaac Harrington, 180, 43, 3600, 20, 87
John M. Oldfield, 200, 70, 5000, 50, 452
Moulton Jacobs, 175, 30, 5000, 40, 315
George T. Morris, 25, -, 500, 5, 125
Josiah Marvil, 115, 20, 3500, 75, 400
Peter Scott, 150, 77, 2200, 35, 232
David Redgester, 150, 77, 2200, 30, 118
James C. Hill, 65, 10, 1000, 20, 240
James Sapp, 110, 40, 6000, 150, 385
John Weekes, 47, 12, 2500, 60, 136
Mary Smithers, 19, 5, 800, 10, 46
M. & J. S. Herrington, 229, 79, 10500, 230, 920
Eliza Adkins, 120, 228, 3500, 50, 275
Henry Vingard, 60, 52, 1500, 10, 195
Asa Dickerson, 60, 30, 1000, 10, 90
Curtis Vinyard, 100, 50, 4500, 100, 300

William Collier, 12, 50, 620, 10, 27
Joshua Johnson, 20, 40, 600, 18, 95
James M. Mosley, 40, 60, 1500, 15, 82
Adam Marvil, 75, 65, 1500, 20, 75
William Marvil, 100, 50, 1800, 50, 308
George Reece, 21, -, 550, 15, 92
Henry Clarkson, 16, 44, 800, 20, 85
____ Marvil, 100, 169, 5000, 25, 288
Emry Marvil, 80, 117, 2000, 45, 270
Clement L. Sharp, 125, 113, 6000, 100, 947
William D. Griffith, 100, 66, 2000, 100, 610
Henry N. Marvil, 75, 145, 2000, 15, 235
Joseph Frazier, 155, 25, 600, 80, 515
Saml. A. Short, 22, 36, 1500, 50, 240
William Benton, 200, 75, 10000, 140, 905
Henderson Collins, 31, -, 3100, 25, 250
James P. Richards, 35, -, 800, 20, 842
John Hill, 400, 100, 3000, 125, 360
William H. Satterfield, 50, 20, 1200, 35, 130
John Duniphen, 80, 160, 3000, 65, 195
James Abbott, 36, -, 400, 24, 45
Elias T. Lofland, 140, 125, 3000, 60, 225
Robert H. Wyatt, 160, 40, 2500, 25, 85
John Harrington, 90, 10, 1200, 25, 187
William M. Harrington, 100, 135, 4000, 145, 512
Nathaniel Minner, 4, -, 300, -, 21
Ephraim Dill, 80, 77, 3300, 25, 424
Charles Warren, 200, 120, 8000, 200, 950
Johnathan Minner, 125, 87, 3000, 30, 250
Samuel S. Harrington, 140, 67, 4200, 100, 475
John Lewis, 100, 60, 2500, 60, 300
George Graham, 200, 65, 4500, 100, 500
Hosea Harris, 40, -, 800, 20, 90
Curtis Calaway, 75, 75, 2000, -, 40
William D. Edwards, 100, 54, 4000, 50, 180
John Minner, 45, 5, 1000, 50, 425
Luff J. McKnatt, 45, 4, 1500, 20, 137
Waitman Clark, 160, 35, 5000, 50, 370
Benjamin Williams, 80, 176, 5000, 75, 522
John Booth, 175, 60, 4700, 75, 480
James A Mooree (Moorse), 180, 100, 4000, 75, 400
John R. Curtis, 40, 64, 2000, 30, 120
John D. Hill, 100, -, 1000, 15, 185
Joseph Booth, 120, 28, 3000, 100, 405
John T. Sapp, 300, 150, 9000, 130, 867
Thomas T. Melvin, 40, 20, 1000, 40, 140
James Anderson, 300, 35, 6000, 130, 550
John Satterfield, 170, 148, 5000, 100, 275
John W. Slayton, 90, 93, 1800, 25, 75
Sarah Harrington, 175, 62, 4700, 30,175
James Scott, 75, 50, 1600, 25, 30
Wm. B. FitzJarrel, 70, 26, 1000, 30, 225
Nicey Williams, 240, 60, 4000, 30, 175
Joseph A. Masten, 100, 50, 2250, 75, 330
John Wyatt, 50, 11, 1800, 65, 364
Samuel Minner, 40, 10, 1000, 40, 140
Elijah Sapp, 80, 20, 2000, 45, 460

Robert Harrington, 100, 40, 2000, 25, 230
Jefferson Fisher, 20, 4, 500, 30, 300
Thomas H. Jester, 130, 70, 4500, 135, 750
James G. Jester, 60, -, 1400, 55, 336
John Whitaker, 125, 75, 3000, 100, 460
Susan Reed, 75, 20, 1200, 40, 545
Jacob Hartzel, 150, 150, 3000, 100, 350
James W. Smith, 125, 25, 2500, 30, 420
John W. Reed, 60, 6, 500, -, -
James Hughes, 75, 81, 1200, 35, 435
James Duniphen, 60, 15, 2000, 75, 385
William Minner, 20, 28, 1200, 30, 135
William Camper, 30, 20, 1000, 35, 333
Thomas H. Laramer, 15, 5, 600, 20, 95
Thomas Laramore, 40, 40, 1800, 30, 256
Samuel Scott, 40, 35, 1000, 20, 45
Soloman Gormer, 50, 50, 8000, 15, -
Mathew Fleming, 25, 75, 1500, 20, 260
Jacob Hickman, 25, 35,800, 30, 160
James B. Fisher, 30, 67, 1400, 25, 235
Joshua Reed, 75, 77, 1500, 25, 33
James Hopkins, 140, 160, 5000, 125, 661
William Scott, 30, 30, 2000, 40, 417
Levi Scott, 20, 30, 1000, 25, 40
William Langrell, 40, 20,800, 25, 56
Joseph Laramore, 20, 30, 1000, 25, 167
Tuff (Luff), Sipple, 35, 35, 1500, 30, 152
James M. Killen (Rillen), 20, 17, 400, 20, 56
Ezekiel S. Cooper, 70, 48, 2000, 50, 290
Jacob Welsh, 20, 60, 800, 25, 75
George W. White, 50, 50, 1000, 12, 46
Samuel Price, 175, 125, 4000, 50, 350
Allen Thomas, 90, 50, 2100, 25, 247
William Wheeler, 60, 40, 1000, 25, 206
James Porter, 90, 10, 2500, 35, 340
John Travis, 40, 40, 500, 20, 225
James H. Price, 40, 70, 1000, 25, 145
William H. Thomas, 40, 60, 900, -, -
William Sapp, 200, 100, 7500, 125, 1437
Ebenezer Heuges, 65, 35, 1500, 35, 300
James McKnatt, 30,70, 1000, 25, 150
George White, 30, 45, 800, -, -
Pheleman Dill, 125, 15, 1680, 60, 257
James H. Smith, 200, 80, 6000, 200, 1115
Asberry Duniphen, 300, 300, 12000, 250, 320
Riley Melsen, 175, 40, 5000, 150, 630
Henry Demby, 100, 30, 800, 15, 35
Isaac Melvin, 54, 8, 1200, 10, 110
Charles Baynard, 75, 25, 1800, 15, 265
James Stafford, 200, 43, 6000, 300, 750
John W. Stafford, 75, 15, 1100, 100, 525
Ferdinand Baynard, 230, 70, 8200, 125, 630
Henry Thrauley, 100, 25, 1500, 30, 240
Samuel Lewis, 90, 28, 1800, 200, 500
John R. Wyatt, 200, 70, 4000, 30, 345
John T. Carter, 84, 76, 2000, 60, 363
John Calaway, 90, 70, 2000, 65, 256
John Cain, 140, 160, 3000, 150, 490

Nathan Smith, 100, 50, 3000, 290, 1230
Nathan Smith Jr., 90, 40, 3000, 50, 335
Elijah Benson, 120, 20, 2000, 50, 400
John Porter Jr., 90, 110, 2000, 75, 135
Nathan Scott, 50, 50, 1000, 15, 25
Richard J. Harrington, 80,60, 1600, 40, 500
William E. Cahall, 100, 40, 1800, 40, 500
Mary Breburne, 70, 40, 2000, 50, 205
John A. Cahall, 120, 60, 2500, 100, 541
James R. Morris, 90, 20, 1500, 30, 136
Oliver Drape (Draper), 90, 40, 2000, 75, 380
James Voss, 30, 20, 1000, 40, 143
George Cally, 92, 34, 2000, 150, 425
Martin W. Maloney, 80, 28, 2000, 100, 445
James Sullivan, 150, 400, 8000, 1100, 355
Elias Turner, 25, -, 375, 15, 100
Daniel Wyatt, 100, 50, 2500, 125, 600
Alexander Baker, 50, 50, 1100, 20, 137
John J. Dansick, 175, 100, 3000, 125, 545
John Brown of T, 80, 120, 2500, 200, 550
Robert Raughley, 200, 154, 5000, 100, 832
John Wyatt, 40, 20, 1000, 125, 630
Zadoc Sipple, 20, 20, 500, 40, 120
Thomas Webster, 60, 45, 1200, 55, 120
William Tavies, 24, 12, 400, 25, 150
William Harrington, 26, 7, 300, 15, 150
Daniel Anthony, 135, 40, 2300, 30, 190
Benjamin Anthony, 50, -, 750, 20, 120
Waitman Hopkins, 50, 150, 2000, 60, 327
William W. Wix, 95, 2, 2500, 100, 475
Elizabeth Wix, 100, 50, 3000, 125, 450
James P. Lister, 65, 35, 1000, 20, 160
James Barwick, 90, 6, 1500, 40, 165
Major Wyatt, 100, 50, 2000, 75, 415
James Thrawley, 100, 200, 3000, 25, 180
James Porter, 100, 180, 2500, 50, 505
George H. McKnatt, 80, 60, 2500, 75, 530
Nathan Morgan, 116, 34, 900, 20, 240
Samuel Harrington, 56, 50, 1500, 100, 366
Henry Ables (Abler), 43, 10, 600, 20, 110
Waitman Hopkins, 150, 154, 3000, 100, 222
Samuel Hopkins, 70, 40, 1000, 75, 165
Thomas Brown, 100,9, 3000, 35, 650
Outten Anderson, 150, 100, 12500, 300, 280
Robert Anthony, 57, 20, 925, 25, 225
James H. Calaway, 50, 103, 1700, 56, 204
David Taylor Jr., 100, 150, 3000, 100,700
David Sylvester, 80, 5, 1500, 30, 231
William Vikley, 75, 12, 1800, 75, 325
Thomas Reynolds, 120, 10, 2600, 100, 414
Levi Bowen, 65, 12, 1500, 35, 256
William Anderson, 106, 65, 2308, 40, 345

S___ A. Graham, 90, 10, 1500, 25, 150
James Manlore, 20, 31, 1000, 15, 45
Wesley Brown, 60, 40, 900, 35, 160
Charles Rekards, 80, 80, 1600, 20, 160
Eli Doherty, 100, 87, 4000, 100, 800
Charles Outten, 60, 10, 2000, 40, 340
William Jones, 45, 40, 1500, 50, 213
Edward Collins, 50, 50, 2000, 20, 35
Ruben Ross, 150, 100, 6250, 150, 80
Joseph Clymer, 50, 20, 1400, 50, 185
Sidney Melvin, 130, 75, 3800, 150, 661
John Porter, 100, 125, 2500, 150, 600
Mary Ferns, 60, 20,800, 30, 108
Elias Turner, 75, 25, 1000, 20, 90
Tamzy Williams, 40, 40, 600, 6, 46
Samuel Carlisle, 150, 50, 1500, 20, 164
Thomas Fearns, 75, 50, 1500, 25, 222
James Layton, 105, 65, 1750, 60, 576
Martin Smith, 150, 44, 2500, 50, 545
Garrison Saulsberry, 170, 80, 4000, 50, 545
Frederick Black, 150, 275, 5000, 60, 430
William J. Layton, 50, 60, 1000, 15, 96
John Feris (Ferns), 100, 25, 800, 30, 260
Bure (Buse) Smith, 80, 60, 2000, 30, 150
John Baker, 50, 60, 1400, 25, 60
Rubin Carpenter, 30, 25, 700, 25, 140
Edandels And___, 70, 50, 1400, 20, 100
Curtis Hopkins, 120, 48, 1700, 25, 128
William Hopkins, -, -, -, -, 150
Nathaniel Pleasanton, 80, 20, 1500, 55, 428
Jesse Ward, 100, 43, 5000, 100, 642
John W. Smith, 120, 60, 3500, 150, 755
Tuff (Luff) Cohall, 165, 60, 5000, 1225, 357
William Turner, 125, 125, 5000, 75, 230
Hooper B. Hopkins, 80, 70, 2000, 100, 5655
George W. Porter, 115, 25, 2000, -, 18
Eli Wooten, 90, 40, 3000, 100, 350
Charles Wooten, 70, 20, 2500, 75, 467
William Ga__er, 40, 25, 1050, 30, 130
John Hopkins, 196, 96, 2500, 100, 713
Green Powtain, 75, 50, 2000, 100, 554
Zebulon Hopkins, 100, 60, 2500, 180, 590
James H. Fisher, 125, 60, 3000, 35, 202
Artemus Smith, 50, 22, 1100, 35, 100
John H. Hardesty, 75, 30, 2000, 40, 250
Henry Saulsbery, 50, 50, 800, 125, 415
John V. Callaway, 80, 60, 2000, 100, 603
Thomas J. Saulsbery, 50, 50, 1200, 15,100
Charles Willliamson, 200, 65, 5300, 100, 658
Willis W. Butler, 60, 44, 1500, 30, 248
John Hickman, 60, 44, 1500, 40, 230
Nicholas O. Smith, 90, 40, 2000, 50, 415
Henry Adkinson, 100, 103, 3300, 100, 340

Johnathan Wilson, 30, 23, 1320, 30, 200
Samuel Porter, 50, 62, 5000, 24, 170
Richard Peters, 60, 90,800, 30, 302
Mathew Parris, 120, 60, 2000, 60, 503
David W. Smith, 100, 40, 1000, 20, 215
Charles N. Adams, 30, 30, 1500, 100, 1083
Thomas Currey, 50, 30, 600, 20, 80
John Lane, 135, 90, 2500, 85, 385
Benjamine Thistlewood, 110, 12, 7500, 85, 210
William Ellensworth, -, -, -, -, 100
Samuel Powel, 30, 10, 400, 15, 20
Henry Downs, 120, 50, 2500, 30, 329
Daniel Harrington, 60, 35, 3000, 50, 440
Mark Cooper, 11, 4, 250, 12, 28
Alexander Harrington, 125, 30, 2500, 33, 265
David Harrington, 30, 55, 3000, 125, 344
Nimrod Harrington, 74, 45, 5000, 125, 342
Thomas Harrington, 100, 60, 2000, 100, 430
William Harrington, 100, 50, 1000, 30, 266
Daniel Bowman, 60, 25, 900, 20, 75
Robert H. Short, 125, 200, 2500,80, 505
Joseph Barker, 30, 100, 2000, 20, 85
Binus Benson, 25, 110, 2000, 25, 154
William Masten, 190, 180, 10000, 125, 1045
William H. Masten, 150, 220, 7000, 100, 760
Isaac Wyatt, 100, -, 1400, 25, 240
Joseph Jarvis, 95, 25, 2200, 20, 145
Joseph Welch, 100, 90, 3000, 100, 260
George Cox, 250, 350, 7000, 100, 691
Samuel Tuttle, 15, 15, 450, 15, 95
Richard Harrington, 60, 70, 2000, 25, 150
Henry N. Clark, 100, 45, 2000, 100, 725
William W. Simpson, 75, -, 1200,75, 335
Mary Morgan, 60, 30, 1000, 30, 340
Samuel Young, 50, 40, 1000, 30, 250
Edward Porter, 25, 50,700, 10, 125
William Jester, 40, 40, 1500, 25, 225
Thomas W. Rawley, 60, 30, 3000, 50, 295
Henry Graham, 70, 8, 1600, 35, 250
William Cain, 40, 5, 800, 75, 221
Lodewick Cain, 60, -, 1000, 80, 355
George W. Redden, 120, 70, 3000, 100, 654
Samuel Clark, 20, -, 400, 6, 90
Uriah Sipple, 125, 30, 4000, 80, 452
William N. Brown, 82, 11, 4700, 120, 535
Lemuel Morris, 25, 4, 750, 25, 135
John Cain of D, 75, 25, 1800, 75, 430
Clement C. Simpson, 120, 180, 8200, 150, 805
Levi Cain, 100, 130, 4000, 150, 530
James Porter, 200, 90, 6000, 100, 613
Mager Bowen, 8, 12, 400, 10, 40
Clement Anderson, 175, 50, 6600, 75, 505
William Salmons, 40, -, 1200, 30, 357
John Tatman, 220, 40, 10000, 150, 830
Jane Bradley, 40, 25, 2500, 50, 423
Alexander Simpson, 110, 55, 2800, 40,380
Peter Calaway Jr., 90, 10, 1200, 175, 450
Peter Calaway, 120, 30, 3000, 140, 651

Beniah Tharp, 120, 50, 4000, 125, 535
Jacob T. Lewis, 150, 20, 5100, 175, 1783
William H. Taylor, -, -, -, -, 222
Emery Spence, 90, 10, 3000, 80, 403
David Spence, 175, -, 3000, 35, 365
William Roe, 140, -, 4000, 200, 470
Henry Calaway, 75, 29, 3500, 200, 475
Richard Merriken, 200, 100, 3000, 100, 405
Albert Jones, 250, 90, 4500, 100, 460
Samuel Hall, 50, 50, 1500, 40, 296
John C. Hall, 50, 50, 1500, 40, 296
John W. Rawley, 95, 30, 1000, 50,300
Nimroe Lewis, 30, 25, 400, 10, 75
Gypson A. Collins, 60, 90, 2000, 40, 437
Peter D. Harrington, 100, 60, 1200, 125, 329
John Williams, 50, 50, 1500, 75, 319
William Edgell, 200, 200, 3000, 100, 702
Eli Wooten Jr., 30, 120, 600, 35,240
Ezekiel J. Jones, 110, 58, 5000, 275, 1100
Robert J. Hopkins, 180, 150, 3500, 60, 635
William B. Smith, 125, 75, 2500, 35, 181
William N. Hopkins, 160, 40, 2400, 100, 580
Thomas N. Morris, 100, 25, 2000, 40, 193
Snow Jones, 25, 10, 600, 10, 67
Samuel O. Jones, 105, 68, 3000, 100, 461
James Raughley, 150, 90, 7600, 120, 304
James Coutey, 80, 10, 1000, 10, 5
William Vingard (Vinyard), 150, 50, 3100, 25, 300
Isaac Friend, 90, 10, 3000, 20, 140
Henry Whitley, 100, 50, 1500, 20, 90
Pemberton Clifton, 90, 30, 4000, 35, 225
Willliam Smith, 150, 80, 3000, 150, 340
Philemer Edwards, 150, 50, 2000, 80, 100
Jones H. Willis, 150, 90, 3500, 50, 269
Ruth Hill, 250, 150, 4000, 50, 360
Charles W. Morris, 40, 47, 870, 30, 109
Eli H. Macklin, 100, 15, 3000, 50, 233
George W. Dorman, 165, 80, 3000, 40, 215
Mathew Sorden,80, 60,1600, 30, 140
James Satterfield, 45, -, 900, 15, 225
Benjamin Harrington, 125, 75, 2000, 25, 225
Curtis Sapp, 30, 35, 1000, 25, 322
Noah Lynch, 62, 8, 1400, 75, 315
William Shaw, 125, 83, 8000, 250, 1000
Washington E. Satterfield, 120, 96, 6000, 10, 60
Eliza Harrington 50, 25, 2400, 40, 360
Noah Cain, 20, 7, 8000, 40, 140
Mary Calaway, 150, 50, 3000, 40,80
Curtis Calaway, 50, 43, 1000, 25, 67
Elias Sapp, 57, 12, 1000, 35, 250
Elie Calaway, 55, 52, 1000, 35, 125
Rhoda A. Lewis, 100, 100, 2000, 30, 235
Tunis Deats, 150, 100, 2500, 75, 560
Elizabeth H. Jones, 83, 35, 3000, 200, 720
William E. Lord, 200, 70, 4000, 100, 285
Covington Messick (Metsik), 50, 40, 1500, 30, 217
Solomon Bradley, 50, -, 750, 10, 35
William Pretyman, 50, 10, 1500, 20, 110

William H. Powel, 70, 40, 5000, 100, 715
Samuel Fisher, 40, 20, 1500, 20, 129
Henry S. Fisher, 80, 20, 2500, 30, 347
Samuel Cubbag, 300,75, 4375, 100, 520
Minus Willey, 15, 90, 1200, 25, 170
Mahlon G. Kesler, 80, 80, 3200, 20, 190
Shadrick Coliston, 100, 30, 3200, 125, 285
Oliver Hammon, 100, -, 2400, 75, 260
Beniah Anderson, 50, 100, 2250, 20, 249
David T. Anderson, 120, 25, 2200, 30, 345
Isaac Jester, 100, 62, 1620, 65, 42
Clement Harrington, 90, 18, 1500, 30, 262
David Taylor, 90, 120, 3000, 150, 515
Johnathan Harrington, 125, 83, 3000, 124, 433
Elias T. Booth, 150,77, 6000, 200, 560
James H. Morgan, 25, -, 1000, 15, 200
James Raughley, 25, -, 500, 20, 145
George Tharp, 35, 10, 700, 35, 120
Jordan Tindel, 50, -, 750, 30, 200
Thomas E. Obier (Oliver), 80, 70, 2500, 100, 311
Charles F. Walker, 35, 7, 400, 8, 68
James Russ, 67, 33, 2000, 40, 263
Solomon Murphy, 120, 125, 3300, 50, 290
William Dawson, 150, 50, 4000, 100, 443
Wesley Scott, 135, 35, 3000, 75, 527
William Hamilton, 100, 25, 2500, 25, 291
Andrew Lord, 163, 14, 8750, 23, 440
Robert Ralston, 150, 150, 6000, 100, 675
Alexander Pretyman, 70, 30, 1200, 25, 136
Lewellen Tharp, 100, 190, 5000, 70, 257
Benjamin H. Tharp, 75, 32, 1700, 75, 230
James Cubbag, 70, 30, 2500, 75, 215
Robert Thomas, 175, 185, 7200, 50, 540
Thomas M. Raiser, 50, -, 1000, 20, 120
Peter Adams, 100, 50, 2000, 50, 325
Stephen Redden Jr., 100, 44, 2000, 75, 518
Luther Scott, 100, 150, 3600, 100, 430
Moses Harrington, 125, 125, 7500, 250, 724
Dennis Minner, 40, 20, 1500, 20, 50
Shermizer Fisher, 150, 150, 12000, 150, 3235
Elizabeth Johnson, 100, 60, 3200, 60, 655
James B. Perttyman, 100, -, 1600,70, 363
Jeremiah P. Cordery, 130, 120, 3000, 80, 532
Jacob Cordery, 200, 130, 7000, 75, 570
Joseph T. Hatfield, 200, 200, 8000, 30, 380
George FitzJarrell, 75, 58, 1500, 50, 240
Thomas Watson, 100, 102, 4000, 100, 360
Nutter Macklin, 30, -, 600, 25, 105
Jame Johnson, 100, 50, 2000, 35, 175
James C. Tatman, 218, 218, 6000, 150, 652
Zadoc Allen, 33, -, 330, 5, 134
Robert C. Ross, 100, 150, 3000, 80, 360
Ziphora Murphy, 90, 50, 1000, 25, 90

John A. Collins, 200, 100, 1800, 100, 700
Nathan Fleming, -, -, -, -, 100
Outene McColley, 100, 126, 2500, 25, 265
Josiah Wolcott, 60, 4, 1500, 75, 333
Sarah A. Jump, 40, 50, 1000, 25, 110
John H. Johnson, 60, 40, 2500, 60, 350
William Messick, 125, 175, 3000, 25, 283
John R. Whitby, 175, 225, 4500, 25, 190
Lewis Turner, 100, 60, 2500, 25, 212
Curtis Dempsey, 110, 69, 4500, 50, 248
Josiah Dickerson, 140, 43, 3000, 80, 555
Alexander Trutt, 120, 160, 5000, 30, 190
Andrew Argoe, 30, 10, 600, 25, 10
Noah Scott, 112, 14, 4500, 75, 572
George W. Shockley, 100, 160, 1800, 60, 307
Ezekiel Walker, 110, 57, 2500, 75, 194
Charles H. Fleming, 200, 169, 5500, 50, 390
Ash Walls, 75, 100, 3000, 75, 266
John S. Rion, 80, 20, 2000, 50, 455
Nathaniel C. Powel, 140, 70, 3000, 75, 445
Mashock Duker(Daker), 90, -, 150, 15, 20
Sarah Herrington, 100, 60, 2400, 85, 351
Nathan Harrington, 100, 131, 3400, 25, 350
James Friend, 40, 40, 1200, 15, 67
Johnathan Herrington, 100, 67, 2500, 30, 358
Joseph W. Willis, 100, 80, 2000, 100, 175
William A. Willis, 17, 8, 600, 20, 105
Nathaniel B. Johnson, 60, 90, 1500, 20, 185
James J. Wood, 133, 66, 8000, 50, 342
James J. Walker, 75, 113, 1400, 25, 101
Robert Wilson O, 185, 15, 7000, 300, 600
Mc___ McIlvaine O, 196, 4, 12000, 400, 1500
Sainana McIlvaine O, 90, 10, 3200, 75, 150
Henry Barnett T, 50, 30, 4000, 100, 400
Thos. Jones O, 300, 100, 15000, 300, 1200
Thomas Barker T, 150, 50, 3000, 75, 250
Wm. Gray O, 25, -, 1000, 25, 100
Danl. J. Jackson T, 105, 75, 7000, 100, 300
Joseph W. Milliams T, 900, 40, 2000, 50, 200
John Sartton O, 175, 75, 8000, 125, 800
John W. Carren (Carrow) T, 100, 100, 5400, 75, 800
Saml. Harrington T, 100, 130, 5000, 50, 700
Thos. Reed O, 225, 400, 10000, 150, 1000
Saml. Shorts O, 17, 3, 2000, 25, 250
Mary H. Camper O, 40, -, 1500, 25, 100
Wm. Lindale T, 60, 10, 2500, 40, 250
Thos. H. Wyatt T, 80, 40, 4000, 40, 600
Burton Donivan T, 150, 150, 9000, 75, 700
Jas. Ward T, 80, 6, 5000, 150, 400
James Lindale T, 150, 20, 6000, 75, 350
James P. Shorts M, 138, 15, 8000, 125, 450

Elias Russel O, 140, 78, 5000, 100, 600
Jas. Bradley T, 140, 20, 5000, 50, 800
James Grier O, 200, 100, 10000, 500, 100
John Luff O, 53, -, 2500, 75, 240
Wm. VanBurkalow T, 40, -, 1600, 100, 200
Ezekiel Jenkins O, 112, -, 5000, 75, 400
James Green O, 75, 25, 9000, 900, 350
John Griffith M, 191, 25, 6000, 80, 700
George Gibbs O, 30, 7, 1200, 75, 150
John Duncan O, 48, 6, 3000, 950, 200
Brincus Townsend T, 275, 200, 8000, 50, 500
Joseph Powell O, 88, 20, 4000, 50, 400
Wm. H. Sapp T, 50, 20, 1500, 50, 300
William Voshell O, 40, 35, 2500, 50, 275
William Holston O, 50, 50, 1500, 25, 175
Elizabeth Swiggett O, 75, 25, 2500, 20, 150
Wm. P. Maloney O, 75, 25, 2700, 50, 325
John R. Thomas O, 25, 50, 900, 40, 250
Cornelius Hinsley T, 180, 45, 7000, 100, 800
Charles Conwell O, 150, 20, 7000, 150, 600
Benedict Gildersleve O, 180, 200, 7000, 150, 900
Susanna Brown O, 38, 4, 1800, 50, 175
William Brown O, 94, 24, 3000, 75, 300
James Anderson O, 78, 28, 4000, 100, 700
James Case O, 40, -, 1200, 30, 300
John Davis O, 52, 5, 1800, 50, 200
Thos. P. Lindale O, 75, 15, 2000, 75, 300
John Young T, 900, 15, 3500, 900, 500
Thos. G. Hopkins M, 182, 2, 3000, 150, 550
Danl. Gildersleve M, 290, 10, 9000, 200, 500
Andrew Ireland M, 120, -, 4000, 120, 350
Elias B. Baker O, 100, 7, 4000, 200, 325
John Benson T, 60, -, 1500, 25, 150
James F. Faulkner T, 60, -, 1600, 28, 250
William Carter T, 300, -, 9000, 300, 700
Edward Collins T, 200, 25, 6000, 50, 400
Robt. W. Carter O, 180, 10, 14000, 500, 900
John Nelson (Melson) O, 18, -, 1600, 100, 150
John W. Collins T, 200, -, 6000, 100, 600
Timothy Terry O, 65 15, 3000, 150, 350
James Jackson O, 75, 15, 3000, 50, 350
Daniel Jackson O, 163, 25, 7000, 50, 100
Mark G. Chambers O, 200, 20, 6000, 150, 500
Charles Baker T, 140, 10, 4000, 125, 600
James Jester T, 80, 10, 2000, 25, 200
Wm. Burke T, 60, 90, 5000,75, 400
Job Coverdale T, 75, 48, 2500, 25, 25
Geo. N. Roberts O, 180, 45, 6500, 150, 400

Wm. S. McIlvaine O, 116, 15, 5000, 250, 325
Joseph Calley T, 125, 25, 6000, 100, 500
James Green T, 140, 15, 6000, 150, 500
George Walker O, 32, 30, 2000, 25, 250
Mary J. Downhan O, 25, 6, 9100, 25, 125
Wm. Herring O, 300, 60, 12000, 500, 1000
Saml. Dewees T, 200, 70, 10000, 200, 800
William H. Ridgway O, 200, 235, 10000, 300, 1000
Thos. H. Howell O, 97, 20, 8000, 50, 350
John M. Rickards O, 36, -, 3000, 100, 300
Helmsly Walls O, 30, -, 2500, 100, 200
Martin Knight T, 100, -, 7500, 125, 600
Thos. L. Temple O, 160, 35, 6000, 20, 145
Hunn Jenkins, 75, -, 7000, 3000, 600
Allen Haines T, 60, 12, 1700, 150, 200
Frisby B. Clark O, 40, 6, 4000, 100, 300
Edwd. Lord O, 28, -, 2800, 50, 400
Danl. McBride T, 45, -, 1500, 75, 200
William Wheatly M. 150, 6, 6000, 75, 550
Wm. Evans O, 100, -, 7000, 150, 500
Isaac Griffith T, 275, 28, 8500,100, 600
Solomon Townsend T, 150, 26, 4500, 100, 800
Joseph Gibbs O, 100, -, 2000, 50, 225
Robt. Lewis O, 21, -, 600, 25, 175
Ezekiel Shaw T, 900, 50, 3000, 50, 225
Charles Postles T, 40, -, 2000, 100, 250
Manuel Highutt T, 100, 10, 2000, 28, 100
Reuben Johnson O, 175, 25, 4000, 125, 252
Nathaniel Bailey T, 200, 25, 2500, 95, 175
Andrew Tilghman M, 100, 25, 3000, 50, 300
Barratt P. Comer O, 80, 25, 3500, 75, 200
Henry J. Anderson O, 6, -, 600, 25, 500
Silas C. Mason M, 100, 8, 5000, 125, 650
Saml. R. Mason T, 40, 10, 2000, 100, 350
John W. Lynch T, 90, 10, 3000, 50, 200
John C. Prettyman T, 25, 5, 2000, 75, 400
James Massy O, 170, 30, 6000, 75, 300
James Black O, 100, 42, 4000, 100, 450
James Vincent O, 25, 9, 1200, 50, 175
Wm. Porter T, 80, 20, 3000, 40, 100
Peter L. Bonvill O, 163, 30, 8000, 1000, 1200
Saml. Black T, 100, -, 4000, 100, 375
Bevins Morris O, 150, -, 5000, 150, 550
Wm. Lober O, 22, -, 600, 30, 160
George Bateman O, 140, 60, 4000, 50, 375
Henry Lofland T, 190, -, 4000, 75, 350
Stansbery Murray O, 39, 6, 1000, 75, 275
Thos. Reynolds O, 216, 60, 7000, 200, 300
Nathan Trebbett T, 80, 50, 2500, 50, 500

John P. Emerson T, 100, 50, 3000, 75, 400
John D. Anderson T, 250, 50, 12000, 125, 500
Erasmus Burton T, 110, -, 4000, 150, 700
John Quillen T, 140, 37, 2500, 150, 650
Risdon Williams O, 80, 35, 3000, 75, 400
William McCauley T, 90, 50, 3500, 75, 350
John G. Melvin T, 120, 35, 4000, 125, 225
Joseph Smith Jr. O, 120, 26, 3000, 100,5 00
Peter Sowber O, 30, 10, 1200, 25, 175
Thomas Bell O, 25, -, 700, 40, 125
Robert McCauley T, 40, 38, 2000, 50, 200
George Smith O, 90, 10, 5000, 100, 550
David Coverdale T, 90, 60, 3500, 60, 350
Jeremiah Postles T, 70, 5, 2000, 25, 250
Samuel Thomas T, 40, 10, 1500, 20, 200
Saml. D. Darby T, 250, 150, 15000, 1500, 1000
Caleb Smithers O, 90, 30, 10000, 200, 400
Caleb Smithers O, 300, 60, 12000, 200, 1000
Robert J. Sowber O, 40, 10, 4000, 200,3 50
Nathan Young O, 100, 100, 6000, 50, 500
John Emerson O, 100, 75, 12000, 300, 1000
John West O, 300, 100, 20000, 500, 1200
Ezekiel Abrams T, 600, 5, 4000, 56, 300
David S. Postles T, 100, 25, 6000, 25, 700
Richard J. Camper T, 112, 20, 3000, 500, 500
Robert H. Stevenson T, 200, 57, 10000, 150, 900
Jehu M. Reed O, 250, 100, 10000, 100, 700
Thomas Vickery O, 150, 30, 8000, 125, 800
John Emory T, 120, 50, 5000, 400, 900
Isaac Godwin T, 180, 120, 12000, 150, 700
Edward Quillen T, 100, 100, 3500, 50, 350
John Young O, 17, 3, 800, 25, 175
William Kersey T, 20, -, 1000, 50, 300
Alex. Tilghman M, 70, 30, 3000, 100, 600
James Prattis O, 25, 5, 1000, 25, 200
Wm. C. Brown O, 100, 35, 3000, 100, 500
James Clymer T, 70, -, 2000, 50, 250
Wm. Spencer T, 250, 150, 6000, 100, 450
Saml. Herring O, 83, 50, 4000, 150, 350
Alfred Clifton T, 140, 60, 100000, 100, 950
Nathl. Harrington O, 40, 20, 1000, 75, 325
Elias Townsend O, 100, 50, 4000, 75, 400
A___ Johnson T, 100, 50, 3000, 50, 340
John Case (Cox) T, 130, 70, 2000, 50, 250
Wm. E. Hillen, T, 70, 40, 1500, 25, 150
Wm. Credick (Cradick) O, 250, 50, 3500, 125, 400
William Case O, 150, 50, 4000, 150, 800

Alex. Fleming T, 225, 75, 6000, 150, 700
Thos. Sipple O, 125, 40, 2000, 50, 300
Nehemiah Walker M, 320, 30, 10000, 200, 1000
George Rathledge M, 100, 60, 4000, 100, 500
Andrew Barratt M, 175, 25, 5000, 900, 450
Danl. Nichols T, 100, 50, 6000, 50, 350
John Patterson O, 200, 50, 8000, 300, 400
Josiah Bradley T, 150, 100, 4000, 75, 325
James H. Evans T, 100, 50, 4000, 50, 450
Purnell Mosley T, 75, 25, 5000, 50, 350
David Needles O, 75, 20, 3500, 40, 200
Jas. R. Needles O, 80, 40, 3500, 50, 350
Henry Marker T, 40, 20, 1000, 25, 200
Perry Knotts O, 50, 30, 2000, 75, 300
Jas. Beauchamp T, 150, 50, 5000, 100, 550
Wm. C. Marker O, 130, 30, 1000, 5, 500
James Anderson T, 200, 200, 7000, 50, 450
James Anderson M, 75, 125, 4000, 50, 200
Jonathan Doronham O, 100, 20, 3000,125, 450
Abner Roe T, 30, 60, 1000, 50, 150
Wm. H. Jarrold T, 80, 25, 2000, 35, 200
Saml. Cook T, 75, 15, 1800, 40, 75
Mathew Benson O, 50, 10, 1500,75, 275
Jerry Prichett T, 50, 30, 1600, 25, 100
Benj. Harris T, 50, -, 1000, 30, 135
Aaron P. Olsmond O, 120, 80, 8000, 75, 200
Robt. Ross O, 50, 1, 1500, 40, 151
John Jarrold O, 120, 60, 4000, 75, 350
Peter D. Hubbard O, 88, 30, 1200, 25, 300
Henry K. Hargadine O, 150, 50, 5000, 200, 800
Sylvester H. Willis T, 160, 100, 7000, 150, 500
Latetia Harrington O, 105, 10, 2500, 50, 250
Wm. H. Jones M, 150, -, 3500, 200, 400
Wm. B. Harrington O, 40, 66, 3000, 40, 250
Nathan Shurry T, 120, 44, 3500, 150, 300
Thos. B. Causey O, 100, 20, 3000, 200, 500
John Harrington T, 85, 15, 4000, 150, 600
Henry Harrington T, 135, 55, 7000, 150, 800
James H. Jeter O, 25, 100, 1500, 50, 300
James Alexander T, 75, 45, 1500, 25, 250
William Satterfield O, 200, 120, 9000, 75, 650
Fanett Curbey T, 250, 100, 12000, 125, 800
Benj. Hurlock T, 150, 50, 2500, 40, 150
Wm. Billings T, 200, -, 6000, 100, 600
Wm. Simpson O, 100, -, 2500, 50, 350
Thos. Kersey T, 70, 20, 2000, 50, 150
George French M, 90, 30, 2000, 50, 200
Wm. K. Adkins, 120, 230, 5000, 50, 275

John Bailey O, 80, 40, 6000, 300, 600
John Jarrold T, 70, 15, 3000, 100, 4000
John Cullen O, 80, 45, 3500, 50, 280
John Harrington O, 125, 5, 3000, 50, 450
Elsberry Richardson O, 75, 25, 2000, 40, 75
Wm. Satterfield O, 200, 50, 4000, 200, 700
George Anderson O, 62, 20, 3000, 100, 650
Chas. C. Case (Cox) O, 54, 10, 1800, 50, 200
Everett Billings T, 50, 25, 1500, 50, 200
Jas. Needles O, 65, 43, 3500, 100, 3500
John Tinley T, 100, 60, 4500, 50, 400
Henry C. Cooper O, 100, 75, 3000, 500, 600
Lewis Chapier M, 100, 100, 4000, 125, 450
Peter D. Taylor M, 70, 130, 3500, 50, 200
Mary Hopkins O, 120, 35, 5000, 75, 400
Major Wyatt T, 100,80, 2000, 50, 300
Uriah Meredith T, 100, 130, 1600, 40, 250
Elijah Morris O, 120, 60, 4000, 200, 350
Benj. Reeves O, 125, 350, 8000, 200, 450
Noah Holding O, 300, 100, 5000, 150, 600
Thos. R. Waters M, 80, 50, 3000, 50, 300
Saml. Herring T, 50, 20, 2500, 40, 225
Wm. Dawson O, 200, 100, 10000, 200, 800
Edwd Cook T, 30, 30, 1200, 20, 200
Thos. Sherwood Jr., T, 75, 25, 2000, 30, 250
John Sherwood Sr. O, 150, 250, 10000, 125, 600
Wm. Graham T, 200, 100, 15000, 200, 1200
Wm. S. Clauk O, 150, 104, 7000, 100, 500
Robt. H. Jones T, 125, 40, 5000, 100, 450
Henry Cowgill O, 300, 300, 20000, 500, 1500
Daniel Jones T, 100, 30, 4000, 50, 400
Saml. Cohee O, 125, 40, 3500, 125, 350
Jas. Downhan O, 900, 100, 4000, 75, 225
John Bennett T, 50, 50, 2000, 40, 175
Jas. L. Billings T, 110, 115, 3000, 125, 400
Alex. Voshell T, 100, 100, 5000, 100, 325
Woodward Stockley T, 100, 100, 5000, 125, 300
Danl. Perry O, 100, 15, 2500, 75, 450
Spencer Perry O, 20, 100, 2000, 45, 300
Thos. Keys O, 25, 5, 500, 30, 130
Elijah Colwell O, 23, 15, 500, 45, 300
Saml. Shutz O, 100, 43, 3000, 150, 600
Saml. K. Burnite T, 200, 100, 9000, 100, 400
Jas. H. Carter T, 20, -, 600, 25, 250
Wm. J. Killen T, 100, 62, 3500, 50, 375
Jacob J. Cook T, 80, 40, 3100, 50, 400
Wm. S. Vickery T, 90, 30, 3000, 60, 150
Wm. Knotts T, 80, 60, 3500, 250, 200

Alexr. Frazier T, 150, 18, 3700, 100, 200
Jonathan Tinley O, 100, 112, 4000, 75, 400
James Johnson T, 40, 166, 4000, 100, 275
Abner Dill O, 120, 100, 4000, 100, 600
Alexr. Hewes O, 150, 100, 4000, 200, 425
Reuben Cain O, 100, 25, 2000, 50, 250
Thos. C. Green O, 280, 40, 10000, 400, 1500
John Foncook(Hancook) T, 100, 25, 3000, 50, 350
Zadock Sipple T, 200, -, 4000, 100, 500
Saml. Hurd T, 250, 50, 4000, 50, 400
Emeline Draper O, 900, 25, 2000, 40, 250
Peter Dill O, 75, 33, 2000, 30, 225
John W. Cooper, 250, 50, 10000, 200, 800
Wm A. Dill T, 175 100, 6000, 100, 450
Wm. Greenley O, 125, 50, 3000, 100, 700
Watson Pickering T, 228, 100, 4000, 100, 275
Garratt Voshell O, 50, 30, 700, 35, 275
Saml. D. Conner O, 75, 125, 2000, 30, 350
John Cooper O, 100, 60, 2000, 75, 418
John W. Cooper T, 100, -, 1500, 45, 300
Thos. Bell T, 50, 50, 1000, 30, 150
John Tribbett T, 100, -, 1000, 40, 250
Caleb Clark O, 70, 20, 1000, 15, 200
George Reynolds O, 900, 50, 2500, 75, 350
Edwd Pindar O, 100, 50, 3000, 75, 450
John Erwin O, 75, 40, 2000, 100, 500
James Hurd T, 180, 120, 5000, 785, 300
Benj. Reed O, 60, 65, 2000, 50, 300
James Clark T, 75, 25, 2000, 40, 300
Alexander Penison T, 100, 160, 4000, 50, 200
Benj. Goodwin O, 160, 100, 5000, 100, 500
Mathew Kemp T, 45, 10, 1000, 50, 225
Lewis Sermons T, 60, 50, 1200, 150, 500
Major Cain O, 60, 43, 1000, 50, 325
Philimon Carter O, 340, 60, 7000, 500, 1200
Benj. Draper T, 50, 13, 1000, 50, 200
Benson Dill T, 100, 100, 2000, 50, 300
Ann Lister O, 100, 30, 2000, 75, 325
John Green O, 100, 50, 4000, 300, 600
Hinson Melvin O, 50, 5, 1000, 75, 400
Lob C. Carson T, 125, 175, 3000, 100, 275
John Scanvenger O, 75, 75, 2000, 57, 328
Thomas Mimnier T, 200, 200, 5000, 125, 650
Thos. Hurd O, 75, 45, 8000, 350, 350
Jonathan Longfellow O, 100, 68, 5000, 150, 450
Thomas W. George T, 100, 68, 3000, 100, 500
Nathan Clark O, 100, 60, 3000, 100, 350
James Fausett M, 75, -, 2500, 25, 300
Sol Sharklin O, 52, 10, 4000, 100, 400
Chaulky Ballinger O, 70, 15, 2000, 125, 250

Jos. Cook T, 225, 85, 3000, 50, 300
Warner Barcus O, 50, -, 1000, 28, 250
Andrew J. Cally O, 200, 40, 4000, 100, 450
John Cleaver O, 38, 10, 1600, 50, 200
John Patterson T, 208, 48, 6000, 300, 650
Mary Lewis O, 90, 5, 2000, 75, 300
William Stubbs O, 100, 200, 4000, 125, 500
Henry Stetson T, 80, 20, 3000, 100, 300
Ezekiel B. Clements O, 110, 50, 6000, 200, 650
Thomas Clements O, 73, 4, 2000, 50, 225
John Rochester O, 60, -, 2000, 60, 300
James B. Bostick T, 100, 100, 4000, 50, 400
John Wright T, 80, 70, 4000, 75, 600
Alexr. Jackson O, 150, 100, 15000, 200, 250
John Goodwin Jr. O, 100, 100, 8000, 300, 630
David McBride T, 95, 75, 2500, 100, 400
James Cooper T, 125, 95, 4000, 75, 600
Henry K. Draper T, 95, 81, 4000, 75, 650
Chas. H. Johnson M, 105, 5, 4000, 100, 400
Louder T. Layton T, 60, 30, 4500, 75, 300
Thos. Downham O, 90, 8, 4500, 50, 125
John Moseley T, 125, 25, 5000, 30, 200
Thos. L. Madden O, 125, 25, 7000, 150, 460
John B. Lewis O, 45, 5, 1000, 50, 150
Wm. Darling T, 900, 100, 5000, 75, 400
Richabald Alliband Jr. O, 900, 50, 6000, 200, 650
Wm. Alliband O, 50, 36, 2000, 125, 350
Jas. D. Voshell T, 120, 10, 4000, 40, 200
Benj. Stradley O, 80, 10, 3000, 125, 400
Wm. K. Dickson M. 80, 34, 5000, 150, 1000
Alfred Massey T, 90, 19, 4500, 75, 450
Wm. H. Dennis T, 140, 10, 3500, 40, 250
Wm. Lewis O, 25, -, 1500, 40, 150
Thos. Lewis T, 150, 50, 6000, 100, 550
Avery Draper O, 120, -, 5000, 150, 600
Wm. Slay O, 140, 37, 7000, 200, 700
John Caulk O, 95, 42, 4000, 50, 300
Nathan R. Douglass O, 89, 32, 4500, 100, 300
Thos B. Cooper O, 100, 15, 4000, 125, 750
Risdon Cook T, 100, 10, 2500, 75, 225
Griffin Moore T, 90, 150, 2500, 10, 500
Thos. Goodwin O, 125, 125, 8000, 300, 1200
Washington Greenley O, 75, 25, 2000, 50, 200
Thos. Downhan T, 35, 40, 1600, 25,175
Benj. Dill O, 85, 18, 1500, 50, 300
Thos. Kemp T, 75, 57, 2000, 30, 200
John Sipple O, 60, 44, 2500, 75, 400
Alex. Holden T, 50, 10, 1500, 40, 125
Wm. Frazier O, 100,70, 4000, 100, 400
Chas. Holding O, 150, 75, 3500, 80, 325

Wm. Pearson O, 300, 60, 11000, 40, 100
Mary Hurd O, 80, 35, 2000, 75, 325
Lemuel Clark, 950, 60, 2000, 100, 400
James Longfellow O, 35, 35, 2500, 50, 350
Andrew Warren T, 150, 50, 2500, 50, 325
James Longfellow O, 700, 50, 2500, 150, 450
Robert Greenly O, 150, 150, 4000, 80, 325
Wm. C. Jump, 75, 25, 1000, 75, 350
Wm. Edwards O, 150, 140, 4000, 100, 600
James Cohee O, 70, 60, 1500, 50, 300
Job Clark O, 180, 70, 3000, 100, 700
John S. Hopkins O, 190, 100, 1500, 50,300
Joseph D. Grewell O, 140, 50, 2000, 60, 450
John Grewell (Gruwell) O, 135, 15, 3500, 75, 650
Benj. L. Cohee O, 110, 90, 4000, 100, 500
Alex. Frazier O, 180, 50, 7000, 300, 500
Ezekiel Frazier T, 95, 85, 6000, 100, 400
Whitely W. Meredith T, 100, 20, 6000, 75, 400
John H. Cook T, 100, 50, 4800, 100, 450
Robt. W. Reynolds O, 120, 80, 7000, 300, 1000
Andrew Slaughter T, 120, 80, 7000, 300, 1000
Wm. Slaughter T, 120, 100, 5000, 80, 1150
Thos. Cubbage O, 200, 100, 3000, 50, 500
Jacob B. Meredith O, 100, 30, 2500, 75, 125
Lemuel Clark T, 80, 20, 1600, 50, 280
Wm. Meredith O, 145, 45, 5000, 125, 800
Alex Dill O, 100, 85, 2500, 75, 600
Mary Grewell O, 120, 40, 3000, 50, 500
Saml. S. Cooper, 150, 30, 3000, 75, 550
Edwd. J. Carter O, 225, 75, 12000, 300, 2000
Wm. Smith T, 300, 60, 10000, 200, 100
Wesley Morris T, 100, 30, 2000, 50, 350
William Burk T, 300, 100, 8000, 125, 400
Peter Raughley O, 100, 100, 2000, 75, 450
Nathaniel Green T, 150, 50, 3500, 100, 500
Elizabeth Jump T, 100, 50, 3000,75, 350
Jonathan Bostick M, 80, 80, 2500, 50, 350
Saml. B. Cooper O, 147, 77, 2000, 290, 500
Peter L. Cooper O, 150, 50, 3500, 125, 550
Elias Perry T, 110, 110, 2500, 50, 450
Shadrack Johnson M, 150, 50, 4000, 100, 500
Wm. Cook T, 150, 50, 4200, 150, 600
Joshua K. Clements O, 200, 60, 8000, 250, 1500
Thos. E. Frazier O, 85, 25, 3000, 150, 300
Wm. Cubbage T, 65, 245, 3000, 150, 400
Wm. A. Adams T, 100, 150, 3000, 150, 600
James Frazier O, 200, 40, 12000, 200, 700

Robt. Frazier O, 100, 40, 5000, 125, 550
Saml. C. Dill T, 100, 80, 3000, 100, 650
Isaac Gooden O, 1250, 65, 5000, 150, 800
Benj. C. Cubbage T, 100, 50, 2500, 75, 250
James Cohee T, 175, 25, 3500, 50, 600
Saml. B. Cooper O, 30, -, 1000, 30, 150
James Godwin T, 150, 50, 4000, 80, 500
Robert Knotts T, 70, 20, 2000, 70, 275
Saml. Cooper M, 135, 15, 3500, 75, 800
Rachel Reed O, 95, 5, 1500, 50, 500
John W. Cullen, 350, 45, 25000, 800, 2500

New Castle County Delaware
1860 Agricultural Census

This agricultural census was filmed from original records in the Delaware State Archives in Wilmington Delaware by the Delaware State Archives Microfilm office.

There are some forty-eight columns of information on each individual. Only the head of household is addressed. I have chosen to use only six columns of the information because I feel that this information best illustrates the wealth of the individuals. These are shown below:

1. Name of Owner
2. Acres of Improved Land
3. Acres of Unimproved Land
4. Cash Value of the Farm
5. Value of Farm Implements and Machinery
13. Value of Livestock

Thus, the numbers following the names represent columns 2, 3, 4, 5, 13.

The following symbol is used to maintain spacing where information in a column is left blank (-). This symbol is used where letters, names or numbers are not legible (_).

Josep Roberts, 170, 57, 11000, 300, 1000
Isaac Gibbs, 410, 200, 35000,500, 1200
Rich. F. Hanson, 175, 25, 1200, 300, 800
John McCune, 270, 80, 20000, 300, 900
James L. David, 200, 40, 8000, 200, 800
William Wilson, 150, 152, 20000, 600, 2000
J. H. Hanson, 130, 10, 8000, 200, 1000
John Grace, 300, 40, 10000, 100, 800
James Kandy(Handy), 200, 70, 14000, 500, 1200
Cyrus Tatman, 130, -, 8000, 200, 500
R. R. R. Rothwell, 100, 20, 9000, 400, 600
James Rothwell, 132, 20, 1200, 200, 800
Lydia Rothwell, 240, 5, 16000, 300, 450
Doct. J. V. Crawford, 250, 5,17000, 400, 1250
William E. Riley, 170, 8, 14000, 300, 1200
Robt. A. Cochran, 170, 20, 14000, 500, 1500
Richard Lockwood, 150, 10, 13000, 200, 1000
A. S. Nandain (Naudain), 100, 20, 8000, 100, 1000
John M. Nandain, 250, 20, 15000, 300, 2000
Robt. A. Cochran Jr., 300, 56, 25000, 300, 2000

Nehemiah Davis, 300, 10, 15000, 300, 1500
Samuel B. Guin, 280, 20, 8000, 150, -
William H. Townsend, 140, 30, 8000, 250, 800
Joseph A. Rickards, 100, 200, 5000, 100, 200
Archibald Finley, 50, 8, 3000, 300, 350
James Lewis, 150, 20, 8000, 200, 3000
James Bird, 175, 10, 8000, 100, 500
William Daniel, 100, 27, 6000, 200, 800
Robt. D. Ratledge, 100, 27, 9000, 200, 800
Joseph West Sr., 160, -, 16000, 200, 1300
Benjamin Gibbs, 200, 5, 18000, 500, 1500
David J. Morgan, 200, 152, 15000, 100, 1000
Henry Davis, 200, 19, 15000, 200, 1000
Henry C. Gott, 150, 20, 10000, 100, 800
John H. Hutchinson, 100, 100, 6000, 100, 500
Mariah Fish, 90, 5, 2000, 50, 200
David H. West, 130, 40, 5000, 40, 500
James H. Short, 200, 20, 6000, 100, 500
John M. Rothwell, 212, 20, 16000, 300, 1000
William Ginn, 175, 50, 8000, 100, 500
H. P. Reading, 125, 5, 8000, 200, 200
Saml. Townsend, 170, 50, 12000, 100, 700
Saml. Townsend Sr., 179, 3, 18000, 900, 2300
John Townsend, 160, 5, 15000, 500, 1400
Levi W. Lattermus, 70, 60, 2500, 200, 900
Jacob Staats, 120, 25, 8000, 100, 800
Thomas Bedlow, 100, 70, 8000, -, 200
Michel Daugherty, 200, 100, 6000, 100, -
John Bedwell, 125, 75, 3000, 100, 400
Reese B. Bostic, 20, 100, 2000, -, 100
John Rash, 30, -, 200, -, 100
William Shaw, 90, 10, 1000, 100, 400
Theodore Sweetman, 60,60, 1000, -, 100
James Walls, 60,70, 2000, -, 200
Henry Holten, 70, 36, 1500, 100, 300
William Francis, 40, 60, 1000, 100, 200
William Pinson, 150, 150, 5000, 100, 100
John Colyer, 192, 25, 20000, 300, 1000
Z. McD. Roberts, 251, 60, 12000, 100, 1100
Jacob Clayton, 25, 46, 1000, -, 200
Ben McCoy, 60, 45, 2000, 100, 300
James McCarter, 60, 80, 3000, 100, 300
Thomas Scaggs, 65, 30, 2500, 200, 800
Cooper Dickson, 250, 150, 6000, 100, 800
William Clayton, 50, 62, 2000, 100, 400
Nathaniel Gausford, 30, 12, 1000, -, -
Timothy H. Clayton, 51, 50, 1000, 100, 200
David Tibbit, 40, 30, 500, -, 100
William Ford, 100, 40, 1000, -, 200
James Carrow (Carven, Carrew), 50, 75, 1000, -, 300
Henry Donel, 100, 10, 1000, 100, 100

Samuel Beck, 60, 132, 2000, 100, 300
Elizabeth Chiffin, 100, 85, 2000, 100, 400
Moses Marshal, 120, 80, 4000, 100, 200
John Scaggs, 70, 70, 4000, 50, 200
William Scaggs, 65, 85, 5000, 50, 400
William Budd, 40, 20, 1000, 50, 100
Daniel Darrell, 42, 2, 1000, -, 200
William Scotten, 25, 75, 800, -, 100
Henry Shafer, 65, 25, 2000, 50, 400
Noah Davenport, 80, 50, 1000, 50, 100
Martin Hoffecker, 80, 120, 2000, 50, 200
James Hodgson, 20, 100, 1000, -, 100
John M. Ford 50, 90, 3000, 50, 300
Samuel Guism, 50, 40, 800, 50, 200
John Garman, 70, 30, 3000, 100, 800
Isaac Clayton, 30, 6, 600, 50, 200
James Pain, 35, 5, 700, 50, 300
Edward Davis, 20, 1, 400, 10, 100
James Fortner, 40, -, 800, 10, -
Nickolas Johnson, 80, 20, 2000, 50, 200
Abram Carny (Carry), 100, 20, 3000, 100, 500
Christopher Nandain, 230, 30, 15000, 100, 800
John S. Silcox, 120, 75, 12000, 200, 300
Nathaniel Williams, 130, 23, 13000, -, 500
Manlore Wilson, 200, 12, 15000, 300, 2000
John Nandain, 150, 50, 10000, 300, 1000
John F. Staats, 100, 100, 6000, 100, 300
Samuel Tomas, 160, 30, 8000, 100, 600
Huey Whitlock, 230, 70, 20000, 200, 1200
Aaron Reynolds, 176, 20, 17000, 200, 1300
James C. Lattemus, 125, 25, 8000, 200, 1200
James R. Collins, 300, 100, 20000, 300, 1800
Marth H. Davis, 180, 23, 10000, 100, 800
David T. Rose, 130, 50, 8000, 100, 600
Edward Loatman, 100, 100, 6000, 50, 300
Joseph Rash, 240, 100, 5000, 50, 600
Ephraim Vandyke, 150, 50, 7000, 100, 800
Moten Rickards, 300, 100, 12000, 100, 1000
John W. Thomas, 56, 14, 2500, 100, 500
Elizabeth Thornton, 50, 14, 800, -, -
William P. Reynolds, 50, 10, 1000, -, 300
James Reynolds, 25, 11, 1200, 100, 400
John Hines, 65, 26, 1500, -, 200
Thomas Marien, 130, 50, 5000, -, 600
Nathan Fursons (Farsons, Parsons), 45, 49, 1000, -, 100
James Powell, 76, 46, 4000, 100, 500
Rebecca Boots, 40, 70, 800, -, 300
Samuel Roberts, 130, 20, 9000, 200, 700
John Brockson, 225, 55, 15000, 200, 1200
William E. Riggs, 185, 65, 12000, 200, 1200
William M. Johnson, 135, 55, 5000, 50, 500
Gideon N. Servison, 149, 20, 6000, 100, 400
John Anderson, 100, 20, 4000, 50, 400
Robt. Wright, 110, 86, 1600, 54, 500
John B. Flaherty, 140, 10, 4000, 54, 600

Ward S. Vandergrift, 200, 41, 10000, 100, 800
Benj. Money, 120, 10, 6000, 100, 1000
Elias Lockerman, 200, 104, 6000, 100, 500
Philip Mattis, 250, 50, 40000, 200, 1200
Eben Olprey, 100, 10, 5000, 100, 400
John Pierce, 160, 40, 6000, 100, 900
Mark Davis, 175, 25, 10000, 100, 800
William E. Appleton, 190, -, 20000, 200, 1100
Francis Matthews, 180, 120, 15000, 100, 800
James V. Moore, 150, 25, 20000, 600, 1500
Thomas M. Moore, 151, -, 20000, 400, 1300
Elias A. Moore, 160, 10, 20000, 400, 1400
F. T. Perry (Penny), 195, -, 20000, 300, 1200
Joshia Fennimore, 200, 200, 20000, 1000, 2200
Edward C. Fannimore, 300, -, 20000, 500, 2000
John Atherly, 180, 120, 18000, 400, 1500
Joseph Higler, 200, 150, 12000, 100, 300
John Thomas, 45, 45, 1000, -, 400
Gideon E. Barlow, 90, 108, 4000, -, 400
Waide Deacon, 40, 20, 1000, -, 400
A. C. Lattamous, 160, 60, 7000, 75, 900
William Nailor, 110, 50, 4000, -, 500
A. P. Porter, 55, 29, 3000, -, 200
Geo. D. Collins, 100, 50, 400, 100, 500
Thos. Middleton, 150, 50, 6000, 100, 600
Joseph Wells, 160, 80, 10000, 200, 900
Samuel Wells, 120, 44, 6000, 200, 700
Jas. Donoho, 140, 10, 7000, 40, 400
G. F. Mason, 60, 22, 8000, -, 300
N. Davis, 190, 50, 15000, 200, 1600
James Johnson, 125, 100, 7000, 100, 600
E. Cooper, 200, 50, 10000, 100, 1200
O. C. Crow, 160, 20, 10000, 100, 700
S. Meredith, 100, 25, 8000, 100, 600
John Stephenson, 230, 50, 15000, 300, 800
Andrew Spear, 200, 75, 7000, 150, 900
Gideon E. Rothwell, 200, 100, 12000, 200, 1000
Samuel Staats, 38, 2, 1500, 100, 500
Geo. A. Deakyne, 80, 16, 3000, 100, 500
Colen Ferguson, 125, 52, 5000, 100, 600
Saml. A. Armstrong, 275, 40, 15000, 200, 1500
John French, 200, 50, 10000, 100, 700
James W. Alexander, 120, 60, 5000, 100, 600
Ezekiel View, 100, 20, 4000, 100, 500
George Sheldon, 125, 175, 8000, 100, 600
Alex. Lee, 38, 2, 2000, 100, 400
John Appleton, 300, 50, 20000, 300, 3000
Edward Price, 100, 60, 30000, 100, 500
Jonathan Brown, 175, 325, 10000, 100, 500
Emery Lambert, 60, 20, 2000, -, 400
John Blackiston, 40, 85, 2000, 50, 300

Alphey O. Hill, 120, 60, 5000, 100, 800
Walter Beck, 180, 25, 8000, 100, 800
Joseph Rauer, 150, 48, 5000, 50, 500
James Carrow, 80, 80, 2000, -, 100
George Lewis, 200, 93, 9000, 300, 1300
Purnel T. Jones, 160, 176, 10000, 50, 700
Car Watson, 60, 35, 1500, -, 150
G. W. Ingram, 25, 8, 1500, -, 100
Isaac Ratliff, 150, 50, 6000, 50, 500
Luel Mannering, 58, 2, 1000, -, 500
Thomas Scott, 132, 75, 10000, 200, 1250
John T. Darham, 75, 6, 2000, 50, 400
Benj. Truah, 240, 20, 8000, 50, 544
Jonathan Jester, 105, 42, 4000, 50, 700
Peter Warren, 100, 60, 3500, 50, 500
Josiah Miller, 196, 20, 10000, 200, 1200
John H. Bennett, 100, 20, 10000, 200, 1200
William Farrell, 150, 40, 10000, 100, 500
Nickolas Vandyke, 100, 80, 6000, -, 400
Isaac Rickard, 160, 20, 8000, 100, 800
Jane Young, 200, 50, 9000, 100, 1200
Benj. Songo (Sango), 210, 30, 12000, 200, 1200
John Collins, 101, -, 200, 50, 1000
James Staats, 181, 90, 18000, 400, 1500
George Deakyne, 80, 25, 5000, 200, 1500
Isaac Staats, 130, 20, 6000, 200, 500
Albert G. Deakyne, 80, 195, 3500, 100, 400
Geo. C. Deakyne, 150, 15, 6000, 200, 500
P. F. Deakyne, 200, 100, 10000, 200, 1500
Henry Cooper, 59, -, 1500, 50, 300
Enoch J. Fleming, 30, 60, 3600, 300, 500
William Townsend, 100, -, 10000, 100, 400
James Reed, 237, 30, 14000, 200, 1000
Thomas Deakyne, 162, 20, 7000, 100, 1200
Robt. Derickson, 144, 147, 8000, 300, 1200
Thomas Hartup (Hartsup), 108, 40, 7000, 150, 800
Samuel R. Martin, 80, 22, 4000, 150, 50
Robt. Gaskin, 50, 117, 3500, 400, 800
Ben. W. Stann (Shann), 45, 20, 2500, 100, 300
W. C. Alston, 160, 40, 10000, 200, 600
Jacob Hill, 250, 150, 8000, 300, 1500
Geo. W. Buchanan, 250, 50, 15000, 200, 1500
James Nowland, 80, 30, 3000, 50, 300
Elizabeth Barnett, 95, 95, 4000, 50, 400
Antony Watson, 80, 20, 3000, 50, 400
Thomas Paris, 295, 150, 8000, 100, 500
John C. Macy, 200, 12, 6000, 50, 500
Richard Cubly, 30, -, 1000, -, 300
Lawrence Doile, 60, 20,1500, 50, 400
John W. Nandain, 120, 30, 5000, 100, 600
Benj. Enos, 56, 45, 1000, 50, 300
Miles T. Jones, 26, 3, 800, 50, 100
Godgrey Harstenton, 60, 120, 3000, 50, 400

Jacob V. Nandain, 180, 70, 8000, 100, 600
James Bell, 170, 20, 8000, 100, 400
Saml. J. Pierce, 250, 50, 10000, 200, 800
Susan Hakill (Hukile), 128, 30, 6000, 100, 700
William A. Hukile, 80, 80, 4000, 50, 500
Samuel Armstrong, 187, 187, 10000, 50, 600
William Matthews, 100, 10, 3000, 50, 200
Daniel R. Johnson, 80, 50, 3000, 50, 300
Miles T. Dickerson, 100, 20, 3000, 50, 400
William F. Staats, 120, 50, 4000, 50, 500
Morris Collins, 100, -, 4000, 100, 600
John Wilds, 130, 30, 6000, 50, 600
John Derickson, 140, 14, 4000, 50, 800
Michel Bryan, 100, 75, 6000, 50, 400
Annias Enos, 171, 45, 12000, 200, 1000
Robt. M. Warner, 100, 50, 6000, 100, 400
Samuel Bartley, 43, 10, 600, 4, 200
William Hobson, 150, 120, 4000, 50, 200
James Jackson, 150, 30, 5000, 50, 400
John Ferris, 60, 60, 1000, -, 200
Wm. M. Gardener, 140, 120, 5000, 50, 300
Nap B. Deakyne, 130, 37, 6000, 100, 500
Aurther Bungy, 60, 20, 1000, 50, 200
Chas. Walker, 175, 5, 8000, 50, 1100
Jerry Gale, 125, 25, 5000, 100, 600
Robt. Powell, 150, 150, 2000, 50, 300
Ben Smith, 47, 5, 800, 50, 200
Perry Hamilton, 36, -, 400, -, 200
John McCoy, 350, 50, 20000, 200, 2000
Andrew J. Collins, 300, 25, 2000, 200, 2000
Abram Hayden, 300, 50, 15000, 300, 200
Wm. S. Deakyne, 125, 25, 4000, 200, 800
Robt. Johnson, 100, 20, 6000, 50, 500
Andrew J. Fortner, 280, 10, 10000, 400, 2000
B. Franklin Davis, 150, 75, 8000, 20, 4000
Archibald McLane, 60, 50, 2000, 50, 400
Delight Gardner, 40, 23, 3000, 50, 400
Jacob Deakyne, 160, 140, 11000, 600, 1000
David Staats, 30, 5, 800, 50, 300
Manlore Davis, 150, 20, 5000, 50, 600
William Weldin, 75, 42, 3000, 50, 500
Eli C. Welch, 80, 70, 3000, 25, 300
John Wood, 50, 70, 2200, 50, 200
John A. Hurlock, 130, 90, 6000, 200, 900
John Winford, 120, 30, 5000, 50, 600
William Buckson, 100, 50, 4000, 50, 400
James Buckson, 175, 75, 10000, 200, 5000
William Craig, 60, 20, 2000, 50, 300
Rob Huggins, 300, 50,10000, 200, 1100
James D. McNatt, 42, 200, 1000, 20, 400
Alley Deakyne, 90, 39, 4000, 100, 600
Henry H. Woodkeeper, 80, 10, 2000, 50, 400

Jeremiah Allen, 56, 10, 800, 20, 200
Philemon R. Dill, 80, 20, 1000, 20, 300
Benj. C. Harris, 100, 50, 2000, 20, 300
Isaac Caulk, 100, 60, 3000, 50, 400
Geo. Daniels, 120, 60, 4000, 50, 400
John Neuson, 60, 20, 1200, 50, 300
Jacob C. Vandyke, 120, 30, 4000, 100, 800
Jeremiah Pryor, 60, -, 3000, 50, 500
Edward Thomas, 450, 50, 40000, 200, 2400
William Harris, 39, 3, 800, 50, 300
Jacob Harris, 30, 10, 600, -, 300
Richard Bostic, 100, 80, 3000, -, 300
Jacob Daniels, 200, 500, 8000, 50, 500
Isaac Allen, 60, 40, 1800, 50, 200
William Burk, 80, 20, 2000, 50, 300
Stinger L. Tinley, 120, 60, 8000, 40, 500
Peter View, 100, 10, 8000, 100, 600
Perry Prettyman, 300, 100, 12000, 100, 800
Jas. W. Crawford, 200, 30, 1000, 200, 100
Richd. McKee, 290, 10, 20000, 300, 2000
Samuel Pennington, 80, -, 8000, 100, 800
Saml. T. Crawford, 90, -, 15000, 200, 600
John C. Corbit, 300, -, 35000, 300, 2300
Chas. Beaston, 75, 20, 112000, 300, 1200
Jas. W. Geary, 240, 10, 15000, 300, 1100
Nehemiah Beamis, 250, 50, 15000, 300, 150
Alfred S. Hudson, 300, 90, 25000, 500, 2000
Lenick F. Shallon (Shalleross), 475, 4, 26000, 50, 3000
Westly Heims, 135, 15, 10000, 300, 600
Laurence Davis, 200, 13, 13000, 100, 900
Israel Townsend, 100, 25, 12000, 400, 1000
Parker Green, 25, 3, 1800, 20, 100
Jacob W. Satterthwait, 100, 23, 8000, 50, 715
Chas. & Jas. Terraddele, 90, 24, 10000, 200, 800
William T. Talley, 100, 33, 8000, 200, 850
Amor H. Chandler, 7, -, 1400, 40, 200
John Chandler, 8, -, 1200, 100, 120
Hayes Chandler, 83, 10, 5000, 200, 330
Lukins Tomlinson, 80, 16, 5000, 200, 375
John A. Jordon, 22, 20, 2500, 10, 145
Joseph Hanbey, 77, 26, 9000, 150, 700
Joseph Kellam, 50, 8, 3000, 150, 475
John E. Grubb, 15, 3, 1700, 25, 145
Joseph D. Talley, 77, 6, 6000, 75, 600
Hiram Talley, 217, 5, 4000, 100, 250
John Mekeever, 36, 25, 4000, 100, 250
James B. Burnet, 50, 1, 3000, 30, 270
William Fourney, 70, 33, 5000, 50, 275
Abram Palmer, 110, 40, 5000, 200, 580
Elizabeth Kennady, 15, 10, 1500, -, 90
Rebecca Gallot, 15, 10, 1000, -, 80
Cornelius Sweeney, 20, 4, 900, -, 50
Thomas Keene, 20, 40, 1200, 50, 100
James Martin, 31, 1, 2000, 40, 190
James Righter & Co., 90, 30, 8000, 200, 765

William F. McKee, 45, 29, 4000, 125, 450
Spencer Robinson, 16, 6, 1000, 75, 95
Isaac S. Thompson, 85, 15, 3500, 200, 690
Oliver H. Parry (Perry), 18, -, 5000, -, 500
P. Miller Taylor, 23, 1, 2000, 50, 200
Hugh Haughley (Raughley), 25, 1, 2000, 30, 260
Mary Memore, 38, 4, 2000, 100, 220
Joseph Mousley, 7, 5, 1000, 10, 300
John W. Day, 80, 38, 9000, 150, 535
Thomas R. Day, 20, 20, 3000, 100, 300
Jacob Backus, 20, 20, 3000, 100, 300
Adolphus Husband, 90, 30, 10000, 150, 1068
John Husband, 165, 50, 12500, 250, 1210
Joseph B. Langley, 118, 7, 12500, 300, 1150
John Proud, 22, 18, 4000, 50, 270
Samuel H. Derrick, 8, 7, 1320, 100, 136
Charles T. duPont, 45, 45, 8000, 200, 400
Francis Petitedamange, 76, 30, 10000, 200, 900
Abram Husbands, 20,7, 3000, 250, 230
William F. Husbands, 30, 28, 4500, 200, 675
Thomas Husbands, 60, 23, 7000, 500, 750
Hugh Sterling, 17, -, 3000, 100, 364
James E. Hornbey, 39, 5, 4000, 100, 427
Joseph Hunter, 94, 12, 8500, 100, 655
Maria Green, 35, 4, 3900, 100, 273
William Sayers, 55, 35, 6500, 100, 500
William Pattison, 10, -, 2000, 100, 250
Thomas F. Mahaffy, 48, 2, 15000, 500, 525
Maria Plan, 15, 13, 1000, 20, 125
Emory Nicholson, 16, -, 1500, 25, 200
John Friel, 43, 30, 4000, 100, 325
William Wilson, 100, 30, 6000, 150, 250
Lewis Wilson, 57, 20, 5000, 50, 600
Thomas E. Peirce, 45, 6, 2000, 50, 150
Charles E. Willbank, 38, 10, 2000, 15, 90
Valentine Forwood, 55, 15, 6000, 150, 445
Jehu P. Pyle, 127, 25, 10000, 400, 859
Samuel M. Tally, 48, 5, 4000, 200, 495
Curtis Mousley, 40, 810, 2500, 100, 260
Hezekiah Talley, 36, 10, 3000, 100, 240
Zachariah Ebright, 34, 1, 1200, 25, 186
Pennell Peterson Jr., 30, 16, 2500, 75, 320
John Sharply, 175, 40, 16000, 100, 500
Jane McBride, 30, -, 4000, 150, 516
David E. Derrickson, 25, -, 5000, 100, 250
John Senott, 40, 22, 4000, 50, 290
William Sharply, 35, 15, 4000, 200, 310
Irwin W. Peirce, 22, -, 2000, 50, 260
Alexander McDonald, 30, 10, 3000, 100, 170
Alfred D. Murphey, 50, 25, 7000, 200, 369
James McGilligin, 78, 20, 10000, 175, 556
Amos Bird, 15, -, 1000, 50, 125
Smith & Blackwell, 190, 60, 25000, 500, 1655

Eli B. Talley, 104, 20, 8600, 600, 1075
John Anderson, 20, 27, 2500, 250, 615
Jessup & Moore, 35, 25, 7000, 300, 418
J. C. Elliott, 90, 7, 15000, 200, 800
George Mously, 70, 20, 8000, 200, 760
Benj. Elliott, 92, 8, 15000, 200, 1000
Isaac S. Elliott, 162, 25, 2500, 500, 2500
Joshua Hotten (Healten), 150, 50, 12000, 350, 1200
George L. Miller, 50, 20, 5600, 400, 600
Mary Stullian, 66, 14, 8000, 50, 262
George S. McKee, 100, 20, 17000, 500, 700
Howard Crosedale, 20, -, 13000, 60, 70
Alex Hand, 25, 5,1200, 20, 100
Isaac _. Lesuderman, 22, -, 1600, 100, 150
John Bowen, 30, 12, 2500, 30, 200
Francis B. McCullem, 40, 1, 14000, 250, 200
Isaac Hand, 30, -, 2000, 50, 150
William Weer, 75, 33, 22750, 150, 950
Joseph Tatnall, 45, 17, 13500, 135, 900
Sebastian Raymond, 50, -, 14000, 230, 490
William Lea, 25, -, 6000, 75, 530
Caleb Taylor, 90, 35, 7500, 150, 850
Robinson Beeson, 56, 10, 5230, 75, 325
William Peters, 32, 8, 3000, 75, 145
Robert Morrison, 40, 8, 5000, 75, 20
Curtis Barlow, 11, -, 1500, 20, 145
Jordon Nicholson, 28, 16, 2000, 100, 325
Thos. J. Closs, 55, 10, 3500, 100, 360
Edward _. Bellah, 160, 6, 29000, 400, 1800
Theopholus Magne, 90, -, 20000, 300, 650
Eli Wilson, 90, 4, 17000, 800, 1800
Joseph Mendenhal, 75, -, 8000, 100, 885
James R. Jefferis, 115, -, 23000, 1000, 2624
William R. Weldin, 69, -, -, 5700, 200, 470
John S. Beeson, 100, 300, 10000, 300, 730
Thomas E. Weer, 145, -, 30000, 380, 1250
James Price, 188, 6, 33000, 1000, 2600
Jacob Zebley, 80, 52, 25000, 500, 1120
William Orr, 100, 60, 25000, 150, 1850
Isaac Lendersman, 65, 5, 14000, 300, 1284
Lewis Bird, 41, -, 4000, 100, 257
George Daugherty, 50, 20, 5000, 200, 1000
David Beeson, 63, 2, 6500, 300, 1055
Joseph Saring, 60, 2, 7700, 50, 100
John Todd, 66, -, 5200, 250, 610
Wm. & Jas. Todd, 127, -, 19000, 800, 2010
Charles Crompton, 100, 10, 20000, 60, 610
Mary Ann Robinson, 140, -, 20000, 275, 2000
Isaac Naylor, 70, 12, 12000, 100, 1000
Henery Webster, 80, 30, 10000, 150, 675
George W. Talley, 350, 10, 21500, 600, 5000
Jacob R. Weldin, 80, 10, 9000, 300, 1185
Beulah Weldin, 40, 23, 8000, 200, 500

John Talley, 60, 16, 6000, 100, 475
George Harriet, 30, 10, 7600, 150, 266
Martin Miller, 67, 8, 12000, 200, 700
Joseph Shipley, 124, 70, 37000, 600, 1340
Henery Paskill, 72, 8, 6500, 200, 715
Stephen G. Weldin, 85, 30, 11500, 500, 748
Joseph Miller, 60, 6, 7000, 300, 525
William L. Wilson, 55, 15, 7000, 300, 368
Curtis Grubb, 17, 10, 1500, 50, 125
Edward Biggers, 25, 3, 2200, 100, 175
Daniel McVey, 35, 25, 3000, 100, 175
Isaac D. Weer, 25, 15, 2000, 50, 225
Hannah Wilson, 40, 25, 3200, 100, 295
Edward Bri_gurst, 34, 13, 6000, 300, 650
John Almond, 35, 6, 2500, 50, 300
John Bradford, 103, 20, 15000, 500, 1500
Phillip Prince, 68, 30, 8000, 200, 520
Charles Simon, 15, -, 1600, 200, 220
John Talley, 25, 15, 2000, 100, 310
Richard Rowland, 40, 42, 25000, 50, 300
John Day, 60, 10, 5000, 200, 660
Marie Jordon, 49, 3, 5000, 250, 685
Jacob Sharpley, 55, 12, 5500, 600, 775
Clark Webster, 72, 75, 7200, 100, 642
Jos. S. Derrickson, 71, 25, 12000, 1000, 1240
Isaac Webster, 42, 6, 4000, 150, 532
Samuel Forwood, 16, 4, 2800, 150, 350
Thomas Talley, 75, 10, 8500, 250, 580
Charles Talley, 20, -, 2000, 75, 245
Lewis Zebley, 64, 4, 6000, 200, 600
Lewis Talley, 35, 30, 5000, 100, 393
William Ball, 25, 27, 4000, 100, 300
Peter Talley, 20, 60,2800, 200, 361
Thomas M. Talley, 75, 7, 6400, 150, 670
George E. Weldin, 24, 1, 3500, 150, 300
Thos. L. Talley, 118, 12, 13000, 400, 1305
David H. Forewood, 39, 20, 5000, 700, 635
Jacob Beeson, 137, 40, 22300, 200, 680
Edward Beeson, 55, 50, 5500, 50, 343
Mrs. Barbara Carr, 241, 100, 23700, 400, 1900
Jesse Sharp Jr., 215, 10, 20000, 300, 2700
Peirce Forwood, 58, 10, 8500, 90, 500
Jehu (John) Forwood, 114, 40, 2000, 50, 1410
Joseph W. Peirce, 13, 1, 1000, 50, 178
John Foulk, 80, 20, 10000, 200, 618
Penrose R. Talley, 77, 33, 10000, 400, 1150
Lewis Weldin, 8, 2, 5000, 200, 300
William P. Weldin, 52, 3, 7000, 125, 560
McIlvain & Lobdell, 89, 1, 13500, 200, 560
Levi M. Weldin, 60, -, 9000, 100, 470
B. & J. H. Guest, 75, 50, 20000, 400, 821
Mary Beeson, 7, -, 1400, 50, 295
Hanson Robinson, 52, 3, 30000, 500, 1000
James Osborn, 122, 4, 17000, 300, 1260
Daniel B. Perkins, 22, 2, 3000, 150, 270
Isaac N. Lodge, 125, 20, 15000, 300, 1327
Robert Orr, 82, 6, 10000, 200, 620

William McCullough, 64, 1, 5000, 50, 353
Thomas Carterwell, 50, 50, 6000, 60, 445
Sarah McSordley, 10, -, 3000, 50, 350
Joseph B. Guest, 34, -, 5000, 150, 382
Michael B. Mahoney, 90, -, 25000, 500, 1625
Clorist _. Perkins, 40, -, 5000, 150, 375
Amer Perkins, 110, 10, 19000, 600, 870
Robert Miller, 80, 50, 7800, 100, 540
Phillips & Derrickson, 100,50, 20000, 150, 1160
Clark Lodge, 58, 12, 8000, 150, 625
Samuel Lodge, 20, -, 7000, 150, 185
William P. Lodge, 100, 14, 21000, 500, 1040
Mary Lodge, 108, 10, 12000, 400, 1375
Casey & Scudders (S_dders), 122, 13, 12000, 300, 1100
Joseph McNamer, 35, 1, 10000, 200, 506
James Grubb, 60, -, 15000, 300, 740
George Valentine, 51, -, 1000, 200, 720
Amos J. Forwood, 17, 8, 3750, 150, 337
Fred Bird, 25, -, 4700, -, 50
Richard Edwards, 36, -, 6000, 100, 300
William C. Lodge, 75, 15, 18000, 400, 1500
George Lodge, 335, 100, 32000, 500, 3405
Tage (Gage) & Myers, 55, 7, 15000, 250, 700
Lot Cloud, 80, 80, 9600, 250, 900
Marshall Hill, 28, 2, 12000, 500, 400
Willis P.H Hazard, 10, -, 6000, 300, 250
J. C. Darley, 10, 2, 5000, 110, 250
J. B. Clemson, 25, -, 13000, 200, 270
Geo. W. Churchman, 78, -, 10000, 500, 700
Geo. Thompson, 25, 5, 5000, 150, 600
William Goodley, 100, -, 14000, 300, 1300
John B. Gray, 50, -, 6000, 150, 600
Abner Vernor (Vernon), 23, 2, 10000, 80, 200
Mercer J. Way, 95, 25, 10000, 100, 760
Thomas Goodly, 41, 3, 5000, 100, 420
Francis Ennis, 20, -, 2000, 100, 135
Jacob Carpenter, 90, 10, 10000, 100, 505
Adam (Odom) Prince, 98, 24, 12000, 700, 955
Valentine R. Ennis, 37, 3, 3000, 70, 275
Mary A. M. Robinson, 40, 14, 4500, 125, 375
Walter Peirce, 110, 50, 10000, 200, 980
George Moore, 90, -, 7000, 150, 670
Thomas Clyde, 220, 10, 25000, 500, 2500
George Williamson, 83, 17, 10000, 200, 820
Saml. J. Lodge, 56, 4, 8000, 150, 337
Lewis B. Harvey, 75, 25, 12000, 200, 880
George Veale, 155, 40, 4000, 500, 950
William Bird, 75, 10, 6000, 250, 468
John B. Smith, 32, 6, 3200, 150, 350
Mary Weer, 5, -, 900, 70, 115
Thomas Rambo, 60, 5, 5000, 200, 610
Isaac N. Grubbs, 1335, 22, 13800, 250, 650
George W. Smith, 80, 20, 10000, 250, 840
Thomas C. Smith, 13, 2, 800, 50, 200

Dr. J. H. Shevers, 65, 8, 10000, 215, 284
Thomas S. Gray, 90, 90, 18000, 200, 950
William Hanby, 80, 20, 8000, 100, 690
Owen Zebley, 45, 1, 1700, 50, 75
Joseph Pyle, 58, 18, 7600, 300, 900
Adam Peirce, 110, 40, 8000, 200, 676
Robert Casey (Cosey), 68, 11, 4000, 150, 674
Uriah Peirce, 26, 10, 3500, 75, 160
Charity Hanbey, 14, 1, 1500, 100, 100
Thomas Zebley, 31, 3, 3600, 75, 430
James G. Hanbey, 31, 2, 4000, 200, 396
Elias Peirce, 54 8, 4700, 100, 522
Hiram Lodge, 81, 10, 9000, 400, 155
McCaulley & Crabb, 110, 53, 13000, 300, 1310
Samuel Hanbey, 110, 40, 9000, 300, 800
James A. B. Smith & Son, 30, 21, 8000, 300, 850
Samuel G. Chandler, 90, 20, 7000, 110, 800
Levi Springer, 50, 30, 5000, 100, 500
Edward Mendenhall, 100, 60, 10000, 500, 800
James Martin, 130, 20, 10000, 500, 1600
Miller Speakman, 25, 15, 4000, 400, 200
Otley Vernon, 100, 20, 9000, 400, 1250
Eli Call, 11, 3, 1000, 100, 120
Jackson Holmes, 85, 7, 8000, 200, 600
Caleb Sharpless, 55, 5, 4500, 100, 451
Samuel Green, 60, 20, 7000, 200, 900
Caleb Sharpless Jr., 43, 9, 4000, 2225, 1500
James M. Bracken, 140, 66, 20000, 600, 1600
Jacob Hanna, 70, 30, 1000, 350, 500
James Jackson, 100, 16, 8000, 300, 1400
Samuel Yeatman, 25, 29, 4000, 75, 200
William P. Passmore, 140, 32, 12000, 600, 1300
Bennet Klear, 80, 16, 7000, 200, 500
James Chandler, 30, 35, 2000, 200, 300
Harman Davis, 25, 15, 2500, 150, 200
Jeremiah Springer, 70, 100, 10000, 100, 300
Samuel Gamble, 55, 5, 7500, 200, 400
Jethro Bugless, 27, -, 1200, 50, 150
Harlan Cloud, 100, 20, 13000, 300, 1300
David Taylor, 60, 10, 5000, 200, 500
Levis Walter, 62, -, 6500, 400, 600
S. Canby Clement, 111, 30, 10000, 400, 400
Thomas Harlan, 63, 10, 6000, 200, 600
Israel Bonsel, 45, 15, 6000, 100, 200
Chalkly Way, 72, 9, 8000, 200, 300
Joseph P. Chandler, 67, -, 7000, 300, 1000
Levi Walker, 41, 20, 5000, 110, 250
James H Gould, 50, 50, 5000, 101, 200
James L. Carpenter, 135, 15, 12000, 300, 1600
James L. Delaplane, 160, 25, 15000, 500, 500
Joel Dalton, 26, 4, 3500, 50, 600
Jeremiah Underwood, 60, 12, 8500, 150, 500
Hugh Passmore, 35, 10, 3000, 51, 200

Chalkley Bullock, 25, 19, 3000, 50, 250
J. Garret Pyle, 70, 5, 5000, 100, 500
Charles Bogan, 30, 3, 2000, 100, 250
Morris Palmer, 180, 40, 14000, 500, 1000
Marshal Chandler, 120, 30, 11000, 225, 1000
Robert Wilson, 25, 5, 3000, 100, 250
Thomas Wilson, 80, 10, 7500, 250, 500
Eli Nichols, 80, 25, 8000, 150, 400
Robert Husbands, 140, 40, 20000, 250, 600
J. Pulson Chandler, 125, 80, 14000, 500, 1200
_. McHullough, 96, 4, 7000, 300, 500
Samuel P. Chandler, 140, 20, 20000, 800, 1200
Francis Hobson, 160, 10, 13000, 300, 900
Henry Swayne, 70, 10, 8000, 250, 700
Hugh Bogan, 65, 15, 7000, 100, 500
William Thompson, 95, 25, 8000, 150, 500
Joseph Cloud, 50, 10, 5000, 410, 500
Benjamin Seal, 100, 12, 12000, 300, 1000
Isaac Cloud, 25, 15, 3500, 50, 150
Reece Pyle, 72, -, 5000, 400, 401
Jeremiah Dodsworth, 16, -, 3000, 50, 250
James Scott, 26, -, 2500, 40, 200
Zeph__ C. Fairel, 60, 75, 7000, 300, 500
Enos Wolter, 45, 15, 6000, 150, 30
Robert Barks (Burks), 85, 4, 6000, 150, 500
Joseph Lynam, 92, 23, 7000, 100, 501
Henry Smith, 70, 40, 5000, 150, 500
Charles L. Casperton, 90, 50, 10000, 200, 800
John Gregg, 60, 20, 10000, 500, 800
Peter W. Gregg, 100, 25, 15000, 500, 1110
Augustin Carnog (Cernog), 72, 16, 9000, 500, 1000
George Proud, 95, -, 7000, 200, 500
James Donohu, 30, 10, 2000, 70, 200
James Starett, 60, 15, 4500, 50, 150
James White, 44, -, 5000, 200, 500
Henry Wolter, 50, -, 10000, 100, 400
Adaline Hendrickson, 90, 22, 12000, 200, 1000
James McGullis, 40, 10, 3000, 100, 200
John Thompson, 180, 20, 13000, 500, 1000
William Wilmet, 125, 75, 13000, 400, 1610
John M. Cralle, 85, 23, 8000, 300, 510
Ellis M. Nichols, 70, 10, 7000, 200, 500
Samuel Armstrong, 35, 6, 3000, 100, 300
William Atkinson, 4, -, 300, -, 5
Ben Brown, 30, -, 1800, 50, 150
Caleb S. Way, 80, 15, 10000, 250, 800
John Pyle, 90, 20, 8000, 100, 1200
WilliamS. Dillworth, 170, 24, 14000, 400, 1800
William Armstrong, 80, 10, 8000, 75, 500
Alexander McCoy, 50, 12, 4500, 70, 400
Thomas Kinsey, 93, 5, 9800, 300, 700
Henry Messimer, 60, 7, 5500, 150, 300
Louis Grave, 120, 25, 12000, 300, 1000
Albert Cummins, 85, 20, 10000, 600, 700
James Mullen, 30, 10, 3100, 75, 250
Pter Fermier, 100, 27, 13000, 800, 900

Moses Loutha, 80, 20, 8000, 500, 1500
Am___ Way, 55, 4, 5000, 400, 600
James Brown, 100, 36, 10000, 300, 700
John Brannan, 60, 10, 5000, 200, 600
Patrick Daugherty, 35, 15, 3000, 150, 350
Abner Woodward, 100, 50, 8000, 500, 800
William Cockson, 55, 11, 4600, 200, 450
William Edwards, 11, -, 1500, 75, 150
Hugh Reed, 50, 16, 4000, 150, 300
Peter McLoughlin, 36, 4, 4000, 50, 200
Joseph Daugherty, 30, 20, 6000, 50, 400
John Nickson, 45, 5, 4000, 75, 200
Lilliam M. Parkens, 9, -, 800, 50, 200
Robert Marom, 75, 5, 8000, 200, 700
Ben Beaty, 50, 44, 8000, 50, 200
Thomas Neals, 30, 10, 3000, 50, 250
Ben Bartram, 65, 20, 7500, 400, 800
Charles Green, 50, 10, 6000, 250, 600
John Cooper, 36, -, 5000, 100, 300
Isaac Hendrickson, 70, 40, 8000, 200, 800
Thomas Lynam, 70, 44, 10000, 100, 600
William Robinson, 120, 20, 11000, 300, 500
John Banning, 80, 36, 10000, 401, 800
Peter Collins, 5, -, 1000, -, -
William S. Fleming, 60, -, 15000, 150, 1800
E. I. duPont & Co., 500, 600, 60000, 500, 7750
Elwood Mitchiner, 75, 10, 9000, 500, 550
Isaac Clough, 60, 13, 4500, 100, 250
Jesse Hallowell, 100, 11, 15000, 500, 700
George Eli, 30, -, 6000, 200, 500
Stephen Scott, 22, 10, 2000, 50, 100
James Smith, 14, -, 3000, 50, 250
Jacob Vanhan, 109, 2, 10000, 300, 500
James H. Heffecker, 100, -, 13000, 500, 300
Moses Journey, 50, 15, 7000, 100, 500
James Arnor, 66, -, 5000, 200, 500
Abram Ford, 65, 5, 20000, 150, 500
Richard Brindley, 100, 80, 12000, 800, 1200
James Dunn, 35, -, 3500, 100, 400
William Pyle, 30, -, 1500, 200, 500
John Brinkley, 30, -, 3000, 200, 350
John Thompson, 30, 30, 3500, 100, 300
William Laws, 30, 5, 3000, 75, 800
Henry Grant, 190, 30, 30000, 300, 2000
John Ball, 45, -, 3000, 100, 200
Thomas Smith, 22, -, 6000, 50, 500
Henry Foulk, 20, -, 2000, 50, 300
Philip M. Howell, 11, -, 2200, 50, 450
Marcillus Price, 130, 40, 25000, 300, 1700
David C. Wilson, 75, -, 30000, 50, 400
James Biddle, 100, 30, 17000, 100, 1850
Charles W. Howland, 33, -, 4000, 50, 900
Henry Lentz, 10, -, 2000, 50, 100
T. M. Rodney, 23, -, 15000, 100, 300
Thomas Seals, 14, -, 3000, 50, 150
R. Fendon Clark, 27, -, 4000, 100, 350
Bailey & Brother, 5, -, 6000, 50, 100
Henry C. Fermick, 35, 5, 12000, 100, 450
Thomas Kellem, 30, 10, 5000, 75, 275

Thos. Lamplaugh, 31, -, 8000, 100, 500
John B. Lynam, 105, -, 5000, 310, 506
John Booth, 100, 28, 1100, 50, 200
Henry Howard, 250, -, 8000, 200, 9000
Robert Flinn, 80, 27, 18000, 200, 1000
Wm. B. Flinn, 120, 14, 15000, 400, 1700
Evan T. Flinn, 35, 5, 15000, 150, 400
Benj. Rothwell, 37, 11, 4000, 100, 350
J. S. Lynam, 100, 30, 10000, 400, 600
Samuel Cranston, 180, 45, 20000, 300, 1560
Francis Flinn, 125, 25, 12000, 500, 1000
John H. Snyder, 20, -, 1500, 50, 300
Joseph Woodward, 125, 15, 10000, 300, 500
Christian Keiffer, 100, 10, 14000, 300, 1000
James Armstrong, 90, 20, 10000, 200, 800
Gilpin P. Stidham, 90, 6, 10000, 100, 600
Abner Hollingsworth, 120, 12, 12000, 400, 500
Robert L. Armstrong, 90, 20, 12000, 300, 900
Samuel Canby, 160, 30, 20000, 400, 2000
J. P. Armstrong, 100, 18, 10000, 200, 700
Thomas Foot, 100, 13, 13000, 200, 800
John Stuart, 95, 8, 5000, 200, 550
John Evans, 42, 12, 4000, 200, 250
Joseph Pritchard, 122, 25, 15000, 100, 800
John Flinn, 75, -, 6000, 300, 800
William Armstrong, 80, 8, 9000, 150, 700
Alfrd Bevell, 25, 5, 3500, 100, 300
Edward V. Paterson, 20, 15, 2500, 75, 150
John R. Tatam, 90, 40, 15000, 200, 800
Samuel Marshal, 100, 55, 15000, 200, 600
William Z. Derrickson, 60, 15, 7000, 200, 700
James Brown, 70, -, 9000, 250, 1800
Samuel Richardson, 80, 30, 10000, 150, 200
William Richardson, 40, 15, 8000, 150, 800
Ashton Richardson, 56, 3, 10000, 200, 550
Joseph Grubb, 105, -, 12000, 200, 1100
Ephraim Magargel, 85, 60, 10000, 150, 900
Abner Lynam, 100, 7, 10000, 100, 1000
John B. Marshal, 125, 15, 20000, 200, 1400
James B. Stidham, 75, -, 7500, 150, 1000
William Voucher, 55, -, 6000, 200, 400
George Bramble, 54, -, 7000, 150, 750
David Lynam, 100, -, 12000, 200, 600
Joseph S. Forman, 60, 60, 15000, 400, 1000
Joseph _. Richardson, 60, 40, 15000, 200, 1200
John Richardson, 50, 41, 15000, 200, 1000
Robt. R. Robinson, 60, 26, 9000, 210, 700
Thos. Welsh, 156, -, 25000, 500, 2500
Edward Thatcher, 95, -, 12000, 200, 1000

James T. Bird, 20, 18, 12000, 100, 400
William E. Harget, 17, -, 3000, 50, 100
Charles Wert, 95, 20, 12000, 200, 400
J. A. Brown, 80, -, 15000, 50, 800
William A. Lynam, 95, 14, 10000, 400, 300
Esther Walker, 77, 20, 7000, 700, 440
John Walker, 160, 50, 16080, -, -
William Walker, -, -, -, 300, 70
Thomas Walker, -, -, -, 250, 500
George Thompson, 100, 25, 11000, 350, 940
Edward Singley, 106, 15, 9000, 150, 500
Evin B. Plumly, 40,-, 3000, 100, 280
Melton Mitchener, 92, 15, 8000, 300, 700
Lukins McVaughn, 73, 3, 6080, 250, 650
T. L. J. Baldwin, 67, 4, 4000, 150, 470
John Springer, 78, 9, 8000, 250, 500
James McDowell, 100, 67, 12000, 150, 500
Beeson Yeatman, 25, 6, 2500, 250, 300
William Wilder, 35, 12, 4000, 100, 325
Laban Pierson, 30, 3, 3000, 100, 250
Thomas Pierson, 93, 7, 7300, 600, 700
Joster H. Dixon, 85, 15, 10000, 400, 600
John G. Jackson, 200, 40, 22000, 250, 3500
William Pierson, 30, -, 3000, 25, 185
James Dixon, 165, 15, 12000, 500, 1000
Ephram Wilson, 128, 8, 12000, 500, 900
Mathew Bunting, 30, -, 3000, 150, 200
Peter Connell, 22, 2, 2000, 100, 236
J. Calvin Hall, 100, 12, 12000, 400, 1000
Levis Lamborn, 20, -, 2000, 30, 100
Caleb Entrikin, 90, 20, 6000, 350, 1000
Samuel Jacobs, 95, 26, 12000, 700, 900
John Heald, 88, 10, 9500, 400, 1000
Howard Flinn, 135, 22, 14000, 500, 1140
Samuel Sharpless, 135, 23, 9000, 300, 800
Jessse Graves, -, -, -, 100, 400
Mary Armtrong, 81, 14, 5000, 100, 800
Abner Marshall, -, -, -, -, 400
Harris Jackson, 22, 4, 5000, 50, 150
Westley Flinn, 419, 15, 15000, 25, 1015
George Collins, -, -, -, -, 218
George Springer, 195, 23, 19000, 500, 1720
James C. Jackson, 135, 15, 15000, 50, 940
Joseph Highfield, 11, -, 1200, 65, 150
William Little, 100, 20, 11000, 350, 945
Pusey Wilson, 100, 23, 9500, 210, 610
Joseph Poole, 220, 39, 12000, 200, 455
Spencer Chandler, 185, 31, 17000, 2000, 1500
Amos Sharpless, 85, 15, 10000, 100, 900
Samuel P. Dixon, 110, 40, 11000, 210, 1350
Isaac T. Hoopes, 47, 5, 3500, 100, 525
Chandler Lamborn, 100, 50, 9500, 600, 700
James Donnell, 38, 2, 2600, 150, 320
John K. Waterman, 111, 30, 8000, 200, 648

Egbert Klair, 115, 10, 10000, 200, 637
Samuel Taylor, 68, -, 3750, 250, 380
James Woods, 21, -, 1700, 200, 50
Joseph Eastburn, 200, -, 18000, 600, 1295
Alben Buckingham, 67, 8, 6000, 200, 590
T. _. Moore, 67, 10, 5000, 150, 500
Lewis Thompson, 96, 15, 11000, 300, 910
Benjamin W Duncan, 140, 20, 12000, 390, 90
Jonathan Catlin, 120, 6, 16000, 800, 885
David Wilson, 150, 30, 10000, 300, 1700
William Howard, -, -, -, -, 250
Phillip C. Wilson, 65, 10, 6500, 100, 425
Thomas Shakespier, -, -, -, -, 240
Bethia Brackin, 38, -, 4000, 50, 220
Thomas Chandler, 29, 4, 2950, -, 300
Abigal Brannen, 35, -, 4500, 50, 170
Richard Fisher, 75, 25, 6000, 300, 625
John Armstrong, 159, 45, 14000, 350, 780
Hiram Pierson, -, -, -, 400, 648
Jacob Chandler, 80, 30, 5000, 400, 612
George Montgomery, 16, 4, 950, 50, 165
Thmas Vandever, 70, 30, 5000, 700
Josh... B. Barker, 55, 10, 7000, 200, 535
David Klair, 123, 20, 10000, 225, 739
B. F. Gebhart, 60, 30, 9000, 600, 885
John Woodward, -, -, -, -, 730
John Hulett, 73, 5, 6440, 300, 520
Ann Lobb, 10, -, 600, 50, 55
William Hobson, 68, 16, 4000, 200 425
Joseph Leach, 106, 25, 9000, 350, 690
James Leach, 36, -, 3600, 300, 495
Wm. Springer, 33, 6, 2700, 70, 320
Jo__ Petitidemong, 61, 15, 000, 300, 710
Francis Ransford, 63, -, 5750, 150, 500
Seyeus Pyle, 47, -, 10000, 350, 800
Thos. Hoops, 52, -, 5400, 200, 250
Thos. Love, 42, 7, 1300, 250, 275
Jo__ Hollingsworth, 100, 16, 9500, 100, 350
Joseph Nies, 50, 7, 5000, 200, 725
John Foust, 50, 12, 5000, 200, 650
Lewis Miller, 97, 6, 8200, 200, 340
Wm. Ralston, 25, 3, 2600, 30, 87
Samuel Graves, 40, 6, 5000, 375, 585
Wm. Robinson, -, -, -, 300, 1080
Edwd. Saunders, -, -, -, 450, 913
Reuben Hall, 25, -, 3300, 50, 155
James Bailey, 65, 5, 7000, 250, 670
Henry Clark, 47, 3, 4500, 25, 665
John Shakespier, -, -, -, 200, 165
Joseph Taylor, 30, 19, 2000, 200, 200
Samuel Nevel, -, -, -, -, 75
John Ridgway, 131, 5, 6500, 150, 325
George Pierson, 100, 13, 8000, 200, 635
Lemuel Graves, 49, 6, 4050, 200, 420
William Peoples, 64, -, 5000, 200, 330
Chs. Furgeson, 50, 20, 5000, 300, 300
Isaac Crossner(Crosser), 90, 12, 6500, 250, 325
Eli Crosser, 44, -, 3600, 250, 420
John Mitchell, 200, 20, 15000, 350, 575
John Bailey, 51, 4, 5000, 150, 460
Jonathan Fell, 65, 35, 10000, 250, 1500
David Graves, 73, -, 7500, 350, 730

Nathan Tearsley, 61, 9, 7000, 200, 320
Peter Coyle, 80, 25, 20000, 1000, 700
Wm. G. Phillips, 10, -, 4000, -, 60
J. B. Robinson, 72, 8, 7000, 100, 450
James Cranston, 250, 38, 19000, 500, 1200
James McCabe, -, -, -, 350, 725
Corn__, Derickson, 77, 8, 5500, 400, 575
Benj. Cranston, 136, -, 13000, 800, 1435
Levi Workeman, 10, -, 3000, 150, 100
Benj. Garrett, 27, -, 2700, 250, 575
Wm. Hazlett, 64, -, 7600, 300, 620
Fred___, Woodward, -, -, -, 300, 600
James Claman, 50, 20, 4000, 200, 500
Mary Gregg, 29, 4, 2850, 150, 375
James Crossen (Crosser), -, -, -, 100, 175
Jane McClure, 50, 20, 3500, 250, 300
James Walker, 33, -, 2600, 200, 375
Garret Johnson, -, -, -, 150, 680
John Ball, 53, 5, 2000, 100, 280
Jos. Cranston, 200, 20, 10000, 100, 1750
Andrew McKee, -, -, -, 400, 1230
Jabez Banks, -, -, -, 436, 1555
Thos. Smith, 39, 3, 5000, 50, 680
Saml. Kelly, 100, 5, 8000, 400, 650
Barton McClure, 137, 15, 9150, 600, 890
Lewis McClure, 60, 6, 4500, 250, 660
Jane McClure, 27, 8, 2200, -, -
James Foot, 59, 12, 5680, 200, 480
Zacharia Derickson, 113, 16, 9800, 300, 1000
Thos. Lynam, 97, 3, 8000, 385, 875
John H. Sanders, 64, 18, 7150, 300, 340
Wm. Sanders, 58, 14, 4300, 75, 240
Saml. Anthony, 50, -, 7000, 300, 450
John Foot, 95, 7, 9000, 200, 560
Mathew Rilett, -, -, -, 350, 775
Walker Denney, 100, 5, 8000, 150, 970
Jos. Woollason, 98, 7, 787, 300, 635
Stephen Mitchell, 90, 11, 9000, 200, 846
Abner Mitchell, -, -, -, 300, 850
Jos. Mitchel, 170, 30, 17500, 300, 825
John Ochletree, 55, 5, 4000, 150, 325
John Thompson, 153, 30, 13700, 400, 780
Levi McCormick, 18, -, 1200, 12, 80
Nathan McCormick, 28, 10, 1700, 55, 200
John McCormick, 18, -, 1200, 100, 200
Henry Taylor, 60, 5, 4200, 150, 350
Abner Woodward, 113, 12, 9000, 400, 700
Thos. Mitchell, 73, 10, 6000, 350, 500
Saml. Little, 93, 7, 4000, 300, 650
James Akin, 54, 10, 2000, 75, 235
James Stinson, 22, -, 2000, 100, 150
M. E. Easttoten, 137, 10, 10000, 350, 645
Benj. W. Shakespear, -, -, -, 250, 825
Chas. Jackson, -, -, -, 300, 780
Jos. Hanna, 54, 8, 4650, 100, 500
James Denison, 75, -, 4700, 100, 450
Robt. Hanna, 28, -, 2000, 50, 275
Jos. Trender, 46, 12, 5500, 150, 325
Edwd. Brayfield, 18, -, 1500, 100, 175
Saml. Montgomery, 65, 5, 3500, -, -
Bennet Wingate, -, -, -, 120, 700
James Denny, 110, 40, 10000, 300, 1070
Jos__ Pritchard, 157, 20, 7700, 300, 935
Lewis Pennock, 100, 15, 5000, 200, 450

Jos. Paxton, 54, 2, 6000, 300, 330
David Mediel, 15, 20, 1220, 150, 700
Jos. Greenwalt, 51, -, 3750, 150, 325
John Greenwalt, 110, -, 2100, 150, 425
Jacob Cuniner, 197, 352, 18000, 300, 950
Wm. Brown, -, -, -, -, 375
Thos. Hunt, -, -, -, 100, 300
Richd. Bonsell, 66, -, 6000, 230, 505
Wm. M. Derickson, 115, 5, 6000, 300, 560
George Harrison, 14, -, 200, 40, 100
Thos. Perse, -, -, -, 500, 940
Aquila Derickson, 332, 40, 29000, 600, 1400
David Chiles, 177, 30, 18670, 800, 635
Uriah Drake, 153, 10, 13000, 500, 800
James Little, -, -, -, 600, 1425
Jas. Phillips, -, -, -, -, 85
Wm. Little, -, -, -, 100, 325
Thos. Higgans, 142, 22, 1000, 400, 995
Jos__ Woorrell, 52, -, 4000, 400, 395
Benj. Morrison, 10, -, 1200, -, 40
Chas. Whiteman, 110, 16, 6500, 300, 630
Isaac Vansant, 150, 30, 7000, 300, 660
Joseph Mote, -, -, -, 250, 450
Oliver Eastbourne, 12, -, 1000, 600, 905
Peter Phillips, -, -, -, 500, 500
John A. Reynolds, 230, 33, 18000, 400, 700
Jas. Maxwell, 10, 1, 700, 50, 100
Randolph Peters, 127, 20, 9000, 500, 980
John Anderson, -, -, -, 300, 340
Ann Mitchell, 52, 10, 3500, 100, 125
John Brown, -, -, -, 101, 325
Edwd Meet___, -, -, -, 125, 265
Wm. Johnson, 112, 20, 8000, 200, 646
Mary Macklin, 13, -, 1200, 100, 135
Sarah Thompson, 50, 4, 2600, -, 275
Jackson Haywood, -, -, -, -, 300
Robt. Fitzsimmon, 40, 40, 4000, 200, 320
Joseph Gutherie, 12, 8, 1200, 100, 150
John Cloud, 115, 25, 8000, 150, 600
William Bell, 192, -, 10000, 800, 1370
Jos. Lindsey, 140, 60, 16000, 600, 1200
Elizabeth Eastbourne, 260, 75, 18900, 250, 800
David Eastbourne, 338, 40, 28940, 300, 534
Saml. Loyd, 40, -, 4000, 300, 580
Robt. Tweed, 87, 20, 7000, 250, 731
Walter Craig, 54, 5, 3400, 300, 470
Jacob Whiteman, 119, 8, 11000, 300, 900
Henry Whiteman, 1145, 6, 7000, 300, 700
Eli Mote, 228, 24, 14600, 400, 633
Saml. Harkness, 40, 5, 4000, 250, 515
Elizabeth Guthrie, 43, 8, 2500, 100, 200
Saml. Vansant, 10, -, 1200, 75, 125
John Elliott, 330, 45, 25000, 500, 1545
Jacob Stillwell, 50, -, 2500, 250, 450
Sarah Jacobs, 75, 15, 4500, 75, 295
Saml. Russell, 31, 5, 2500, 75, 175
Wm. Flinn, -, -, -, -, -
David Lynam, -, -, -, -, 1146
Jno. Rhubencome, 112, 7, 8000, 500, 650
Ja___ W. Ball, 125, 15, 12500, 300, 800
Aaron Klair, 115, 20, 12600, 400, 800
Saml. Adams, 5, -, -, 400, 900

Robt. McGonigal, 50, 2, 3500, 100, 300
Nathl. Pritchard, 35, 8, 3500, 150, 328
Mansel Tweed, -, -, -, 300, 400
Jas. Mitchell, 70, 14, 6150, 350, 800
James Brown, -, -, -, 250, 715
Lewis Fell, 203, 40, 19400, 700, 1840
John McCartney, -, -, -, 350, 700
Wm. Carlile, 53, 8, 4500, 400, 500
John Haspen (Hasper), 75, 4, 4800, 400, 645
Levi B. Moore, 110, 10, 7000, 250, 800
Wm. Harkness, 170, 30, 13000, 700, 775
Jacob J. Fendley, 75, 25, 8000, 150, 780
Eli Davis, 80, 20, 8000, 500, 940
Barton Hayes, 169, 10, 17900, 1000, 1955
Thos. Rankin, 110, 400, 12750, 600, 1170
Jos. Rankin, 110, 50, 12500, 580, 855
David G. Niven, -, -, -, 150, 660
John D. Forest, 81, 20, 8550, 1200, 990
Wm. Baldwin, 105, 5, 7000, 350, 605
Reuben Ball, 150, 50, 11700, 400, 1500
John Peach, 96, 20, 9000, 450, 700
Elizabeth Chambers, 26, -, 2300, 100, 200
John Chambers, 101, 2, 8000, 550, 650
Robt. Walker, 100, 28, 9000, 500, 566
Rebecca Walker, 100, 12, 10000, 400, 1052
Wm. Brackin, 36, -, 3600, 200, 225
Benj. Gregg, -, -, -, 300, 885
Stephen Springer, 93, 7, 6000, 300, 598
Isaac Mendelhall, 80, 20, 6000, 300, 512
Wm. Foot, 73, 4, 5300, 150, 310
Jas. Springer, 85, 15, 7500, 300, 832
Givens Kendle, 36, 4, 2700, 75, 270
George Story, -, -, -, 100, 325
Alex. Guthrie, 160, 40, 12000, 500, 465
John Harper, 24, -, 1500, 125, 165
David Moore, 14, -, 1000, 50, 65
Emma McDenise, 185, 5, 20000, 300, 1791
George Klair, 108, 12, 9000, 300, 705
John Husband, 98, 10, 8000, -, -
Saml. Denison, 85, 12, 7200, 150, 485
George Jamison, -, -, -, 75, 1100
Arnold Naudain (Nandain), -, -, -, 600, 1090
M. B. Ocheltree, 100, 12, 10000, 500, 1035
John Holand, 100, 4, 6000, 250, 415
John Gregg, 190, 7, 17750, 550, 900
Benj. Chambers, 18, -, 115, 75, 225
Sylvester Townsend, 200, -, 28000, 400, 2000
Henry Quinn, 10, -, 4000, 50, 150
Zadoc Townsend, 65, -, 20000, 300, 700
James Bennett, 20, -, 10000, 175, 400
Jesse Alexander, 10, -, 4000, 200, 271
Lewis Peugh, 10, -, 4000, 150, 100
Jacob Brown, 50, -, 8000, 200, 500
Anna Moore, 198, -, 38000, 500, 2000
Rebecca Gettze, 20, - 5000, 200, 150
John Peugh, 119, -, 10000, 500, 1500
Wm. Gibbons, 22, -, 4000, 150, 350
Ephrem Sutter (Sutton), 5, -, 350, 125, 200
Patrick McMahan, 35, -, 400, 200, 200

Benjamin Townsend, 145, -, 15500, 200, 1500
Francis Carm, 127, -, 15500, 500, 1500
Lucus Alrich, 135, -, 12000, 350, 1000
George Townsend, 215, -, 25000, 800, 4000
Peter Alrich, 148, -, 12000, 500, 1500
Robert Hicks, 26, -, 12000, 400, 500
Edward Broodford, 17, -, 12000, 400, 500
Zadoc Townsend Jr., 325, -, 22000, 400, 2500
Brannough Derringer, 222, -, 11200, 500, 1500
Jason Davis, 300, 20, 25000, 500, 2000
Alexander Eves, 200, -, 20000, 400, 1200
Wm. Edmundson, 120, 20, 14000, 200, 750
Wm. White, 30, -, 50000, 500, 2000
John LeFebre, 180, -, 18000, 500, 2000
John Peach, 125, 20, 8000, 300, 800
George Hairson, 84, -, 15000, 400, 1000
John A. Webster, 15, -, 8000, 250, 300
Theodore Rogers, 60, -, 50000, 1000, 1300
John S. Smith, 150, -, 15000, 200, 700
Giles Lambson, 188, -, 20000, 500, 2500
James Russell, 165, -, 23000, 500, 2000
Albert G. Stutton, 28, -, 10000, 300, 350
John Morrison, 172, -, 25000, 500, 2000
Albert Sutton, 70, -, 7000, 200, 500
John Echels, 166, -, 20000, 500, 1500
James Califf, 165, -, 20000, 300, 1000
James Hurst, 84, -, 12000, 500, 1000
Edward Downing, 101, -, 10000, 300, 600
Wm. Hope, 111, -, 10000, 200, 1000
James Cavender, 175, -, 20000, 500, 1200
Wm. Stroup, 132, -, 14000, 500, 1000
Wm. Banks, 150, -, 20000, 400, 1000
Mandlove McFarlin, 112, -, 15000, 200, 300
Wm. Smith, 80, -, 10000, 200, 1000
Giles Lambson Jr., 230, -, 30000, 400, 800
Charles Hal, 20, -, 10000, 200, 325
Edward McCaferty, 125, -, 15000, 200, 1200
Jacob Walten, 210, -, 35000, 400, 2500
Jahu Clark, 270, 40, 3700, 800, 2500
John Teiler, 70, -, 7000, 150, 300
Bryan Jackson, 210, -, 30000, 200, 4000
Thomas Lynam, 130, -, 14000, 200, 3400
Isaac Grubb, 130, -, 10000, 320, 800
John McFarlin, 150, -, 20000, 500, 1000
Edward Hanse, 375, -, 40000, 500, 2000
Frank Halcomb, 120, -, 20000, 1000, 2000
Michel Callahan, 140, -, 15000, 300, 1200
Thomas M. Ogle, 84, -, 10000, 400, 650
John C. Morrison, 135, -, 15000, 500, 1000
George Morrison, 100, -, 1000, 500, 1000
Phillip R. Clark, 101, -, 7000, 250, 1750

Louisa Marley, 133, -, 7000, 400, 600
Benjamin S. Booth, 330, 33, 24000, 1000, 2500
John B. Smith, 200, 100, 20000, 300, 1000
James Bolton, 100, 5, 10000, 400, 1000
George Allen, 50, -, 4000, 200, 300
Lewis Ashton, 115, -, 10000, 200, 600
James Caulk, 115, -, 10000, 100, 500
George Topper, 100, -, 7000, 200, 500
John T. Simmons, 190, -, 18000, 580, 1500
John Smith, 170, -, 17000, 500, 1100
Lewis Reynolds, 96, -, 6000, 200, 500
Andrew Groves, 200, 100, 20000, 300, 700
Myers Hitchens, 130, 50, 12000, 500, 1000
Wm. McCallister, 190, 30, 17000, 500, 1500
Joseph T. Brown, 130, -, 15000, 600, 1700
George Parmer, 36, -, 4000, 100, 500
David W. Gemmill, 120, -, 12000, 300, 2500
Richard Jackson, 180, -, 20000, 1000, 1925
George Lopdel, 182, -, 25000, 600, 5000
Josiah Lewder, 90, 60, 7000, 350, 900
Reuben Janviel, 160, 30, 14000, 200, 900
Henry C. Burton, 325, -, 35000, 1500, 2500
George McCrone, 163, -, 10000, 400, 1500
Wm. Gray, 150, -, 1500, 300, 1000
George Bartholomew, 80, 20, 6000, 300, 500
Edward Edwards, 320, 50, 25000, 1500, 3000
John McCoy, 130, -, 10000, 120, 1200
George Bowl, 116, -, 8000, 400, 1300
John Diehl, 250, -, 28000, 1000, 2450
Eugene Stroup, 90, 72, 5000, 150, 450
Lewis E. Pennington, 130, 20, 10000, 500, 1200
Daniel Rodney, 206, -, 15000, 500, 1500
Wm. Gibbons, 125, 70, 10000, 200, 600
Edward Stroud, 105, 45, 8000, 200, 600
Wm. Thompson, 125, 25, 7000, 200, 500
Abraham Moore, 40, 50, 6000, 100, 400
John C. Bradley, 130, -, 6000, 200, 500
John Biddle, 120, 22, 6000, 200, 900
Charles McNamer, 100, 50, 5000, 150, 500
Roderick Sutherland, 25, -, 3000, 400, 350
Wm. Silver, 230, -, 23000, 500, 1500
John Baker, 200, 30, 10000, 200, 800
Philip Baker, 75, 37, 4000, 100, 300
Walter Turner, 200, 50, 20000, 500, 1000
George Lee, 250, 15, 25000, 600, 2400
John Lee, 125, 25, 15000, 400, 1300
James McCoy, 150, 30, 14000, 200, 10000
Henry McMullen, 200, -, 20000, 500, 1500
Wm. G. Caulk, 54, 5, 3000, 50, 200
Wessel Robinson, 120, -, 12000, 200, 1000

James C. Douglas, 330, -, 45000, 100, 3300
John L. Deputy, 80, -, 10000, 200, 1000
James Clark, 200, 20, 10000, 200, 900
John Burnham, 135, 25, 12000, 400, 500
Adam Miller, 15, -, 750, 50, 150
David McCallister, 100, 40, 6000, 200, 400
James R. Price, 100, 19, 6000, 200, 500
Amos Sharpless, 269, -, 22000, 350, 3500
John Everson, 265, 15, 25000, 600, 3000
Joseph Brooks, 160, 35, 15000, 250, 800
Adam Turnbull, 131, -, 800, 300, 700
Thomas Smith, 76, -, 6000, 100, 500
Sarah McCrone, 76,-, 6000, 100, 500
James Mumsfield, 80, -, 8000, 150, 600
James Massey, 120, -, 10000, 300, 10000
Grantham Nevin, 180, 10, 20000, 600, 1200
Edward Nevin, 100, 3, 12000, 100, 400
Wm. Bolton, 100, -, 7500, 150, 400
John Swan, 125, 35, 12000, 400, 1000
George Bolten, 100, -, 7500, 150, 1000
David Gemmill Jr., 430, 70, 17000, 200, 2000
George Z. Tybous, 540, -, 50000, 700, 9000
Joseph Barnaby, 140, -, 9000, 200, 70
Thomas Appleby, 200, -, 20000, 400, 500
Wm. White, 80, -, 12000, 500, 1000
Wm. Bennett, 10, -, 1800, 50, 40
Robert Grimes, 10, -, 2000, 150, 350
George Hurd, 175, 195, 25000, 350, 2000
Robert Moore, 160, 40, 20000, 400, 1700
J. & S. Truss, 110, -, 55000, 200, 1000
John Bideman, 10, -, 2000, 50, 50
James Johns, 17, -, 5000, 50,500
Solomon Townsend, 90, -, 1000, 200, 1200
John Titler, 90, 10, 4000, 300, 465
James Short, 70, 20, 5000, 175, 306
George Moore, 27, -, 2500, 100, 240
Mayland Batten, 64, 40, 4000, 200, 860
James Ogram, 100, 97, 8850, 350, 850
Geo. Painter, 100, -, 5000, 175, 200
Abram Moore, 80, 20, 6000, 424, 840
George Clark, 100, 130, 11500, 418, 640
Saml. Porter, 110, 40, 7500, 230, 635
John Calhoun, 40, 3, 2200, 300, 200
John Racine, 45, 5, 3000, 236, 320
John Frazier, 110, 15, 12300, 350, 610
William Bartlett, 40, 10, 2000, 150, 200
John Scott, 80, 10, 5000, 220, 530
Andrew Harner, 80, 15, 7000, 335, 705
Geo. Norris, 100, 40, 7000, 250, 430
Geo. Graves, 90, 10, 5500, 108, 304
Thomas Slack, 90, 3, 5015, 175, 500
Saml. Wright, 100, 50, 12000, 100, 7740
Cantwell Clark, 900, 520, 85200, 1200, 4827
John Davis, 41, -, 3076, 80, 204
David McConaughey, 175, 151, 10000, 250, 425
Wm. McConaughey, 125, 30, 6000, 482, 1140
James Murry, 72, 12, 5000, 185, 405

Wm. Bower, 200, 100, 20000, 1000, 2250
Wm. Reynolds, 150, 25, 11550, 215, 405
Saml. Rambo, 70, 30, 6000, 242, 625
Jacob Scott, 60, 42, 6000, 160, 258
Geo. Heritage, 335, 15, 1800, 100, 162
Wm. Slack, 75, 75, 12500, 336, 782
Isaac Yocum, 75, 61, 6000, 195, 410
Saml. Silvers, 30, 20, 2000, 94, 110
Peter Miggett, 60, 10, 2800, 200, 346
Robert Melvill, 40, 10, 2000, 200, 379
Elias Cole, 11, 4, 1300, 64, 191
William Mote, 170, 40, 24000, 250, 95
Mrs. Bradley, 104, 22, 9450, 186, 430
John Lockard, 43, -, 3440, 175, 161
Thomas Lecke, 45, 8, 3500, 216, 372
Michael Keeley, 17, 90, 1200, 175, 175
Andrew Fisher, 45, 21, 2000, 160, 360
Saml. Clendenin, 14, 29, 1500, 140, 240
Patrick O'Rusick, 18, -, 940, 50, 155
Elijah Walton, 35, 15, 2000, 166, 315
Timothy O'Rusick, 18, 6, 1200, 56, 150
Saml. Green, 25, 15, 2000, 122, 350
Stephen Ott, 70, 15, 6000, 248, 555
Joseph Pennington, 23, 27, 2000, 155, 304
Geo. Lewis, 32, 22, 200, 135, 380
James Stewart, 70, 25, 2880, 1444, 303
George Veach, 84, 19, 3000, 236, 301
Marbin Komether, 50, 6, 1960, 212, 299
Levy Green, 25, -, 1550, 185, 2100
Wm. Irons, 18, 20, 1000, 100, 250
John Welden, 80, 40, 4000, 125, 200
John Ash, 100, 39, 5560, 325, 565
Isaac Stanton, 60, 20, 5600, 212, 219
Levi Gooch, 300, 200, 25000, 525, 1320
Francis Livingston, 42, 10, 2800, 66, 252
Thomas Kendall, 40, 20, 1200, 112, 213
Patrick Sullivan, 12, -, 1000, 100, 177
Alexander Simpson, 90, 44, 8000, 390, 545
Eli Ridgley, 25, 42, 2375, 50, 257
Saml. McIntire, 100, 70, 10000, 340, 345
James McIntire, 100, 50, 10000, 366, 490
Joseph Veach, 760, 20, 10000, 400, 1656
Frank Moore, 95, 15, 3100, 96, 182
James Stewart, 150, 50, 16000, 400, 865
Isaac Jesture, 120, 96, 8000, 150, 455
Robert King, 10, 2, 1200, 125, 242
William Stewart, 100, 61, 12880, 352, 756
Saml, Stewart, 80, 60, 10315, 490, 900
George Grubb, 140, 20, 12000, 400, 600
Wm. Calhoun, 100, 62, 8000, 1000, 712
John McCray, 70, 24, 3000, 309, 550
Wm. Huggins, 100, 60, 6400, 250, 580
Messet (Merret) Paxson, 380, 20, 40000, 2067, 6950
Joseph Sapp, 200, 97, 10000, 526, 833
Richard Groves, 130, 25, 8000, 300, 1060
Henry Gray, 150, 37, 8500, 250, 865
John Crumpton, 72, 58, 6500, 300, 509

John Hentton (Heartlan), 160, 30, 7000, 250, 450
James Veazey, 90, 50, 9500, 421, 925
Richard Gray, 160, 33, 11500, 240, 590
James Coage, 35, 3, 1500, 150, 260
Peter Wright, 19, -, 1140, 125, 200
John Dempsey, 75, 25, 4500, 150, 369
Isaiah Stanton, 70, 10, 5600, 200, 465
James Loans, 70, -, 800, 75, 295
Wiliam Ford, 150, 50, 12000, 150, 760
David Ford, 150, 50, 12000, 405, 1110
Curtis B. Ellison, 156, 10, 12450, 350, 1052
Isaac F. Jesture, 90, 70, 10000, 200, 600
Jesse Boulden, 166, 38, 11280, 300, 884
Wm. Gutherie, 80, 28, 6480, 250, 640
Daniel Cason, 120, 30, 9200, 300, 1011
James Gray, 160, 40, 12000, 455, 780
James Cason, 125, 25, 9000, 500, 1040
Benjamin Reed, 211, 26, 10550, 225, 452
Charles Alrich, 120, 55, 7550, 166, 578
James McMullen, 50, 30, 3000, 175, 197
Mrs. Baynard, 80, 43, 6150 12, 695
Robert Black, 200, 70, 16000, 320, 980
Saml. Frazier, 150, 50, 12000, 332, 1075
Thomas Lindell, 100, 43, 7150, 399, 495
Augustus Smith, 85, 42, 5040, 73, 470
Davison Pierce, 40, 35, 2000, 200, 230
Nathaniel Jaquett, 60, 70, 5000, 155, 560
Benjamin Nicole (Nicolds), 70, 50, 5000, 150, 705
Isaac Nicolds, 26, 7, 1320, 75, 245
Charles Nicolds, 50, 20, 2100, 100, 266
David (Daniel) Shaw, 50, 20, 2800, 200, 400
Geo. Boulder, 400, 215, 30750, 735, 1867
Ruben Roy, 100, 12, 4000, 300, 580
John Jordan, 230, 60, 18000, 600, 1910
James McCoy, 120, -, 7200, 200, 784
James Boulden, 300, 50, 21000, 550, 1390
Causden Johnston, 16, 2, 720, 75, 170
Robert Price, 15, -, 600, 70, 190
John Davidson, 204, 8, 18800, 1038, 1736
Charles Lecompt, 100, -, 9000, 250, 580
Daniel Stephenson, 14, -, 800, 100, 175
Wm. Dayett, 200, 25, 14000, 600, 1720
Wm. Cavesider, 375, 31, 30000, 590, 2175
Thomas McCoy, 340, 60, 35000, 630, 1690
Thomas Scott, 170, 7, 10000, 150, 681
Benjamin Denney, 427, 35, 27720, 765, 2688
Wm. Conlyn, 140, 6, 12000, 500, 798
Henry Appleton, 170, 30, 14000, 680, 1620
Wm. Price, 250, 10, 15000, 350, 1000

Jonathan Ellison, 250, 40, 18000, 655, 2520
Thomas Fisher, 350, 50, 32000, 685, 2520
John W. Kane, 68, 12, 5000, 550, 600
Geo. Herbart, 340, 40, 27200, 800, 2605
John Price, 200, -, 16000, 850, 2237
Thomas McIntire, 180, 35, 15000, 950, 1055
Lewel Biggs, 150, 47, 11820, 500, 887
Edward Armstrong, 170, 30, 16000, 429, 819
Benjamin Biggs, 160, 27, 11220, 525, 1030
Pierce Biggs, 175, 46, 11050, 505, 669
Charles Lum, 212, 41, 20400, 712, 2664
Mrs. McCracken, 230, 12, 20000, 655, 1690
Charles Clement, 70, 8, 5460, 150, 325
Thomas Boston, 40, 60, 3500, 300, 175
Melton Hammond, 50, 20, 5000, 150, 310
Mathew Mills, 7, -, 700, 150, 50
There appears to be several pages missing between the end of Pencader Hundred and Red Lion Hundred and between the end of Red Lion Hundred and St. Georges Hundred.
Thomas Belville, 121, -, 9400, 400, 1000
Samuel McCuel, 130, -, 10000, 200, 700
Joseph Blackburn, 92, -, 8000, 200, 700
John Morris, 150, 32, 8000, 100, 300
Asher Cluyston, 100, 18, 6000, 100, 700
Thomas Longfellow, 300, -, 20000, 300, 1000
John C. Clark Jr., 263, -, 25000, 400, 1500
Peter Clever, 180, -, 18000, 300, 1000
John Havelow, 360, 20, 30000, 500, 2000
Thomas Clark, 228, -, 22800, 500, 2000
John P. Belville, 196, -, 20000, 500, 1500
Joab Alster, 111, -, 9000, 200, 700
Henry Taylor, 100, 4, 10000, 300, 1000
Daniel Stewart, 190, 10, 17000, 320, 1200
Laurence Aspril, 170, -, 14000, 200, 1500
John Hurlock, 240, 10, 30000, 300, 1500
Samuel Bogs, 240, -, 25000, 500, 1500
Wm. Hurlock, 310, -, 35000, 400, 3000
Anthony M. Higgins, 350, 12, 35000, 500, 2000
Thomas Jamison, 24, -, 4000, 150, 500
Wm. Stuchart, 280, 20, 22000, 500, 3000
Joseph Cassell, 100, -, 8000, 200, 700
John Ocheltree, 100, -, 8000, 100, 275
Wm. Jones, 72, -, 6000, 100, 500
Purnel Lynch, 212, -, 12000, 300, 1500
Wm. Casperson, 250, -, 14000, 200, 1000
Benj. Pennington, 140, 10, 15000, 400, 1500
John McCall, 100, -, 9000, 200, 600
Jonathan Draper, 370, 30, 25000, 500, 1500
Wm. Simkins, 200, -, 25000, 200, 1000

Robert Ocheltree, 200, 34, 20000, 400, 900
Barney Raybold, 67, -, 20000, 300, 1000
George Clark, 290, -, 290000, 500, 2000
Solomon Deputy, 260, -, 29000, 500, 2000
James Asprile, 200, 40, 20000, 300, 1300
Wm. W. Stewart, 220, 40, 25000, 500, 1500
James Ridham, 160, -, 17000, 250, 800
Wm. Beck, 175, 75, 17000, 250, 800
John Higgins, 225, 50, 27000, 200, 1500
Alexander Biddle, 112, 28, 8000, 250, 1200
Alexander Davidson, 140, 80, 22000, 300, 1000
Adam E. King, 276, -, 22000, 580, 2000
Clement Reeves, 200, 79, 25000, 400, 1700
John J. Henry, 160, -, 30000, 500, 700
Wm. D. Clark, 265, -, 35000, 800, 4500
John Exton, 200, 100, 31000, 300, 1000
Levi Clark, 160, 40, 20000, 300, 1200
Isaac Aspril, 250, 50, 25000, 400, 650
John C. Clark, 600, -, 60000, 1000, 5785
John Aspril, 150, -, 15000, 300, 750
Anthony Raybold, 400, -, 40000, 500, 3000
Wm. Raybold, 450, -, 60000, 3000, 8500
Israel Urin, 400, 300, 30000, 500, 1500
James Husten, 200, -, 20000, 500, 2400
John Dugan (Dagar), 131, -, 17000, 200, 1160
Samuel Sutton, 210, -, 20000, 600, 1500
John Raybold, 750, 100, 80000, 5000, 12700
William Cochran, 175, 25, 16000, 200, 1200
Alex Vail, 150, 20, 20000, 200, 1200
Richd. G. Hayes, 105, 11, 6000, 100, 600
Andrew Eliason, 220, 20, 25000, 500, 1500
Samuel Rothwell, 310, 40, 28000, 200, 1350
John D. Eliason, 132, 16, 12000, 300, 1000
Edwd. Boulden, 100, -, 8000, 100, 800
Alfred Loftland, 250, 25, 10000, 200, 1200
Thos. McCracken, 300, 80, 16000, 200, 2000
Wm. Lore, 200, 50, 20000, 100, 1750
Jon E. George (Genge), 175, 10, 10000, 150, 1000
James Lecompt, 200, 3, 10000, 200, 1000
Andrew P. Armstrong, 97, 10, 8000, 50, 800
John Brady, 75, -, 7000, 100, 600
David Stephens, 250, 50, 12000, 200, 1200
Chas. Foards, 200, 10, 12000, 200, 800
Wm. S. Cleaver, 130, 10, 10000, 200, 800
Edward Townsend, 163, 24, 8000, 100, 800
Saml. Bayard, 110, 30, 6000, 100, 600
Saml. Segars, 100, 20, 7000, 300, 700
Geo. Segars, 115, 14, 5000, 100, 700

Saml. Bendler, 150, 50, 8000, 100, 500
Geo. P. F. Woods, 280, 120, 12000, 200, 1000
Joseph T. Sarchet, 700, 300, 50000, 1500, 3000
William Deihl, 140, 10, 8000, 200, 1000
Richard Eaton, 100, 100, 8000, 100, 900
Byard Allridge, 80, 12, 9000, 400, 800
James Price, 100, -, 6000, 100, 500
John Allridge, 200, 182, 20000, 100, 1000
William Davis, 200, 50, 17000, 100, 1400
William Cleaver, 5, 25, 6000, 100, 1000
William Kennedy, 220, 60, 28000, 300, 1100
James Frazier, 50, 120, 8000, 100, 1200
James Carpenter, 1115, 80, 20000, 200, 2000
John D. Dillworth, 260, 61, 14000, 400, 1400
William McMullen, 180, 90, 20000, 200, 10000
Joseph Cleaver, 380, 20, 20000, 200, 2000
Thos. S. Merritt, 58, -, 9000, 100, 1100
Levin Catts, 180, 7, 18000, 60, 1200
John C. Spear, 315, 2, 31000, 500, 2000
Robt. T. Cochran, 500, 30, 50000, 1500, 3700
James Burnham, 160, 10, 50000, 200, 1200
Hay Templeman, 150, 20, 8000, 200, 800
John Templeman, 120, 20, 6000, 100, 600
Henry H. McVay, 52, 3, 5000, 50, 400
Thos. Houston, 140, 10, 15000, 100, 700
John Houston, 54, -, 5400, 100, 400
David Vail, 200, -, 15000, 200, 800
James Money, 80, 2, 5000, 50, 600
Samuel Brady, 350, 150, 35000, 300, 2000
Edward Attwell, 90, 80, 6000, 100, 400
Saml. Bayard, 90, 10, 5000, 50, 500
Robt. Morrison, 466, 19, 47000, 400, 2000
Wilson Greer 98, 5, 10000, 100, 800
Hay Walker, 80, 5, 4000, 50, 400
John Titus, 120, 20, 6000, 50, 600
Thos. J. Foard, 375, 153, 35000, 1000, 2500
Eli J. Foard, 130, 22, 12000, 200, 900
James J. Janvier, 230, 9, 18000, 300, 1800
Gasaway Walthies, 125, 5, 12000, 200, 800
Geo. W. Kassnur, 375, 45, 50000, 1000, 350
Jase Higgins, 275, 60, 20000,300, 2000
Wm. M. Vandergrift, 150, 50, 8000, 200, 1700
Saml. Jefferson, 60, 30, 10000, 200, 750
Wm. J. Cleaver, 130, 70, 14000, 400, 1200
James E. Eagle, 100, 25, 8000, 100, 600
Charles Cleaver, 100, -, 8000, 100, 600
Geo. Cleaver, 200, 75, 18000, 300, 2900
James Grass (Gross), 100, 50, 10000, 100, 800
Joseph Ellison, 165, 60, 18000, 200, 1000
Philip J. Grass (Gross), 85, 80, 7000, 200, 600

William C. Camp, 225, 25, 18000, 500, 2000
John S. Burnett, 85, 15, 6500, 100, 600
William Bennett, 230, 40, 30000, 300, 2000
John Harman, 160, 10, 13000, 200, 800
Isaac Cleaver, 160, 10, 10000, 200, 900
A. J. Vandergrift, 200, 40, 20000, 400, 1000
Chris Vandergrift, 100, 40, 10000, 200, 700
James Pont, 90, 15, 10000, 200, 800
James McMullen, 113, -, 12000, 200, 800
John Riley, 350, 100, 40000, 200, 3000
Jas. M. Vandergrift, 190, -, 23000, 300, 1200
John R. Boyd, 140, 11, 1500, 100, 500
C. H. Smith, 195, 5, 20000, 200, 1300
Thos. S. Crasen (Craven), 238, 2, 21000, 200, 1200
James Gray, 175, 25, 12000, 300, 1200
Saml. Riley, 125, 26, 11000, 200, 500
Job Townsend, 150, 20, 12000, 100, 600
William Vail, 278, 72, 25000, 1000, 1600
Samuel Atkison, 230, 20, 20000, 300, 1200
Thos. S. McQuarter, 175, 5, 18000, 400, 800
Lelm McQuarter, 218, -, 20000, 2500, 1500
John P. Hudson, 160, 10, 12000, 200, 1400
Robt. Polk, 300, 20, 20000, 300, 1500
James Pogue, 105, -, 10000, 400, 1300
John McQuarter, 260, 40, 25000, 500, 1500
Frank Gray, 115, 10, 7000, 200, 800
Adam Diehl, 100, -, 10000, 100, 500
Geo. W. Sparks, 175, 25, 12000, 200, 700
Boes Webb, 175, -, 12000, 100, 800
Isaac Woods, 160, 35, 18000, 400, 1300
Isaac Vandergrift, 350, 54, 20000, 300, 1800
John Vail, 152, -, 10000, 200, 1000
Thos. P. Riley, 200, 20, 10000, 200, 1000
L. G. Vandergrift, 200, 125, 20000, 200, 800
John Alston, 130, 25, 20000, 300, 700
Richd. W. Cochran, 150, -, 20000, 300, 1500
Wm. A. Cochran, 200, 10, 25000, 300, 1700
James. P. Hoffecker, 300, 150, 30000, 500, 2000
John L. Clothier, 217, 87, 18000, 400, 1600
Jonathan Williams, 296, 30, 30000, 500, 3200
Ephraim Beaston, 327, 20, 28000, 650, 2200
Wm. R. Cochran, 310, 40, 30000, 400, 2500
James Reading, 125, 5, 8000, 200, 600
Wilson E. Vandergrift, 85, 51, 5000, 100, 800
John Stewart, 110, 5, 8000, 100, 700
Eli Todd, 230, 25, 18000, 300, 1200
Isaac W. Vandergrift, 300, -, 25000, 300, 1800
John Asprile (Asprill), 200, 20, 12000, 200, 800
John Loper (Losser), 100, 10, 9000, 100, 800

John Lynch, 125, -, 12000, 150, 800
Henry Jones, 196, -, 15000, 300, 1000
Asher B. Giles, 200, 10, 16000, 300, 1400
Amos W. Lynch, 300, 25, 30000, 500, 3000
Geo. Reynolds, 125, 75, 10000, 150, 1000
Benj. F. Hanson, 160, -, 16000, 350, 1700
Lydia A. Price, 240, -, 12000, 500, 1500
Peter Holstein, 150, 20, 12000, 250, 1600
Chas. P. Cochran, 400, 50, 40000, 500, 2500
Fredus Vanhekle, 325, 75, 15000, 350, 1800
Rich. R. Cochran, 200, 20, 16500, 400, 800
Horatio G. Loyd, 135, 75, 12000, 300, 800
Samuel Dale, 20, 10, 600, 50, 500
Thos. A. Jones, 275, 27, 17000, 300, 1000
Jas. J. Brown, 170, 24, 16000, 300, 1500
Joshua Clayton, 200, 10, 20000, 200, 1000
Thos. Clayton, 240, 20, 18000, 500, 600
Thos. R. Hayes, 303, -, 21000, 500, 1500
John A. Jones, 344, -, 30000, 500, 1200
Chas. Kane, 165, -, 12000, 200, 1000
Saml. McVay, 96, -, 10000, 200, 800
William Haukie, 435, 15, 30000, 500, 2100
William H. Houston, 365, 35, 30000, 500, 2500
John Wilby, 60, 60, 6000, 150, 400
Zacha. Jones, 280, 20, 20000, 300, 2000
Spencer Holton, 150, 15, 8500, 200, 900
James Davis, 100, 17, 8000, 100, 700
William Stoops, 220, 30, 17500, 300, 1800
John A. Moody, 125, 61, 11000, 200, 700
Smith H. Burris, 100, 45, 7000, 600, 650
Edwin R. Cochran, 210, 40, 22000, 150, 1400
Amer Bell, 75, 59, 5000, 200, 500
Wm. C. Parker, 150, 50, 8000, 300, 1200
Stephen Wiley, 142, 5, 7000, 200, 600
Benj. Armstrong, 140, 10, 10000, 300, 1000
Mary Armstrong, 78, -, 5000, 100, 600
Wm. S. Long, 175, 25, 14000, 200, 1500
Thos. Cavender, 390, 62, 25000, 500, 2000
James Dodson, 240, -, 19000, 200, 2000
Thos. Murphey, 1445, 12, 12000, 500, 1100
Thos. Gould, 100, 20, 5000, 40, 600
John Wilson, 210, 20, 12000, 200, 1100
Jowath Johnson, 196, 15, 10000, 200, 1000
John Thomas, 159, 23, 9000, 150, 900
Peter Merideth, 125, 25, 8000, 100, 600
Jeremiah Reynolds, 140, 14, 11000, 300, 1050
Wm. Wood, 330, 100, 21500, 350, 2500
Jas. R. Hoffecker, 158, 5, 10000, 200, 900
Geo. Derrickson, 444, 20, 37000, 500, 3000

Manlore (Manlon) Davis, 180, 30, 20000, 300, 1000
Obadiah Clark, 36, 4, 2800, 100, 480
Mathew Wright, 5, -, 500, 25, 105
Robert Sergent, 25, -, 1200, 60, 210
Aaron Traverse, 22, 3, 1250, 20, 235
Andrew Mercer, 65, 5, 5600, 205, 435
Joel Thompson, 110, 13, 12300, 500, 2195
Lamborn Pyle, 18, 12, 3100, 180, 340
Margaret Mote, 85, 19, 6000, 292, 800
Albert Stoops, 65, 30, 5000, 137, 380
Wm. H. Robinson, 60, 15, 5000, 145, 498
William Olliver, 9, 3, 1000, 75, 150
James Trimble, 18, 4, 200, 60, 190
John McGlaughlin, 30, 14, 3520, 75, 210
William Palmer, 30, 10, 2800, 30, 180
John Jones, 25, 9, 2040, 75, 135
Samuel McMullen, 42, 8, 3640, 75, 258
Samuel Lindsey, 250, 48, 22350, 680, 1916
William Smith, 150, 50, 18000, 460, 2400
Aaron Baker, 85, 15, 7000, 145, 574
Abram Cornog (Cornoy), 155, 20, 16000, 822, 1785
Alexandter Coulter, 40, 10, 3000, 130, 230
Clayton Platt, 96, 4, 8000, 500, 1585
Joshua Robinson, 75, 9, 6400, 500, 780
Thomas Steel, 100, 40, 16000, 500, 1035
Robert Steel, 90, 6, 7000, 602, 1120
James Crow, 56, 4, 4000, 200, 481
Thomas McKeevon, 38, -, 1520, 50, 236
Peter Stuart, 34, -, 2000, 100, 505
John Morrison, 100, 100, 1000, 125, 823
Andrew Donnell, 132, 18, 10000, 500, 552
Daniel Thompson, 100, 30, 13000, 250, 1635
John Chambers, 100, 50, 12000, 436, 657
John Tweed, 100, 25, 10000, 250, 1041
William McClelland, 80, 28, 10000, 300, 910
Joseph Hossinger, 125, 35, 11200, 312, 905
William Evans, 120, 20, 15000, 350, 800
Levi Fisher, 43, 3, 1200, 100, 447
George Casho, 96, -, 4365, 330, 675
Andrew Kerr, 90, 60, 12000, 435, 975
Jonathan Conkey (Coskey), 6, -, 600, 50,2 00
George J. Smith, 240, 20, 3500, 535, 1300
Edmund Lewis, 125, -, 10000, 150, 288
Charles Whiteman, 125, -, 10000, 400, 1550
Edward Porter, 56, -, 8000, 200, 560
Eliza Motherall, 272, 20, 24000, 425, 1600
Thomas Pennington, 135, 5, 13000, 500, 585
Casper Sheppard, 65, 8, 5000, 312, 400
Joseph Maxwell, 30, 10, 5000, 125, 503
Hiram Moore, 80, 50, 12000, 250, 985
Ann Naudain, 60, 49, 5000, 100, 290
Thomas Simpress, 17, 1, 1000, 50, 46
Robert Furgerson, 120, 24, 11520, 500, 1079
Alexander Crawford, 100, 31, 10480, 150, 715

John Cavsender, 15, 13, 1120, 50, 115
James Hammon, 100, 187, 11480, 425, 790
Jediah Allen, 104, -, 8600, 500, 700
William Ruth, 70, 11, 8080, 500, 735
Levi Ruth, 40, 38, 4000, 210, 300
William Hawthorn, 68, -, 5090, 175, 520
Mary Camby (Carnby), 200, 90, 20000, 500, 1300
Abram Rothwell, 80, 20, 6500, 220, 698
Sarah Alrich, 85, 17, 5000, 1500, 740
Samuel Stroud, 100, 20, 9600, 515, 1220
Margaret Maree (Marel), 30, 25, 4000, 125, 710
William Morrison, 140, 20, 12000, 375, 811
Samuel Scott, 32, 6, 3000, 150, 212
Lewis Morrison, 42, 10, 4000, 155, 400
Jephtha Ayres, 350, 150, 27000, 400, 2734
Joseph Petiete, 140, 55, 12000, 300, 1335
Wm. M. Hawthorn, 95, 16, 6100, 245, 758
Thomas Lynam, 120, 20, 6200, 179, 1035
James Wright, 130, 8, 900, 500, 905
William Jones, 160, 20, 10000, 250, 1498
George Johnston, 66, 44, 5500, 125, 580
Evan Statsonbery, 165, 75, 10000, 250, 1616
James Morrison, 140, 40, 18000, 370, 800
Amos Warrick, 50, 50, 5000, 175, 730
Mary Bustick, 19, 1, 1200, 90, 150
Benjamin Groves, 70, -, 2100, 275, 640
Josiah Fesmire (Fermire), 125, 75, 15000, 400, 1110
David Simpson, 17, -, 2600, 50, 220
Fred Cavsinder, 14, 8, 1320, 33, 116
William Ward, 50, 30, 4000, 175, 115
Thomas Ocheltree, 38, -, 2500, 67, 279
Charles Anderson, 50, 38, 4500, 40, 500
Stephen Cunningham, 121, -, 9860, 200, 650
Samuel Biddle, 135, -, 8500, 125, 531
Sylvester Townsend, 50, 6, 6000, 300, 690
Henry Peters, 70, -, 2000, 100, 200
Benjamin Peters, 40, 120, 4800, 100, 310
John Megginson, 115, 25, 15000, 481, 1089
James A. Groves, 150, 16, 8000, 400, 977
Thomas Lamb, 115, 16, 11700, 300, 732
George Morton, 20, 5, 1500, 125, 315
William Brooks, 130, 35, 9000, 500, 1125
David Pogue, 100, 81, 7500, 315, 875
John Dempsey, 55, 21, 3000, 150, 340
John Ragen, 75, 75, 6000, 150, 435
Lorenza Hammiell, 116, 40, 10000, 275, 900
Thomas Brooks, 113, 40, 7000, 363, 1074
John Hall, 18, 3, 1200, 40, 90
James Lee, 16, 4, 1500, 83, 218
Wm. R. Lynam, 160, 45, 12000, 500, 1095
Robert Stuart, 145, 15, 10000, 250, 700
Jonathan Groves, 75, 20, 5000, 212, 610

Edward McElwee, 80, 90, 8000, 160, 949
Silas Dodd, 100, 52, 8000, 175, 541
James Smalley, 65, 10, 6000, 200, 975
James Morris, 75, 59, 6000, 100, 400
William Batten, 40, 18, 2900, 166, 297
George Johnston, 25, 37, 2500, 75, 185
James B. Groves, 50, 50, 3000, 50, 260
Samuel Roberts, 126, 114, 17000, 2500, 1082
Charles Loudenslayet, 100, -, 7000, 470, 610
Thomas Reece, 52, 59, 3650, 100, 200
Henry Curdman, 150, 50, 16000, 820, 1602
Geo. Colemerny, 150, 10, 15000, 450, 670
Samuel Fenley, 64, -, 6400, 125, 781
Robert Armstrong, 130, 40, 11000, 345, 918
Harry Massin (Marrin), 40, 50, 3500, 150, 180
Palmer Chamberlain, 37, -, 2590, 50, 200
Andrew Rando (Rands), 18, -, 1800, 100, 30
Charles Blandy, 40, -, 4000, 156, 310
Thomas Whitten, 144, 126, 13000, 465, 956
Robert Hawthorn, 100, 25, 250, 255, 675
James M. Woodnall, 100, 23, 8000, 475, 950
Joseph Griffith, 148, 12, 12880, 610, 930
William Cornog, 150, 15, 15000, 1000, 2620
Hannah Bradley, 133, -, 12000, 350, 1140
John R. Reece, 125, 27, 16000, 12, 1320
Calvin Jones, 130, 70, 10000, 275, 640
Thomas Jaggard (Haggard, Hazzard), 75, 25, 5000, 200, 500
William Howard, 3, -, 1200, 75, 100
Benjamin Champiere, 20, 4, 1500, 25, 102
Robert Armstrong, 100, 50, 9000, 424, 545
James Livingston, 190, 30, 20000, 700, 2111
William Frazier, 75, 30, 7950, 150, 400
George Dewees, 66, 10, 6000, 195, 364
Samuel Coneley, 130, 33, 10000, 500, 830
William Couch, 40, 40, 4000, 28, 600
Elias Lofland, 100, 50, 7000, 175, 900
William Vansant, 110, 40, 12000, 225, 517
Adam Dayett, 50, 60, 4200, 150, 364
Frederick Racine, 100, 30, 5000, 100, 255
John Price, 100, 100, 6000, 150, 317
Charles Morrison, 70, 10, 5000, 300, 484
William Knotts, 80, 65, 4000, 75, 570
D. B. Ferris, 175, 41, 12980, 400, 1080
Anthony Deputy, 150, 30, 10000, 226, 520
Wm. H. Broton, 17, -, 1500, 75, 113
Ferdinand Janvier, 180, 40, 12000, 850, 1300

Sussex County Delaware
1860 Agricultural Census

This agricultural census was filmed from original records in the Delaware State Archives in Wilmington Delaware by the Delaware State Archives Microfilm office.

There are some forty-eight columns of information on each individual. Only the head of household is addressed. I have chosen to use only six columns of the information because I feel that this information best illustrates the wealth of the individuals. These are shown below:

1. Name of Owner
2. Acres of Improved Land
3. Acres of Unimproved Land
4. Cash Value of the Farm
5. Value of Farm Implements and Machinery
13. Value of Livestock

Thus, the numbers following the names represent columns 2, 3, 4, 5, 13.

The following symbol is used to maintain spacing where information in a column is left blank (-). This symbol is used where letters, names or numbers are not legible (_).

Some of the pages are out of sequence at the beginning or there are a number of missing pages as the handwriting and location did not appear again in the county (approximately 9 pages). There are other missing pages based on numbering sequence near the end of the county (approximately 15 pages). And pages in the last portion of the county are out of sequence—pgs. 21-30 followed by pgs. 1-20.

John Bennett, 40, 40, 1200, 20, 80
Wm. T. Brasures, 4, 7, 1500, -, 200
P. D. Derrickson, 75, 115, 2500, 35, 100
Jacob Lynch, 50, 25, 1000, 30, 50
Thomas Wilgus, 40, 40, 1800, 50, 100
Ebe Megee, 25, 25, 1000, 20, 200
Henry J. Williams, 20, 20, 1000, 30, 150
Garrison McCabe, 11, 45, 1000, 25, 75
Richard Hickman, 50, 10, 20000, 100, 270
Jacob Evans, 30, 40, 150, 200, 370
Ebenezer Holleway, 50, 44, 3500, 200, 250
Josiah Lynch, 60, 30, 2500, 150, 150
Ezekiel Lynch, 60, 30, 2000, 150, 250
Caleb Lynch, 65, 25, 2000, 50, 200
Ebe A. Daisy, 40, 25, 1500, 100, 220
Lambert Murry, 45, 30, 1700, 20, 30
Nath. Hickman, 35, 17, 1500, 20, 60
Ebe McCabe, 20, 30, 1200, 50, 150
Josiah Timmons, 30, 70, 2000, 12, 200
Richard Hickman, 35, 7, 1000, 30, 60

James Daily, 35, 15, 1000, 50, 50
John S. Bennett, 25, 20, 1000, 150, 330
Gengo (George) Collins, 19, 40, 1700, 10, -
Gustlin Timmons, 70, 85, 3000, 200, 125
Peter Banks, 18, 16, 1500, 15, 150
Alfred Hudson, 60, 70, 2000, 20, 200
John Timmons, 19, 25, 600, 20, 400
Eli P. West, 125, 75, 1500, 100, 200
Henry R. Johnson, 100, 150, 2500, 150, 31
Justus Andrews, 15, 20, 500, 10, 50
Peter Holleway, 100, 75, 6000, 50, 300
Reuben Stephens, 75, 50, 4000, 75, 175
John Ball, 45, 45, 1300, 75, 190
Joshua Medwill, 5, 40, 700, 20, 50
Mitchel Rickards, 40, 30, 800, 50, 180
Stephen Melson, 30, 40, 800, 20, 40
James Cherbres, 25, 45, 1000, 20, 200
J. W. Bennett, 100, 75, 2000, 75, 175
B. T. McCabe, 45, 300, 2000, 75, 372
Benj. Truitt, 50, 150, 2000, 25, 50
George Campbell, 30, 35, 1500, 25, 80
Zachariah Hudson, 40, 33, 2500, 20, 125
William Hudson, 25, 30, 1000, 20, 100
Benj. Long (Ling), 50, 50, 1000, 50, 225
Thomas Stephens, 25, 25, 700, 20, 50
Johnathan Baker, 60, 85, 2500, 50, 125
Joseph Harrison, 100, 67, 3000, 75, 280
Elijah Carey, 120, 200, 15000, 40, 200
John Murrey, 40, 20, 1400, 25, 275
Elijah Johnum (Thonum, Thomas), 45, 50, 5000, 20, 125
Elijah Mccabe, 15, 35, 2000, 20, 200
F__ Theroughood, 50, 50, 800, 20, 50
Thos. Rogus, 150, 50, 5000, 150, 450
Isaac McCabe, 200, 85, 6000, 200, 800
Elisha McCabe, 1, 400, -, -, 40
Joshua Murry, 55, 15, 3500, 80, 400
Joshua Long, 35, 15, 1000, 20, 125
Joshua Hudson, 100, 150, 6000, 50, 100
Zeddoc Collins, 12, -, 400, 15, 125
Moses McCabe, 60, 25, 500, 50, 300
John Dell McCabe, 60, 20, 650, 25, 300
George Murry, 25, 40, 2500, 75, 300
Lott. Murry, 70, 60, 1000, 25, 240
Jacob Johnson, 20, 30, 600, 20, 75
Sacar Murry, 60, 20, 1000, 25, 150
Milburn Murry, 30, 20, 1000, 75, 330
John Andrew, 20, 30, 100, 20,-
Zino Long, 35, 15, 1500, 100, 125
Zino Long Jr., 30, 20, 1000, 20, -
James Campbell, 40, 35, 600, 25, 75
Eliza Long, 20, 40, 500, 20, -
Caleb Murry, 20, 7, 1400, 20, 75
Joseph Murry, 70, 100, 1200, 25, 100
Cannon Wells, 125, 100, 2000, 30, 100
Elizabeth Townsend, 125, 125, 3000, 25, 125
Peter Johnson, 60, 25, 2000, 25, 500
Ebe Walter, 150, 200, 8450, 200, 500
Jediah Evans, 150, 75, 5500, 150, 616
John Steel, 80, 100, 2500, 100, 350
Seri Lynch, 30, 40, 1000, 25, 50
John Jacobs, 20, 32, 1500, -, -
Joshua J. Johnson, 7, -, 500, -, 125
A. E. Hill, 40, 60, 3000,75, 225

Nathaniel Clark, 50, 40, 1400, 20, 40
Ann Thomas, 40, 10, 1000, -, -
Mary Aydolothe, 50, 30, 700, 10, 40
Annasias D. Gray, 35, 30, 1000, 25, 50
Ketura Pool, 100, 150, 3000, 100, 250
Jester Andrews, 45, 25, 1000, 20, 20
Silas Records, 30, 10, 600, -, 100
Clement Hill, 40, 50, 1500, 20, 125
Nancy C. Howard, 25, 28, 1000, 20, 175
Israel Townsend, 5, -, 500, -, -
Ebe Townsend, 35, 120, 2500, 75, 200
David E. Hudson, 45, 75, 3000, 50, 250
Stephen C. Aydolett, 60, 105, 3000, 100, 200
Ann C. Townsend, 75, 75, 2000, 50, 125
James Chericks, 10, 35, 1500, 25, 75
John Calhoon, 60, 150, 2500, 100, 250
Nath Tunnell, 150, 190, 6000, 200, 600
Burton Johnson, 25, 25, 500, 40, 40
George West, 30, 30, 600, 5, 25
Wm. J. Daisy, 15, 35, 800, 25, 100
Stephen Collins, 40, 67, 2000, 50, 140
Henry Tunnell, 60, 140, 2000, 75, 180
Jacob Banks, 30, 150, 2000, 25, 50
John Mitchell, 25, 100, 1100, 25, 40
Nancy W. West, 25, 25, 2000, 25, 125
Levin Records, 20, 50, 500, 10, 75
Joshua C. Townsend, 50, 60, 3000, 150, 300
Major Townsend, 40, 40, 1500, -, 100
Isaac Hall, 50, 100, 1500, 25, 120
James Dale, 70, 35, 2000, 100, 880
John McGee, 20, 25, 100, 20, 75
Jacob Hickman, 35, 40, 900, 100, 250
B. W. Brasure, 25, 25, 700, 20, 125
Thos. Johnson, 30, 100, 200, 25, 500
Ezekiel Dukes, 40, 60, 1500, 25, 100
Thomas Gerd___, 10, 25, 400, 10, -
James Hudson, 15, 20, 1000, 20, 15
A. R. Tingle, 11, -, 500, -, -
Nath. Hudson, 60, 50, 1000, 50, 200
John McCollum, 10, 140, 500, 10, 175
Jacob Brasure, 30, 35, 1500, 125, 425
Joshua Brasure, 3, 6, 400, -, 225
Ananias Rogers (Rogus), 75, 75, 1000, 25, 40
John W. Williams, 60, 60, 2000, 30, 150
Ebenezer Williams, 40, 30, 1500, 50, 125
Jacob Collins, 40, 60, 1000, -, 50
Aaron Holloway, 5, 4, 500, 10, 40
David Holloway, 30, 75, 1000, 20, 40
T. H. Atkins, 30, 34, 1200, 20, 30
Charles Collins, 55, 50, 1500, 30, 150
James Brasure, 30, 50, 1200, 50, 150
James Tingle, 7, 3, 500, 50, 200
Henry Johnson, 40, 30, 1000, 40, 160
Mitchel Derrickson, 150, 150, 4000, 50, 125
James Collins, -, -, -, 3, 50
Charles Melburn, -, -, -, -, 50
David Godwin, 100, 80, 1500, 1, 500
G. T. Hudson, 65, 20, 1000, 50, 250
W. W. Long, 50, 50, 3500, 50, 300
Nath. Evans, 35, 35, 2500, 150, 250
A. D. Derrickson, 75, 50, 4000, 200, 650
Wm. L. W. Williams, 60, 75, 2000, 50, 250
Thos. Taylor, 40, 60, 1000, 20, 50
Alfred Lynch, 40,70, 2000, 100, 160
Isiah Bunting, 30, 30, 1000, 20, 100

Charles Bunting, 20, 30, 1000, 20, 150
Henry Bishop, 60, 60, 3000, 50, 150
Elijah Warington, 60, 50, 5000, 100, 500
Richard Warington, 40, 30, 1500, 50, 180
Mitchel Morris, 50, 56, 2000, 100, 350
James Timmons, 30, 70, 1500, 25, 240
John McCabe, 40, 60, 1500, 50, 100
Levin Morris, 15, 45, 1000, 30, 200
Merril Bunting, 60, 40, 3000, 100, 250
Milby Bunting, 20, 10, 500, -, 50
David H. Hudson, 25, 75, 2000, 25, -
Laban H. Murry, 50, 25, 2000, 25, 225
Stephen W. Murry, 40, 3, 2500, 75, 225
Robert Layton, 25, 25, 400, 20, 145
Ebe Layton, 30, 20, 650, 20, 300
Eli Campbell, 20, 30, 700, 25, 50
Wm. O. McCabe, 77, 75, 2500, 25, 100
James Hickman, 45, 35, 2000, 50, 100
Wm. L. Hudson, -, -, -, 20, 190
Ebe Gray, 35, 10, 1500, 30, 100
Edwd. Townsend, 50, -, 800, 50, -
John Chambers, 20, 15, 1000, 10, 200
James Anderson, 40, 45, 2000, 55, 250
Joshua Collins, 60, 50, 800, 25, 250
Benj. Cooper, 50, 50, 1000, 75, 150
John L. B. Carey, 90, 15, 1200, -, 200
Marshal Bunting, 40, 15, 1000, 50, 15
Joseph N. Daisy, 35, 10, 1000, 25, 48
Levin McCabe, 3, -, 1000, 20, 150
Wm. B. Hickman, 40, 10, 1500, 50, 275
Henry H. Watson, 4, -, 900, -, 150
Levin Hudson, 45, 4, 1000, 75, 175
Wm. R. Tubbs, 30, 70, 1000, -, 125
James Gray, 18, 25, 1000, 50, 75
G. B.Holloway, 70, 41, 4500, 100, 400
John Tunnell, 50, 50, 2000, 125, 428
Henry Hudson, 30, -, 120, 100, 250
Albert Lynch, 30, 30, 1000, 25, 100
Jacob Milgus (Wilgus), 10, -, 2000, 150, 475
Mary W. Lynch, 30, 40, 2000, -, -
Samuel Lynch, 50, 70, 3500, 100, 300
James Lynch, 40, 20, 1200, 150, 250
John W. Evans, 4, 20, 1000, 20, 30
E. W. McCabe, 25, 5, 1000, 150, 150
John Lockwood, 25, 35, 1500, 25, 200
F. M. Howard, 5, 5, 400, -, -
Wm. Cooper, 50, 50, 1200, 25, 100
Mitchell Grey, 26, 27, 1200, 25, 125
I. D. Lynch, 56, 25, 3000, 125, 250
Capt. I. M. Taylor, 30, 22, 1500, 75, 200
Andrew Simpler, 25, 30, 800, 20, 150
Isaac McCabe, 40, 30, 700, 25, 100
Wm. D. Layton, 25, 35, 1000, 75, 200
James Tunnell, 50, 20, 3000, 75, 175
Henry Tunnell, 25, 45, 1200, 25, 100
Henry Hickman, 40, 50, 4500, 200, 875
J. H. Lynch, 45, 5, 1500, 75, 75
John Bunting, 50, 20, 3500, -, 50
Benj. Grey, 30, 37, 3000, 150, 450
Levin Moore, 25, 36, 1500, 100, 250
Burton Tubbs, 35, 30, 2000, 20, 50
Benj. Derrickson, 5, 20, 1000, 20, 50
Elijah Lynch, 30, 50, 800, 100, 100
Wm. Derrickson, 35, 25, 3000, 50, 150
Hester Derrickson, 30, 20, 1500, 40, 200

Isiah Derrickson, 40, 15, 2000, 75, 150
John B. Derrickson, 50, 25, 2300, 75, 300
Jeremiah Hudson, 35, 15, 2500, 150, 400
Levy Layton, 50, 50, 1500, 25, 250
John Dosey, 50, 90, 5000, 50, 150
Eli R. Daisy, 25, 30, 600, -, 100
Joseph Daisy, 20, 40, 500, 20, 50
Nath. M. Derrickson, 33, 40, 1200, 75,180
George T. Williams, 40, 65, 4000, 150, 250
E. E. West, 30, -, 1400, 50, 150
Mary Williams, 50, 70, 2000, 100, 400
J. C. West, 70, 60, 3000, 150, 200
James Derrickson, 45, 50, 3000, 50, 125
David H. Derrickson, 30, 40, 800, 50, 150
Peter Moore, 4, -, 300, -, 50
John Bennett, 90, 20, 2000, 800, 250
S. D. Bennett, 65, 65, 4500, 150, 850
Matilda Derrickson, 65, 65, 2000, 50, 150
Isaac Derrickson, 35, 45, 2500, 50, 250
Hiram Rickards, 1, 5, 400, 20, 200
Wilson Cammel, 60, 40, 2350, 75, 500
Mah Atkins, 50, 55, 3000, 75, 150
John Hickman, 30, 35, 1500, 50, 150
G. L. Hudson, 60, 40, 2000, 50, 150
George Johnson, 20, 30, 1000, 20, 125
Lemuel Hudson, 75, 60, 300, 10, 200
Isaac H. W. James, 60, 50, 2500, 60, 380
Danl. Walker, 30, 20, 1000, 50, 150
Wolsey Carey, 40, 20, 1000, 40, 175
Robt.Rickards, 20, 30, 700, 30, 200
Capt. J. Firmer (Firmen), 45, 30, 2000, 50, 175
Jno. S. Derrickson, 60, 40, 700, 75, 150
P. W. Helm, 30, 95, 1500, 60, 150
Peter Johnson, 40, 30, 2500, 100, 200
Joshua Bishop, 60, 35, 1000, 75, 200
Henry s. Birdage, 25, 28, 500, 20, 100
Isaac W. Wharton, 60, 100, 2000, 75, 240
David W. Evans, 20, 40,500, 30, 150
Henry J. Evans, 50, 40, 2000, 75, 150
Lebby H. Evans, 30, 35, 800, 50, 175
Isiah Ellis, 50, 60, 2000, 75, 125
Eli Davis, 35, 35, 500, 75, 250
William S. Hall, 100, 80, 3000, 80, 350
T. B. Short, 100, 50, 3000, 100, 400
Arther McGee, 30, 20, 800, 50, 175
William Wharton, 45, 50, 1000, 10, 80
M. Derrickson, 60, 40,2000, 75, 150
J. B. Collins, 30, 20, 800, 25, 200
Wm. Williams, 60, 40, 1500, 80, 475
James Melson, 75, 40, 2000, 75, 200
John C. Goodwin, 75, 60, 300, 50, 175
Joshua E. Taylor, 30, 40, 700, 35, 400
Wm. S. Evans, 50, 40, 1000, 75, 140
John Melson, 60, 30, 2000, 80, 150
William A. Rickards, 40, 60, 1500, 50, 150
John M. Betts, 75, 30, 1500, 50, 350
John W. Messick, 60, 40, 2500, 50, 175
Jesse Firmen, 60, 40, 2500, 50, 175
Wm. T. Grey, 35, 30, 900, 50, 150
C. M. Wharton, 40, 65, 3500, 50, 400
John James, 60, 65, 3500, 125, 300
Thos. Dukes, 65, 85, 3500, 150, 170
S. H. Derrickson, 48, 30, 1000, 50, 170
Jerry Hudson, 35, 60, 3000, 100, 200

John W. Derrickson, 65, 75, 3000, 100, 150
Parker Nichols, 50, 50, 300, 50, 200
Benj. Gray, 35, 15, 1500, 50, 160
James Roberts, 30, 20, 700, 25, 100
Elizabeth Hall, 45, 60, 1200, 50, 150
J. S. Lynch, 75, 30, 3500, 75, 300
James M. Rickards, 40, 40, 1000, 30, 123
J. F. Hold, 40, 30, 1500, 50, 150
Thomas Daisey, 40, 35, 3000, 100, 250
Kendal Rickards, 70, 25, 2000, 100, 250
G. H. Wise, 35, 40, 1200, 40, 150
John Burhn, 60, 40, 3000, 100, 125
John Layton, 25, 20, 500, 30, 150
Peter Quillen, 40, 10, 700, 50, 200
Ezekiel Milliams (Williams), 40, 30, 2500, 50, 400
Albert S. Evans, 25, 20, 500, 25, 150
Miers O'Stead (O'Staat), 35, 20, 1000, 50, 170
Mary Johnson, 45, 45, 2000, 50, 150
Wm. Roberts, 40, 60, 2000, 50, 120
Elisha Evens, 75, 40, 2000, 95, 300
Philip Short, 35, 30, 710, 30, 150
Clement Evans, 60, 70, 3000, 50, 200
Aaron Lynch, 70, 40, 2000, 75, 300
Robert Tracey, 30, 20, 500, 25, 100
Curtis Jacobs, 20, 30, 500, 20, 500
Robt. Quillen, 50, 40, 1000, 50, 150
S.H. Evans, 25, 10, 800, 50, 200
Morris Messick, 75, 40, 1600, 75, 400
Lemuel Evans, 50, 60, 1600, 50, 300
John Steel, 80, 90, 1500, 100, 350
Ephraim Calhoun, 20, 30, 500, 25, 500
Wolsey Carey, 40, 40, 900,20, 100
Eliza Rickards, 35, 50, 950, -, 50
Capt. Jno. Steel, 80, 100, 2500, 100, 350
John McCabe, 30, 40, 600, 30, 200
Peter Murry, 40, 30, 900, 40, 150
George Rogers, 30, 20, 600, 35, 200
George Roach, 40,75, 800, 40, 150
Prettyman Daisy, 60, 30, 1000, 60, 200
J. Chamberlaine, 40, 30, 700, 30, 150
M. Blades, 30, 25, 500, 20, 10
John Carey, 40, 70, 600, 40, 200
Saml. Vickers, 30, 20, 500, 20, 75
P. W. Daisey, 30, 40, 600, 30, 200
Margt. Elliott, 80, 120, 8000, 100, 275
Wilson Knowles, 60, 20, 3000, 20, 120
Jas. Chipman, 120, 100, 5000, 25, 460
Joseph Windsor, 100, 60, 1000, 40, 390
Wm. S. Moore, 120, 56, 4500, 30, 200
John W. Waller, 160, 25, 6000, 100, 400
Danl. S. Baker, 150, 25, 6000, 100, 400
Henry S. Moore, 100, 40, 2000, 40, 160
Jonathan Moore, 30, 10, 2000, 50, 170
Thos. H. Chipman, 50, 56, 2020, 25, 175
Jeremiah Wright, 50, 103, 1500, 40, 80
Wm. H. Hearn, 115, 20, 4000, 30, 300
Mary Grottee, 120, 80, 1000, 25, 120
Thos. L. Cannon, 400, 300, 7000, 200, 700
Wm. S. Bell, 90, 110, 1500, 50, 250
Boas Bell, 50, 167, 1100, 10, 150
Edwd. B. Bell, 30, 147, 2000, 5, 25
Jas. L. Knowles, 150, 113, 2600, 15, 100
Esther Bell, 25, 92, 1000, 10, 100
Elisha Huston, 100, 50, 1500, 50, 300
Josiah Moore, 60, 24, 1000, 30, 75
Elijah Owens, 40, 20, 600, 10, 40

Wm. Wheatly, 140, 160, 3000, 100, 500
Wilson A. Moore, 250, 50, 4500, 100, 350
Dennis Phillips, 200, 100, 4500, 50, 300
Martin M. Husten, 60, 40, 600, 25, 160
Isaac H. Knowles, 70, 44, 2400, -, 30
Thos. Larmer, 100, 140, 2400, 50, 200
Miles S. Holt, 45, 25, 1000, 40, 80
Danl. Phillips, 75, 25, 1500, 30, 150
Henry W. Phillips, 50, 57, 1600, 20, 100
Wm. G. Elliott, 50, 23, 1400, 50, 175
Jas. A. Marvel, 50, 106, 1200, 30, 150
Jeremiah Eskridge, 200, 110, 4000, 40, 540
Isaac Morris, 75, 30, 1300, 5, 130
Elijah Morgan, 160, 140, 3000, 50, 260
John Wright, 120, 80, 2500, 50, 275
Levin Lauk, 75, 75, 1000, 40, 225
Sallie A. Marvel, 60, 54, 600, -, 25
Stephen F. Waller, 210, 75, 4000, 25, 200
Tilghman L. Spicer, 80,105, 2800, 30, 300
Jacob B. Baker, 90, 21, 1500, 30, 2500
Eccleston Moore, 200, 125, 5000, 150, 500
Lewis T. Moore, 111, 100, 4200, 10, 150
Purnel W. Riggin, 60, 24, 1500, 40, 150
Saml. J. Elliott of J, 180, 40, 6000, 15, 125
Geo. J. Waller, 210, 174, 8000, 30, 200
Wm. Chipman, 75, 125, 1000, 50, 200
Wm. T. Benson, 60, 40, 1500, 30, 100
James Riggin, 100, 50, 3000, 30, 160
Francis A. Lloyd, 60, 200, 2000, 30, 60
Wm. Taylor, 45, 23, 1400, 40, 150
John H. Spicer, 95, 206, 1900, 20, 200
Charles Buck, 60, 40, 1000, 25, 100
Wm. H. Scott, 100, 127, 2000, 75, 300
Jas. M. Morgan, 100, 88, 1500, 50,100
Asa Culver, 35, 46, 1200, 25, 2500
Lorenzo D. Morgan, 80, 29, 600, 25, 2500
Jacob Morgan, 50, 151, 1700, 75, 380
Robt. T. Phillips, 110, 70, 1800, 125, 470
Thos. A. Outten, 100, 102, 2000, 40, 125
James Shiles, 110, 70, 2000, 50, 250
Jas. H. Scott, 60, 40, 1000, 100, 40
David H. Boyer, 100, 50, 2000, 100, 350
Jas. L. Wainwright, 22, 5, 500, 20, 150
Curtis Spicer, 88, 81, 2400, 100, 290
Margaret Wainwright, 70, 20, 1500, 30, 140
Wm. O'Neal, 125, 130, 2500, 50, 300
Philip Graham, 240, 100, 5500, 75, 400
Wm. W. Phillips, 120, 60, 1800, 10, 900
Priscilla Tubbs, 100, 150, 1200, 40, 225
Geo. W. Green, 40, -, 800, 30, 200
Wm. Graham, 40, -, 800, 30, 200
Lazarus Turner, 75, 70, 2000, 50, 256
Jas. Wiley, 70, 65, 1200, 50, 260
Miles Tindal, 160, 140, 2300, 60, 310
Asbury Baker, 100, 60, 3000, 30, 140

Asbury Hastings, 150, 70, 4400, 20, 150
Stephen Green, 160, 20, 3000, 150, 300
John Collins, 116, 44, 2500, 40, 175
Purnell J. Baker, 65, 35, 1000, 40, 240
Warren J. Phillips, 45, 20, 1300, 40, 200
Philip Short, 200, 100, 5000, 60, 450
Geo. W. Gorman, 110, 290, 6000, 45, 140
Belithe Salmons, 210, 110, 6000, 55, 400
Washington Vincent, 50, 30, 1000, 50, 190
Jacob Knowles, 250, 120, 5500, 250, 625
Eli Clifton, 75, 30, 1500, 50, 100
Elias Taylor Jr., 24, 24, 700, 5, 300
J. B. Bailey, 150, 70, 3000, 140, 450
Geo. E. Wiley, 150, 500, 12000, 75, 450
Solomon Short, 150, 105, 6000, 200, 800
Minos B. Lingo, 90, 50, 1400, 100, 110
Jacob L. Messick, 30, 15, 400, 40,75
Elizabeth Workman, 150, 50, 2000, 40, 60
Edwd. Taylor, 125, 55, 1000, 100, 500
Henry Hudson, 80, 25, 2000, 50, 300
John H. Sholes, 69, 11, 1000, 50, 175
Saml. Calaway, 90, 210, 4500, 60, 150
Hudson D. Plumer, 80, 70, 1000, 30, 200
John H. Elliott, 70, 60, 1800, 30, 150
John Hitchens of S, 95, 35, 1400, 35, 160
Edwd. T. Ford, 100, 100, 2000, 40, 350
Edmond Hitchens, 70, 38, 1500, 100, 500
Saml. A. Chipman, 180, 70, 4000, 150, 380
Elias Taylor Sr., 150, 10, 2000, 100, 500
Joseph Vaughan, 200, 100, 4500, 100, 400
John W. Callaway, 125, 125, 5000, 150, 600
Danl. Brown, 80, 120, 2000, 25, 250
John J. Short, 30, 120, 2000, 50, 180
Wm. Giles, 130, 60, 2000, 100, 500
John Isaacs, 80, 60, 1200, 30, 100
Jesse Clifton, 25, 6, 300, 50, 40
David O'Neal, 70, 40, 1000, 150, 300
Jas. H. Boyer, 120, 38, 2500, 50, 325
Thos. Workman, 140, 38, 3500, 95, 450
Jas. H. Whaley, 120, 190, 4500, 40, 100
Anderson S. Pusey, 60, 38, 1000, 40, 100
Caldwell Gibbons, 150, 59, 4000, 25, 140
Isaac N. Fooks, 250, 150, 6000, 450, 650
Wm. W. Morgan, 150, 100, 6200, 200, 1000
Isaac Dolby, 80, 70, 1600, 100, 375
Nehemiah Messick, 30, 20, 600, 20, 80
Jno. W. Messick, 54, 50, 1000, 30, 100
Johnson Cannon, 75, 100, 2000, 100, 340
Joel Messick, 80, 80, 1600, 40, 300
Wm. W. Messick, 85, 25, 1500, 75, 500
Chas. Messick, 220, 170, 6000, 100, 1100
Wilson Messick 80, 95, 1500, 25, 140
Silas Calhoon, 80, 30, 1200, 30, 90
Waitman Bryan, 150, 130, 2000, 100, 700
Mary L. Jones, 60, 68, 1000, 20, 175

Elzey Spicer, 75, 105, 2000, 50, 300
Branson D. James, 150, 61, 3000, 300,700
Mary R. Short, 75, 25, 1500, 100, 700
Peter W. Donoho, 75, 25, 1500, 50, 450
Curtis A. Conoway, 250, 160, 7000, 200, 1500
John C. Collins, 106, 206, 3000, 300, 325
Selby Conoway, 160, 140, 5000, 100, 700
Albert West, 40, 60, 1000, 10, 100
Miles Hitchens, 30, 14, 600, 30, 125
Smith Hitchens, 90, 60, 1500, 100, 400
Nathl. Hitchens, 40, 20, 800, 50, 250
Noah Hitchens, 60, 40, 1000, 15, 175
Minos Hitchens, 80, 60, 1400, 40, 300
Joshua J. Lambden, 150, 170, 3750, 125, 550
Wm. Mitchell, 100, 19, 1700, 75, 300
John Chipman, 100, 150, 3750, 100, 600
Jos. A. More, 140, 50, 1500, 100, 200
Risden Miller, 80, 140, 2500, 75, 150
W.P. Torbert, 100, 47, 1500, 75, 500
George Game, 106, 120, 1800, 100, 250
Edwd. Hitchens, 70, 70, 800, 20, 200
Hamilton Torbert, 120, 120, 1800, 75, 175
Wingate Cannon, 45, 45, 700, 50, 40
Isaac Cannon, 53, 50, 1000, 30, 140
James Cannon, 80, 120, 2000, 150, 300
Wingate Matthews, 100, 105, 2000, 100, 300
Burton Cannon, 150, 171, 3000, 250, 775
Geo. W. McGee, 75, 73, 1000, 50, 200
Saml. H. Hitchens, 50, 95, 800, 30, 100
John Rodney of D, 70,67, 1000, 50, 225
Robt. M. Rodney, 90, 50, 1500, 100, 220
John H. Gordy, 120, 100, 2000, 100, 350
John S. Pusey, 75, 6, 600, 50, 125
Robert Trader, 80, 60, 1000, 10, 65
Nathl. Melson, 100, 71, 2700, 100, 400
Mitchell Hitchens, 70, 80, 1600, 20, 65
Mary A. Melson (Melser), 90, 60, 1500, 30, 140
Saml.P. Dukes, 30, 19, 1000, 25, 125
Joseph Gunby (Gemby), 45, 25, 800, 5, 60
Wm. D. Richards, 60, 25, 1000, 100, 500
Wm. Bryan, 50, 100, 1000, 40, 100
John R. Melser, 70, 29, 1000, 40, 175
Robinson Ellingsworth, 60, 80, 2500, 30, 100
Eleanor Pusey, 70, 95, 1300, 20, 210
Wm. W. Otwell, 10, 60, 1200, 50, 175
Jas. S. Truitt, 100, 60, 1200, 10, 120
James Truitt, 120, 125, 2400, 100, 500
Nutter L. Matthews, 30, 7, 350, 15, 125
Henry B. Truitt, 100, 170, 2000, 50, 300
Eliza Pusey, 70, 38, 800, 30, 110
Lavinia English, 50, 50, 1000, 50, 300
Jane Prettyman, 50, 20, 700, 25, 150
Geo. M. Truitt, 50, 88, 2500, 75, 400
Burton P. Truitt, 60, 38, 1200, 50, 400
Edwd. Phillips, 55, 42, 800, 10, 100

Abel West, 40, 23, 600, 35, 200
Edwd. Warrington, 60, 90, 1200, 3, 100
Jehu N. Phillips, 109, 100, 1700, 50, 300
Lambeson Collins, 60, 30, 800, 100, 225
James Lowe, 40, 35, 800, 150, 400
Benj. Cordry, 20, 49, 700, 30, 100
Robt. K. Smith, 70, 32, 1200, 20, 200
Minos T. Collins, 60, 33, 1000, 25, 100
Elisha T. English, 40, 60, 1000, 25, 300
James W. English, 40, 60, 1000, 150, 600
Philip W. Matthews, 100, 100, 1600, 70, 400
Jacob W. Cannon, 75, 75, 1500, 100, 250
Jas. H. Boyce, 35, 65, 1000, 50, 400
Henry R. Pepper, 75, 25, 1200, 100, 700
John Purnell, 125, 15, 1500, 20, 100
John Carmean, 145, 60, 3000, 100, 500
Wingate Matthews of H. 18, 60, 1500, 50, 150
Mary A. West, 100, 125, 3000, 60, 400
Danl. B. Cannon, 65, 170, 2500, 5, 275
John Gordy of A, 50, 50, 1000, 40, 250
Joseph Wells Sr., 15, 15, 400, 5, 75
Wingate Downs, 105, 180, 2000, 75, 200
Josiah Collins, 130, 120, 2000, 100, 300
Ephraim Collins, 130, 120, 2000, 100, 300
Jos. S. Collins, 25, 25, 500, 3, 125
Wm. P. Gray, 30, 20, 500, 5, 175
John W. Smith, 30, 20, 500, 60, 200
Jas. W. Smith, 30, 64, 1000, 100, 400
Jos. Wells Jr., 50, 19, 600, 5, 50
John S. Matthews, 60, 90, 2200, 60, 350
Henry Davis, 36, 4, 400, 5, 75
John Oney, 30, 65, 1000, 20,50
Thos. Melson, 24, 39, 1000, 8, 250
Isaac B. Short, 35, 99, 1500, 100, 600
Jos. H. Hitchens, 20, 21, 800, 30, 175
Joshua G. West, 50, 51, 1500, 10, 250
Dennis Collins, 45, 130, 2000, 20, 150
Ann M. Hastings, 40, 32, 500, 15, 200
Cornelius D. West, 82, 125, 2000, 50, 500
Peter West, 90, 85, 2000, 75, 300
Wm. S. West, 30, 70, 1000, 20, 175
Burton West, 50, 50, 1500, 50, 200
Lemuel Baker, 10, 28, 650, 5, 200
Smal. S. Truitt, 35, 15, 1000, 30, 200
Jacob Gemby, 20, 50, 700, 25, 100
Sampson B. Smith, 20, 30, 500, 30, 200
Philip S. West, 40, 39, 1100, 75, 300
Wm. Wells, 25, 45, 1500, 30, 150
John Jones, 65, 65, 1500, 40, 250
Elijah R. Parsons, 90, 58, 2000, 100, 400
Silas Pennell, 30, 60, 1000, 50, 150
John Gemby, 25, 50, 800, 75, 400
Mary Pennell, 25, 25, 400, 5, 150
Isaac T. Gorman, 50, 49, 1000, 25, 150
John Beacham, 40, 60, 1000, 25, 75
David Gemby, 75, 75, 1000, 50, 300
John P. West, 30, 82, 2000, 25, 300
Wm. Downs, 30, 70, 1000, 5, 100
Peter Wilkinson, 40, 10, 750, 30, 250
Elisha Mitchell, 50, 25, 1200, 30, 200

Thos. Mitchell Sr., 35, 30, 1300, 20, 175
Robert Smith, 75, 75, 7250, 10, 200
B. B. Banton, 30, 36, 1000, 5, 150
John W. Hickman, 80, 45, 1500, 100, 400
Saml. Mitchell, 20, 30, 500, 10, 50
Joseph Betts, 30, 20, 500, 20, 125
Elzey Collins, 35, 15, 800, 20, 220
Elias T. Collins, 30 10, 500, 15, 50
Urias Short, 60, 39, 2000, 50, 250
Myers Short, 40, 20, 1200, 30, 400
Shadrach Short, 60, 50, 3000, 100, 500
David Long, 25, 25, 1500, 10, 200
Wingate West, 35,15, 700, 25, 300
Peter Hitchens, 30, 80, 1100, 5, 120
James P. Workman, 15, 85, 1000, 10, 60
John Baughton, 10, 90, 1000, 20, 70
Noah Downs, 30, 170, 2500, 25, 175
Ephraim West, 13, 12, 500, 15, 100
Isaac S. Jones, 40, 36, 1500, 100, 450
Wm. A. Truitt, 75, 25, 2000, 50, 300
Geo. W. Jones, 75, 42, 2000, 50, 300
Elizabeth Short, 26, 24, 500, 50, 300
Wm. Whaley, 50, 50, 1000, 75, 350
Rufus J. Elliott, 90, 60, 1500, 20, 150
Wm. Cordry, 100, 56, 1500, 50, 175
Zedekiah P. Hastings, 70, 30, 1500, 50, 320
Mary B. Truitt, 40, 68, 800, 10, 75
Wm. Truitt, 200, 550, 7500, 100, 700
Priscilla A. Gordy, 180, 118, 3000, 40, 175
Louisa W. Gordy, 45, 15, 800, 15, 180
Hezekiah Matthews, 100, 80, 1500, 75, 400
Peter B. Gordy, 75, 19, 1000, 60, 200
Wm. A. Calaway, 37, 38, 1300, 50, 250
Cyrus Carmean, 110, 90, 3000, 75, 300
Noah Timmons, 90, 67, 1200, 30, 175
James Andrews, 40, 83, 1000, 25, 225
Sallie McGee, 70, 60, 1300, 30, 275
Ezekiel Timmons, 79, 71, 1500, 50, 400
James Downs, 20, 30, 1000, 5, 75
Eunicy A. King, 60, 39, 800, 20, 160
Peter Parker, 34, 1, 400, 50, 250
Henry Dunoway, 50, 48, 1000, 50, 275
Sallie Baker, 15, 16, 300, 15, 110
Rufus Mitchell, 22, 64, 1000, 40, 250
Thos. Truitt, 36, 63, 1200, 30, 250
Wesly Mitchell, 50, 25, 750, 30, 200
Sallie Baker of A, 35, 5, 400, 5, 125
Matthew Truitt, 25, 40, 650, 25, 200
Henry Moore, 15, 20, 350, 10, 90
Robt. Mitchell, 21, 9, 300, 25, 125
Richard T. T. Mitchell, 50, 50, 1000, 10, 150
Nathan Mitchell, 25, 25, 500, 20, 90
Wm. B. Parsons, 60, 20, 1200, 50, 500
Henry A. Parsons, 70, 20, 1000, 20, 250
E. B. Hadder, 60, 50, 2000, 20, 125
Joshua Workman, 70, 120, 2000, 30, 300
E. B. Cannon, 50, 22, 1300, 45, 225
James Gorman, 30, 35, 600, 10, 120
Thos. Ake, 55, 45, 1000, 20, 150
John Bromby (Bromley), 20, 75, 2000, 10, 100
Robt. Mitchell Jr., 65, 60, 2000, 10, 140
John Ake, 40, 52, 900, 20, 150
Isaac Wootten, 75, 25, 2000, 125, 600
Ebenezar Gray, 150, 150, 3000, 100, 500
Giles Jones, 40, 10, 500, 10, 150

Thos. Page, 31, 11, 2500, 10, 80
Joshua Gray, 25, 15, 400, 10, 50
Jeremiah Jones, 55, 30, 1100, 75, 200
Noble J. King, 50, 40, 1200, 20, 150
Lemuel Pennell, 75, 50, 1000, 30, 150
Kendall Boughton, 30, 20, 750, 25, 225
Elijah Baker, 30, 20, 500, 20, 150
Saml. Warrington, 80, 120, 3000, 125, 250
Wm. E. Cannon, 60, 91, 1500, 80, 500
James D. Cannon, 25, 125, 1500, 20, 100
Jos. H.B. Cannon, 60, 100, 1500, 30, 125
Wm. B. White, 32, 70, 1300, 3, 175
Jacob Jones, 60, 70, 1200, 40, 350
Benj. Hearn, 100, 95, 3000, 100, 700
Isaac T. Hearn, 30, 70, 1500, 30, 300
Clement C. Hearn, 65, 50, 1000, 450, 70
Ebenezar Taylor, 36, 39, 900, 10, 75
Louden N. Acorn, 100, 50, 3750, 150, 1000
Wm. H. Bets (Botts), 35, 65, 800, 10, 200
Geo. T. West, 30, 66,1200, 25, 325
Aaron Cordry, 75, 75, 1800, 100, 459
John C. Wingate, 40, 60, 800, 10, 200
Andrew Selby, 30, 20, 500, 5, 130
Isaac Selby, 20, 15, 400, 10, 140
Thomas Hearn, 40, 40, 1500, 30, 350
Geo. Phillips, 60, 134, 4000, 30, 175
Thos. T. Neal, 150, 350, 5000, 150, 325
Kendal M. Lewis, 150, 158, 10000, 50, 400
Major Riggin, 60, 38, 2000, 30, 150
Burrows Riggin, 100, 85, 3000, 30, 100
Josiah A. Melser (Melson), 100, 90, 2000, 60, 600
Wm. S. Warrington, 70, 30, 1200, 75, 350
Edward Dill, 25, 80, 800, 25, -
P. K. Steel, 150, 50, 3000, 50,400
B. Macklin, 25, 75, 1000, 40, 150
Burton Dickinson, 35, 25, 600, 50, 200
G. Dickerson, 50, 100, 2000, 50, 500
Wright King, 10, 10, 150, 10, 20
S. Reed, 50, 32, 1000, 50, 30
Russel Dorand (Donovan), 70, 60, 1000, 10, 40
T. E. Cooper, 20, 25, 1000, 10, 100
Mary Reynolds, 75, 225, 3000, 10, 50
Wingate Salwers, 20, 50, 900, 15, 75
Thos. Lindal, 50, 60, 1000, 12, 75
Jas. Shockly, 30, 5, 400, 15, 50
R. Reynolds, 20, 5, 300, 20, 250
Jas. Redden, 100, 103, 2000, 25, 100
M. B. Atkins, 50, 50, 1000, 25, 50
John Sharp, 100, 300, 400, 50, 300
Reuben Donovan, 75, 100, 3000, 50, 400
Foster Donovan, 75, 6, 800, 75, 300
W. J. Donovan, 114, 135, 2500, 100, 340
M. Wilson, 56, 20, 1400, 70, 170
G. Lynch, 46, 5, 300, 15, 325
L. Pepper, 10, 60, 1000, 30, 150
John David, 50, 50, 2000, 100, 300
K. Sharp, 75, 75, 2000, 75, 250
D. H. McColly, 40, 160, 3000, 30, 300
Thos. Connoway, 40, 7, 2000, 15, 200
J. H. Swain, 50, 22, 3000, 20, 250
D. H. Warrington, 40, 110, 3400, 25, 150
N. W. Vaughan, 30, 200, 1500, 75, 200
W. C. Melson (Melser), 40, -, 400, 80, 300
Stockly West, 60, 24, 1200, 10, 120

S. M. Vaughan, 20, 30, 500, 60, 150
Wm. Carpenter, 90, 10, 2000, 20, 110
L. B. Day, 100, 125, 4500, 200, 100
P. D. Smith, 85, 26, 1600, 75, 600
Robert West, 100, 50, 200, 50, 150
Jas. Martin, 100, 100, 6000, 200, 1000
Eli___ Prettyman, 25, 30, 1200, -, 25
Daker Parker, 50, 50, 2000, 40, 200
David Pepper, 600, 400, 14000, 250, 2270
Thos. R. Marvel, 100, 50, 3000, 40, 200
Eli W. Wilson, 30, 75, 1000, 15, 60
A. Carey, 50, 550, 3000, 20, 150
Hester Davis, 70, 30, 1000, 25, 200
Kendal S. Warren, 170, 225, 5000, 45, 250
John Piper, 50, 40, 800, 20,60
John Maclin, 150, 50, 6000, 100, 1500
Richd. Layton, 50, 25, 800, 20, 100
H. Brown, 200, 63, 3000, 100, 400
S. Lofland, 100, 25, 1000, 15, 200
Riley Abbott, 25, 75, 800, 15, 75
Wm.Pettyjohn, 30, 60,750, 20, 40
Isaac Clendaniel, 60, 100, 1500, 30, 300
P. P. Walls, 50, 40, 3000, 15, 125
Z. Donovan, 25, 75, 700, 25, 100
John P. Donovan, 25, 75, 700, 25, 100
Jos. Robbins, 75, 100, 1000, 50, 500
D. H. Robbins, 50, 40, 600, 12, 150
N. Donovan, 50, 50, 500, 20, 100
Geo. Donovan, 50, 50, 1500, 25, 150
Peter Donovan Sr., 80, 120, 1000, 20, 30
Riley Donovan, 40, 60, 1000, 25, 200
Jehu Lynch, 50, 80, 2000, 40, 300
George Torbert, 30, 66,3000, 30, 125
Wm. Donovan, 150, 150, 8000, 100, 300
Geo. Messick, 30, 20, 1200, 20, 80
Jesse Dutton, 30, 77, 500, 20, 40
Peter Donovan Jr., 25, 75, 500, 20, 60
Robert Donovan, 40, 35, 1500, 30, 200
L. Dutton, 70, 70, 1000, 40, 250
Job Donovan, 80, 80, 500, 20, 150
James Donovan of E, 100, 215, 2000, 75, 500
E. Messick, 75, 150, 1500, 40, 200
R. Dutton, 75, 125, 1500, 30, 200
Wm. Walker, 100, 15, 500, 25, 60
C. Fowler, 50, 100, 2500, 20, 150
Elzy Mosely, 100, 400, 2600, 20, 12
Moses Brittingham, 25, 28, 600, 20, 95
James Holston, 75, 200, 3000, 30, 100
Eli Donovan, 75, 150, 2000, 40, 300
Jonathan Donovan, 80, 100, 1400, 40, 300
John Hevaloe, 125, 175, 3000, 40, 350
Kendal Messick, 30, 60, 800, 45, 150
Sarah A. Messick, 40, 110, 1000, 18, 60
Mithel Piper, 50, 45, 700, 35, 70
W. Hopkins, 75, 50, 1000, 100, 300
Phillip Workman, 300, 50, 3000, 100, 600
Thomas S. Brittingham, 15, 25, 80, 50, 100
Elisha Hart, 15, 3, 300, 10, 50
Josiah Sharp, 60, 60, 1200, 60, 200
Wm. Dalton, 50, 50, 800, 40, 150
James R. Chase, 40, 60, 850, 25, 50
Mary Reed, 50, 40, 1000, 20, 150
Warren Torbert, 70, 30, 200, 30, 125
Elisha E. Greenly, 75, 49, 900, 25, 200
Samuel H. Layton, 100, -, 4000, 25, 225
Joseph B. Vaughan, 110, 120, 8000, 30, 100
James H. McColly, 20, 2, 150, 25, 50
Cornelius Dodd, 40, 60, 1000, 25, 50

Isaac W. Short, 75, 60, 2000, 40, 250
Allen Short, 10, -, 1200, 20, 120
David H. Johnson, 25, -, 500, 25, 135
Wingate Morris, 100, 40, 3000, 100, 40
Benj. V. Morris, 8, 8, 32, 12, 80
James P. Milson (Wilson), 40, -, 700, 30, 150
Eli Walls, 90, 122, 8000, 100, 600
Gideon Joseph, 30, 50, 800, 20, 175
Major W. Milser (Milson), 60, 40, 2000, 40, 400
Wm. D. Adkins, 60, 40, 2000, 30, 300
Greenbery H. Pepper, 70, 40, 1000, 12, 300
Robt. Warren, 55, 35, 2000, 30, 125
Saml. White, 45, 45, 1200, 25, 100
Isaac M. Adams, 35, 6, 500, 35, 150
Aaron Marvel, 80, 70, 2000, 50, 150
Edward Marvel, 80, 70, 200, 50, 200
C. C. Hart, 21, -, 800, -, 600
Sussex Co. Alms House, 255, 68, 12800, 300, 1500
Noah Johnson, 60, 90, 2000, 15, 80
George W. Walls, 100, 70, 1500, 30, 200
George Wilson, 75, 50, 1000, 31, 225
Abram Johnson, 70, 50, 1200, 20, 75
John Messick, 60, 40, 1000, 50, 200
Pinky Pettyjohn, 70, 60, 2000, 30, 250
Phillip West, 60, 40, 1500, 20, 200
Perry P. Johnston, 20, 60, 600, 20, 150
David M. Reynolds, 45, 125, 3000, 30, 300
Elsy Milson, 50, 40, 2000, 30, 350
Peter Rust, 100, 25, 2000, 75, 500
Zachariah Joseph, 60, 40, 800, 30, 100
James Johnson, 40, 60, 1000, 35, 150
Peter P. Dodd 50, 60, 1000, 35, 200
Aaron Dodd, 45, 50, 1000, 25, 100
John T. Carey, 75, 60, 1500, 30, 175
Joseph Burriss, 80, 45, 800, 75, 270
Edward Adkins, 50, 113, 1500, 25, 50
James L. Collins, 50, 50, 1000, 30, 175
Catharine Adkins, 100, 120, 2500, 35, 125
Henry O'Bennum, 60, 65, 2000, 25, 250
William Hancock, 100, 140, 3000, 38, 250
Moses Magee, 150, 56, 2000, 35, 300
James A. Collins, 20, 80, 1000, 31, 250
Miers Collins, 50, 50, 1000, 30, 200
Virden Macklin, 100, 100, 3000, 35, 500
James B. Joseph, 40, -, 650, 30, -
Benj. Agin, 50, 32, 1000, 35, 110
Geo. T. Calhoon, 60, 75, 2000, 50, 500
John Wilson, 75, 50, 1500, 35, 75
Edwd. Dickerson, 35, 40, 2000, 15, 150
Henry Milson, 20, 30, 500, 15, 60
Matthew Wilson, 50, 50, 2000, 50, 400
James Workman, 50, 50, 2000, 30, 250
Robert Greenly, 19, 50, 1200, 25, 500
Eli Roach, 40, 40, 600, 30, 150
Joseph Milson, 70, 57, 800, 35, 125
Levin Mayo, 70, 30, 1200, 35, 175
Mary Chase, 60, -, 700, 30, 100
Chas. Tunnell, 150, 50,7000, 125, 420
David Stuart, 50, 50, 2500, 40, 500
Wm. Short, 30, 50, 2000, 80, 150
Maria Dutton, 35, 50, 1000, 25, 20
John Workman, 40, 40, 1500, 45, 300
Geo. Pettyjohn, 60, 40, 600, 30, 100

James Dickerson, 100, 100, 2000, 25, 150
Wm. Spicer, 60, 100, 1200, 50, 200
Wm. W. Wilson, 50, 60, 1500, 40, 300
William Dickerson, 25, 15, 300, 25, 60
Theophilus Solmons, 75, 10, 800, 40, 75
James Duton, 75, 125, 2000, 25, 150
George Chase, 35, 175, 2000, 35, 200
Wingate Milson, 100, 200, 3000, 10, 350
Levi Moseley, 14, -, 300, 25, 200
Charles Wilson, 60, 40, 1000, 30, 250
John Dodd, 60, 7, 1500, 25, 300
James Roach, 40, 75, 1000, 25, 61
Peter Wilson, 70, 10, 800, 20, 100
Nancy Donovan, 25, 45, 700, 25, 125
Isaac Chase, 100, 200, 2000, 30, 200
Zachariah Pettyjohn, 65, 36, 1000, 35, 200
William Pettyjohn, 75, 30, 1000, 40, 150
Abram Donovan, 40, 60, 800, 15, 75
John Blissuret, 50, 150, 2000, 20, 400
Nathaniel Betts, 50, 150, 2000, 20, 15
Jesse Ennis, 100, 100, 1400, 450, 1000
Nathaniel Veasey, 200, 170, 4000, 160, 450
Wm. Veasey, 50, 25, 1000, 50, 150
John Dodd, 100, 25, 1000, 40, 150
Alex Johnson, 45, 50, 1000, 25, 150
Wm. Johnson, 75, 150, 2200, 25, 150
Stephen Norwood, 75, 25, 700, 15, 45
Elisha Holland, 300, 200, 4000, 200, 400
Henry Sharp, 80, 20, 1000, 30, 300
James S. Lauk, 30, 6, 300, 25, 200
Josiah Davison, 40, 12, 500, 25, 125
Nathaniel Sharp, 75, 125, 2000, 30, 250
Miers Clark, 30, 100, 1300, 15, 15
Bo__ Harmon, 125, 275, 3020, 25, 125
Capt. Henry Hudson, 50, 150, 2500, 35, 200
Nathaniel Clark, 150, 150, 3000, 30, 200
Wm. Blissuret, 40, 100, 1000, 15, 75
Phillip N. Hood, 52, 40, 820, 20, 100
Stephen Carpenter, 130, 70, 2000, 35, 250
John Lindale, 60, 70, 1000, 20, 100
Elias O. Day, 50, 30, 1200, 25,150
Purnal J. Maull, 30, 10, 800, 30, 200
Jacob N. Coffin, 130, 15, 1000, 25, 250
Sylvester P. Joseph, 50, 50, 1000, 30, 125
Israel Purnal, 25, 10, 300, 15, 50
Nathaniel Warrington, 20, -, 200, 15, 75
Saml. Martin Jr., 200, 85, 3000, 30, 300
Sylvester W. Palmer, 12, 7, 500, 20, 175
Robt. Brian, 100, 50, 2000, 20, 150
Andrew J. Holland, 100, 100, 2500, 40, 400
Joseph Casey of W, 40, 60, 1500, 40, 150
James M. Lauk, 100, 100, 1200, 20, 75
Mary Simpler, 60, 25, 1000, 25, 175
Sally Shannon, 90, 40, 800, 20, 70
John Shannon, 75, 300, 2500, 25, 130
John N. Joseph, 110, 290, 3000, 35, 300
James H. Hopkins, 50, 70, 1000, 30, 150
Lydia A. Perry, 50, 100, 1500, 35, 150

William Hasel, 50, 75, 1200, 40, 300
Thos. J. Perry, 75, 100, 2000, 60, 300
Henry White, 75, 25, 1200, 50, 250
Jesse White, 10, -, 200, 15, 30
Peter J. Hopkins, 100, 120, 2000, 40, 300
Burton Fowler, 80, 20, 900, 35, 300
Samuel Hudson, 60, 140, 2000, 35, 200
Reuben T. Wilson, 60, 140, 2000, 25, 150
Peter Lauk, 75, 25, 1200, 35, 300
Reuben Milson, 80, 125, 2500, 25, 100
Wm. H. Simpler, 100, 200, 3000, 35, 300
Thomas Milson(Milser), 25, -, 300, 15, 150
Barak Richards, 20, 5, 600, 25, 45
Nehemiah Dorman, 55, 80, 1000, 50, 300
Benjamin White, 80, 20, 2000, 60, 450
Robert White, 150, 200, 4000, 60, 500
William W. White, 200, 50, 2000, 50, 400
James A. King, 148, 40, 2000, 40, 800
William C. Milby, 78, -, 600, 30, 200
Hugh King, 150, 20, 2000, 35, 600
Russel Dickerson, 65, 135, 2000, 35, 400
Jackson Palmer, 100, 100, 2500, 35, 300
Sally King, 60, 30, 1000, 25, 200
George Burton, 16, 2, 300, 20, 125
David M. Russell, 75, 75, 1500, 30, 200
Henry S. Bennum, 60, 40, 1000, 25, 85
Absalom Rust, 67, 30, 2000, 40, 200
Miers R. Fisher, 30, 100, 2000, 40, 150
James M. Hudson, 38, 5, 1500, 40, 200
Hiram Wilson, 50, 75, 800, 30, 100
David J. Ennis, 75, 50, 1000, 40, 250
Josiah Veasey, 75, 75, 2000, 45, 400
Robinson Barker, 80, 80, 2500, 85, 500
John Simpler, 75, 100, 1500, 34, 150
Peter Warrington, 75, 75, 1500, 35, 175
John Martin, 75, 50, 1000, 35, 250
M___ Johnson, 50, 125, 1000, 30, 300
Nancy Dickerson, 75, 150, 1600, 25, 250
Ellen Virden, 80, 20, 900, 35, 200
Wm. Martin, 40, 25, 625, 30, 150
Jos. B. Virden, 46, 54, 1000, 40, 300
Joseph Wilson, 50, 125, 1500, 25, 200
William Prettyman, 60, 42, 800, 35, 400
James Coulter, 20, 80, 900, 35, 300
Lody Adkins, 60, 40, 1000, 35, 325
William S. Vent, 35, 45, 800, 30, 200
Thomas Walker, 100, 40, 1500, 50, 350
Milliam Mason, 50, 50, 2000, 40, 500
Lydia Burris, 20, -, 200, 20, 60
Benton H. Carpenter, 125, 75, 3000, 60, 400
Geo. W. Bennum, 60, 40, 1200, 40, 300
James Carey, 75, 25, 1500, 35, 250
Wm. P. Wilson, 60, 35, 1000, 35, 300
Henry S. Marshall, 125, 45, 1500, 25, 150
James Anderson, 400, 425, 22450, 400, 1750
A. C. Pepper, 150, 72, 5000, 250, 1800
Geo. S. Davis, 100, 100, 300, 50, 300

Ebenezer P. Warren, 114, 100, 3000, 50, 275
James Tull, 150, 100, 7000, 75, 350
James Reed, 100, -, 2000, 35, 275
Abel Pettyjohn, 75, 100, 1500, 25, 175
William Morris, 50, 50, 800, 25, 100
Bartley Macklin, 50, 25, 1000, 25, 125
John Wilkins, 75, 25, 1000, 35, 175
Robert Pettyjohn, 50, 55, 1000, 25, 125
Jehu Rigward (Ragland), 50, 55, 1000, 35, 250
Thomas Morris, 50, 50, 1000, 40, 300
Robert Morris, 40, 44, 800, 20, 125
Thomas P. Wilson, 60,75, 1500, 35, 300
William Wilkins, 75, 57, 1500, 45, 225
Silas M. Reynolds, 6, -, 75, 25, 150
Wm. S. Walls, 75, 175, 2500, 35, 300
James Jones, 251, 50, 5000, 50, 600
William Clyter, 75, 50, 1200, 30, 250
Samuel Walls, 80, 100, 1800, 35, 300
Daniel Burton, 80, 100, 1800, 35, 300
Isaac S. Morris, 32, 15, 300, 20, 100
James Spicer, 75, 50, 1000, 25, 150
Purnal Johnson, 125, 125, 2000, 30, 150
Wm. W. Donovan, 75, 75, 1500, 35, 275
Joseph Lindal, 75, 50, 1200, 25, 100
Robert Warren, 50, 25, 800, 25, 150
Job Sharp, 50, 25, 800, 20, 150
George Abbot, 60, 50, 900, 25, 100
Eunice Abbott, 100, -, 1000, 25, 150
Geo. G. Abbott, 50, 50, 500, 20, 125
Elisabeth Abbot, 125, 75, 1500, 35, 200
Robert Donovan, 45, 45, 800, 35, 150
Donovan Reed, 50, 50, 1000, 35, 175
John Benson, 110, 290, 4000, 35, 300
John Warren, 50, 10, 400, 30, 200
Wm. Roach, 50, 50, 1200, 30, 300
Henry Pepper, 100, 50, 2500, 35, 300
Thomas W. Wilson (Milson), 100, 100, 3000, 40, 350
B. H. Johnson, 65, 8, 2500, 50, 350
Elex. Young, 15, -, 300, 30, 150
Eli Collins, 50, 50, 2000, 35, 350
Eli L. Collins, 50, 50, 1500, 30, 225
Asa Conwell, 50, 150, 2000, 30, 250
Robt. H. Roach, 50, 50, 1500, 35, 340
James R. Roach, 50, 25, 800, 30, 150
Horatio N. Heathens, 40, 10, 100, 30, 75
Robert Conwell, 50, 50, 1000, 30, 100
James Simpler, 50, 50,1000, 35, 200
Joshua Hevalor, 125, 125, 2000, 35, 300
Miles Saunders, 50, 25, 1000, 25, 150
John A. Ca___, 50, 25, 1000, 25, 150
John C. Morris, 75, 50, 1200, 40, 350
David Johnson, 75, 50, 1200, 35, 225
Thos. Williams, 100, 100, 1500, 40, 350
John J. Morris, 150, 40, 3000, 200, 700
Sam Lecatts, 40, 60, 1000, 40, 400
Lemuel Cary, 60, 18, 1200, 35, 70
Robt. W. Betts, 120, 25,2000, 50, 400
Greensberry Betts, 50, 30, 1000, 35, 300
Isaac S. Betts, 75, 45, 800, 30, 275
Wm. A. Conwell, 75, 30, 1200, 40, 400

David Robins, 125, 75, 2000, 50, 250
Joe T. Conwell, 130, 40, 5000, 60, 600
Hevalor Morris, 75, 50, 2000, 60, 300
Alfred Macklin, 75, 50, 2000, 50, 200
Jas. Stephenson, 40, 20, 600, 30, 175
Joseph Conwell, 75, 60, 4000, 60, 350
James J. White, 30, -, 900, 30, 175
Alfred Russell, 95, 40, 1500, 45, 500
David Russel, 50, 28, 2000, 40, 300
Jno. H. Willbank, 80, 23, 4000, 45, 500
David Nailor, 80, -, 1500, 30, 300
Thos. Robinson, 90, 90, 1800, 50, 400
David Willbank, 100, 200, 3000, 50, 350
James Reed, 100, 50, 2000, 50, 450
Abram Reed, 100, 25, 1500, 40, 500
Bennett Johnson, 60, 25, 1000, 30, 250
Peter Reed, 60, 25, 800, 35, 200
Tilney Stephenson, 75, 50, 900, 40, 275
Luke Clendaniel, 50, 40, 700, 30,75
Emiline Simpler, 40, 22, 800, 35, 200
James Robbins, 100, 25, 1500, 40, 350
James Wilson, 75, 25, 1000, 30, 275
James T. Roach, 100, 25,1500, 30, 250
John Stockly, 150, 650, 5000, 35, 450
Paris Willbank, 75, 25, 800, 30, 80
James Donovan, 60, 87, 2000, 40, 250
Sylvester H. Rust, 60, 80, 2500, 45, 450
Henry Johnson, 50, 40,1000, 35, 150
Dr. D. H. Houston, 300, 150, 10000, 300, 100
James Gordon, 39, 5, 800, 35, 200
Jacob Richards, 148, 20, 3000, 40, 250
Cord Warrington, 100, 75, 4000, 160, 700
Cyrus Holland, 30, -, 300, 40, 300
Elzy Holland, 30, -, 300, 25, 30
Jacob Holland, 30, -, 300, 25, 75
John Fisher, 137, 15, 2000, 40, 300
James Fisher, 130, 25, 2000, 40, 250
Paynter Johnson, 40, 25, 800, 30, 150
John Cathorn, 75, 50, 1500, 30, 200
Ebe Holland, 180, 150, 3000, 45, 400
Ebe Martin, 100, 50, 1000, 40, 150
Jay Lecatts, 50, 75, 1000, 25, 75
Thomas Black, 75, 50, 1500, 30, 300
James B. Downing, 60, 50, 1000, 25, 100
G. H. Wright, 60, 15, 7000, 150, 600
Thos. Pepper, 90, 85, 5000, 50, 400
Wm. Elligood, 88, -, 5000, 150, 700
Wm. A. Hazzard, 80, 60, 7000, 125, 450
Robt. W. Carey, 40, -, 1000, 40, 350
James M. Casey, 180, -, 9000, 40, 200
Peter B. Jackson, 75, 30, 9000, 60, 250
Noble C. Ellingsworth, 30, -, 700, 60, 350
Aaron Marshall, 36, 23, 1200, 40, 275
John C. Haggard, 34, -, 2000, 40, 300
Samuel Martin Sr., 40, -, 2000, 60, 600
John S. Holland, 50, 50, 1000, 40, 150
Samuel Parker, 18, 10, 1550, 40, 250
Nehemiah D. Welsh, 50, 50, 1000, 40, 200
David Lofland, 100, 100, 3000, 100, 500

Nathan V. Messick, 30, -, 2000, 40, 125
David Pepper Jr., 100, 50, 2500, 50, 400
Daniel H. Messick, 50, 50, 1000, 25, 125
John Sorden, 50, 130, 3600, 75, 450
Wesly Wolfe, 80, 40, 5000, 75, 500
Andrew V. Truitt, 30, 60, 500, 30,100
James Wilkins, 100, 40, 1500, 60, 206
Wm. P. Jefferson, 80, 20, 1000, 45, 400
Elias Shockly of W, 75, 45, 2000, 50, 400
Wm. V. Shockly, 50, 50, 500, 40, 300
Wm. H. Betts, 50, 80, 1000, 40, 300
David Lindall, 163, 300, 2800, 35, 300
George Owens, 10, 10, 500, 25, 75
M. R. Daniel, 90, 45, 3000, 50, 500
D. P. Coffin, 80, 60, 2000, 35, 75
Lot W. Davis, 85, 150, 2000, 45, 800
George R. Fisher, 150, 160, 300, 150, 600
Amos Staton, 45, 100, 3000, 100, 400
R. K. Johnson, 60, 120, 3000, 100, 300
Wm. Shockly Jr., 80, 20, 3000, 75, 550
Jonathan Dickerson, 60, 75,1500, 30, -
John D. Swain, 160, -, 4000, 45, 400
David Staton, 40, 10, 1000, 40, 300
Wm. Short, 40, 95, 2000, 55, 400
Thos. W. Donovan, 38, 25, 1000, 45, 300
Chas. Young, 80, 7, 1000, 30, 150
Wm. H. Fountain, 100, 35, 2000, 45, 400
Wm. B. Walls, 50, 100, 1200, 35, 500
John Webb, 100, 50, 2500, 45, 600
S. D. Jester, 100, 25, 2000, 40, 700
B. F. Taylor, 75, 100, 5000, 150, 800
John Hays, 132, 60, 3000, 60, 300
Joseph Lingo, 75, 25, 1500, 40, 500
Elias T. Bennett, 100, 110, 2500, 150, 1000
Agail (Abigail) Stevens, 30, 10, 300, 30, 200
Isaac Betts, 250, 150, 5000, 150, 1500
S. P. Collins, 30, 15, 1500, 50, 400
Geo. R.Swain, 100, 100, 3080, 45, 300
Wm. Peirce, 100, 100, 3000, 45, 250
Sally Bennett, 72, 25, 1000, 40, 200
Joshua Richard, 75, 25, 1000, 35, 125
Thos. Steel, 100, 50, 1200, 35, 56
David Young, 50, 75, 1000, 25, 75
Robinson Shockly, 30, -, 400, 25, 200
Clement Pettyjohn, 50, 50, 1000, 25, -
George Shockly, 50, 50, 1200, 25, 150
Miers C. Draper, 500, 200, 20000, 450, 2000
Thos. Roach, 100, 50, 4000, 60, 500
Kinsey Jones, 75, 25, 1500, 40, 200
Robert Davis, 85, 100, 2000, 60, 400
Georg Hevalor, 50, 25, 800, 25, 35
Thomas Roach, 80, 200, 1300, 35, 300
Robt. Pepper, 100, 75, 2000, 30, 200
Edward Joseph, 75, 50, 1200, 30, 75
Jno. Shockly, 30, -, 400, 25, 60
Jas. Cain, 125, 42, 7000, 150, 665
Edward Roach, 75, 100, 4000, 100, 400
Sal. Brittingham, 100, 100, 2500, 60, 75
John Roach, 100, 100, 2500, 100, 300
David Coffin, 100, 100, 2500, 50, 200

David Dickerson, 100, 100, 3000, 50, 150
Miers Reynolds, 80, 50, 2000, 75, 500
Jesse Dodd, 60, 20, 1200, 25, 100
E. C. Abbott, 50, 30, 1200, 100, 200
David Casmors, 40, 6, 500, 25, 150
B. F. Wapes, 170, 240, 8000, 150, 1500
Nehemiah Messick, 50, 46, 1000, 60, 250
Jesse Shockly, 50, 10, 800, 35, 275
Charles Shockly, 50, 50, 1000, 45, 300
David Truitt, 125, 125, 2000, 65, 400
Joseph M. Davis, 60, 15, 1000, 100, 400
Edith Davis, 60, 15, 800, 60, 325
Riley W. Bennett, 125, 200, 5000, 70, 1200
Robert Young, 100, -, 2000, 25, 100
Solomon Midca___, 21, -, 600, 25, 200
William McDowell, 50, 50, 1000, 25, 200
Albert H. Argo, 100, 25, 1500, 35, 250
Henry R. Draper, 100, 150, 2000, 50, 500
Maud Draper, 120, 80, 3000, 100, 700
Thos. J. Davis, 90, 50, 3000, 150, 1000
Jerry Young, 60, 40, 1000, 25, 100
Daniel Deputy, 150, 30, 2000, 60, 500
Solomon Cervilhin, 30, -, 500, 25, 100
James Draper, 50, 10, 600, 20,10
Geo. Wilson, 75, 25, 2000, 75, 520
John Walls, 75, 50, 1200, 30, 125
Robert Roack (Roach) of T, 80, 50, 1200, 35, 200
Joseph Groves, 50, 50, 1000, 40, 100
Sam. R. Tindal, 20, -, 500, 25, 150
Alfred Reed, 50, 10, 700, 25, 50
John Abbott, 80, 80, 1500, 50, 150
Erasmus Abbott, 90, 110, 950, 45, 350
Henry Vaun, 50, 60, 1000, 35, 250
Sam. Clendaniel, 100, 100, 1500, 65, 700
Silas Butler, 100, 50, 1000, 30, 150
Wm. W Smith, 300, 500, 8000, 150, 800
Saml. Davis, 50, 150, 2000, 25, 75
Purnal Sharp, 75, 170, 2000, 45, 500
Selathiel B. Macklin, 60, 60, 2000, 45, 400
Jonathan T. Betts, 50, 50, 1000, 35, 500
Joseph Fisher, 30, 40, 400, 25, 200
Desolem Pettyjohn, 30, 30, 500, 25, 100
Alfred Abbott, 100, 100, 2500, 35, 250
John Clendaniel, 75, 125, 1800, 40, 225
Jesse Walls, 75, 75, 1500, 30, 250
William Benson, 75, 85, 1500, 25, 200
William Pierce, 125, 75, 2000, 35, 200
Elias Reed, 70, 40, 1800, 45, 450
Isaac F. Warren, 125, 175, 3000, 60, 400
Geo. H. Welch, 50, 60, 1200, 25, 250
Littleton M. Lofland, 50, 110, 1600, 20, 100
Jonathan Milman Jr., 50, 75, 1000, 30, 200
Wm. Burris, 50, 75, 1000, 25, 80
Peter Milby, 75, 125, 3000, 15, 150
James B. Reed, 60, 44, 1500, 60, 10
Geo. Clendaniel, 150, 200, 3500, 80, 500
Boaz Passwaters, 80, 90, 1500, 40, 400
Francis A. Warren, 200, 150, 3500, 85, 1100
Jonathan Milmer, 150, 150, 2500, 60, 500

Jesse Tindal, 75, 25, 800, 30, 300
Stephen Warren of S, 100, 200, 300, 65, 400
Jos. S. Watten, 50, 50, 150, 50, 300
Job Tucker, 35, 15, 400, 30, 100
Cloudsbury Sharp, 75, 25, 1000, 35, 200
Nehemiah Abbott, 50, 50, 800, 35, 150
Isaac Russel, 100, 50, 1500, 35, 200
John R. Day, 100, 75, 1800, 35, 300
Jehu Clendaniel, 125, 50, 2500, 50, 400
Wm. H. Clendaniel, 75, 57, 1500, 30, 200
Henry Deputy, 125, 100, 2500, 60, 800
Bourbon Walls, 50, 95, 800, 30, 150
Rachel Clendaniel, 75, 100, 1800, 50, 350
Nehemiah Truitt, 80, 1200, 1000, 50, 300
Jonathan Williams, 50, 25, 400, 35, 200
Jacob Heller, 75, 25, 1000, 35, 400
Thomas C. Baning, 60, 60, 1200, 40, 375
Wm. H. Smith, 65, 60, 1000, 30, 250
Wm. Pettyjohn, 50, 50, 800, 25, 100
Joshua Webb, 80, 60, 1500, 100, 500
Jas. M. Webb, 100, 75, 1500, 30, 225
Sol. Deputy, 100, 13, 800, 45, 500
Elizabeth Deputy, 80, 20, 500, 25, 100
Sam. H. Clendaniel, 75, 25, 1000, 35, 100
Beniah Tucker, 50, 50, 800, 25, 75
Stephen Warren, 60, 120, 15000, 45, 300
Henry Macklin, 60, 40, 1000, 45, 300
Asbury Sharp, 50, 25, 800, 25, 80
Jerry Warren, 50, 20, 700, 25, 75
Danl. H. Marvel, 75, 57, 1500, 25, 150
Clement H. Hudson, 225, 225, 4500, 100, 600
Whittington Williams, 200, 100, 4000, 80, 500
James Deputy, 100, 100, 2500, 75, 300
Alfred Short, 110, 110, 4000, 200, 800
Nehemiah Miller, 50, 75, 800, 25, 65
Purnal Shockly, 50, 20, 750, 25, 50
Alex Johnson, 100, 100, 2500, 60, 300
Isaac Murphy, 50,75, 800, 30, 125
Salathiel Ellingsworth, 75, 50, 1200, 25, 125
John Dennan, 75, 50, 1300, 25, 175
Wm. B. Webb, 65, 35, 1000, 40, 250
James Wheeler, 65, 15, 600, 30, 200
Wm. B. Walls, 75, 25, 1200, 400, 400
Jacob Clendaniel, 50, 25, 700, 35, 150
James H. Hudson, 100, 68, 2000, 50, 400
Henry Austin, 75, 50, 1500, 50, 500
Wm. Vess, 40, 40, 800, 25, 150
Ben E. Jester, 90, 45, 2000, 75, 600
John _. Macklin, 96, 110, 3000, 60, 450
George Prettyman, 140, 60, 3000, 60, 500
George Hays, 50, 25, 1000, 35, 125
Manlove Johnson, 70, 50, 5000, 50, 100
Daniel I. Jones, 12, -, 250, 30, 150
Silas Plummer, 50, 10, 400, 25, 75
John Plummer, 50, 15, 450, 35, 125
Mathias Harmond, 75, 25, 1200, 40, 250
John W. Clifton, 250, 150, 6000, 150,700
Samuel Carilse, 75, 25, 1000, 20, 200
John Dawson, 150, 75, 4000, 50, 350
George Macklin, 50, 40, 1000, 45, 250

Curtis Macklin, 50, 40, 1000, 40, 250
James Wootten, 50, 30, 600, 30, 800
John R. Whayly, 150, 50, 3500, 35, 300
Eliza Stayton (Slayton), 50, 20, 600, 25, 100
Isaac Simpson, 150, 180, 3000, 75, 600
Martha Ross, 100, 100, 2000, 30, 150
Parker Lofland, 100, 60, 4000, 50, 400
Henry Stuart, 80, 200, 1500, 40, 250
Nathaniel H. Johnson, 125, 70, 7000, 100, 350
Wm. Brown, 75, 25, 2000, 50, 400
Samuel Hannon, 75, 75, 1500, 50, 200
Henry Abbott, 100, 50, 2000, 40, 150
Wm. H. Stayton, 150, 40, 7000, 200, 800
Samuel Wilkinson, 100, 25, 2000, 40, 100
Aaron Vanderbilt, 75, 20, 1500, 40, 300
Ingham Smith, 200, 100, 6000, 75, 300
Nicholas Plummer, 30, 150, 1500, 25, 100
James Plummer, 30, 70, 1000, 25, 250
Joshua Truitt, 75, 75, 1000, 45, 350
Kensey Truitt, 40, 40, 800, 30, 175
Nathaniel Derrickson, 75, 40, 1100, 40, 300
Levin W. Carmean, 75, 25, 2000, 30, 150
Wm. Davis, 200, 50, 8000, 50, 300
Wm. W. Simpson, 300, 100, 1000, 50, 350
Clemant Heuston Jr., 60, 152, 4000, 50, 350
Elisha Tharp, 150, 50, 4000, 50, 500
George Tharp, 120, 60, 3000, 40, 150
Nemiah Sharp, 100, 100, 2500, 45, 700
Jame Heuston, 100, 1000, 300, 150, 700
Winlick H. Collins, 80, 137, 2000, 75, 700
Thomas R. Moore, 80, 20, 1000, 34, 200
Solomon D. Jester, 80, 50, 2000, 40, 200
James Tatman, 40, 40, 600, 25, 120
Levin D. Vaughan, 60, 40, 800, 40, 150
Bivins Morris, 20, 30, 400, 30, 100
Grace Warren, 60, 40, 800, 25, 150
William Warren Sr., 90, 60, 2000, 100, 400
William Carpenter Jr., 54, -, 600, 30, 80
William Warren Jr., 100, 100, 3000, 35, 275
Robert J. Walls, 50, 30, 600, 20, 75
Joshua Lofland, 100, 60, 1500, 45,450
Charles Shockly, 30, 5, 250, 20,75
David N. Shockly, 50, 22, 800, 40, 250
Nancy Shockly, 50, 20, 600, 30, 100
Samuel B. Jefferson, 70, 30, 2000, 160, 400
Peter Dickerson, 100, 260, 4000, 60, 350
R. W. S. Davis, 70, 90, 2500, 100, 400
Levin Hopkins, 150, 125, 3500, 40, 300
Lydia A. Jefferson, 60, 10, 800, 30, 150
Elisha Dennan, 100, 100, 2000, 65, 450
Joseph Calhoon, 100, -, 2000, 45, 750
Peter Calhoon, 100, 60, 3000, 100, 750

Wm. Magee, 100, 100, 1200, 45, 200
John Ingram, 70, 20, 1500, 35, 250
Wm. W. Abbott, 50, 50, 1000, 25, 150
Spencer Warren, 60, 50, 1000, 40, 125
Robt. Hudson, 50, -, 800, 30, 85
Weighmiah Groors, 40, 20, 500, 25, 60
Luke Ellingsworth, 50, 5, 500, 25,60
Jesse Lofland, 50, 5, 500, 25, 75
Henry C. Hudson, 50, 10, 500, 35, 150
David Watson, 100, 100, 2000, 60, 250
Wm. Shepherd, 85, 15, 1000, 45, 400
Minos Linch, 60, 80, 1200, 60, 300
David Surch, 40, 40, 800, 40, 200
David Roach, 60, 40, 1000, 60, 275
Elias Shockly of S, 25, 25, 1000, 60, 300
John S. Truitt, 200, 100, 5000, 100, 600
Purnell Griffith, 100, 50, 3000, 50, 300
David Watson, 100, 25, 1500, 50, 300
James Warren, 75, 25, 1000, 40, 150
Mark H. Davis, 40, -, 200, 25, 200
Geo. W. Warren, 60, 40, 1000, 25, 250
W. O. Wilson, 100, 50, 1500, 40, 350
John C. Wilson, 75, 25, 800, 30, 150
Wm. Hill, 60, 100, 800, 25, 50
Peter Calhoon, 100, 100, 3000, 45, 300
Burton Carpenter, 150, 50, 5000, 45, 350
Wm. Watson, 100, 75, 3000, 40, 350
Joseph Mason, 100, 50, 3000, 45, 400
Jas. H. Reed, 250, 62, 11000, 250, 500
John Hollis, 85, 15, 3000, 150, 900
Benj. Reed, 100, -, 4000, 60, 200
Mason D. Webb, 60, 20, 1200, 40, 150
Rev. T. P. McColly, 200, 32, 16000, 240, 1430
Stephen M. Ennis, 60, 40, 800, 45, 300
Jos. Watson, 50, 148, 3000, 40, 150
Jas. Reed of J, 80, 53, 2500, 40, 350
Joshua Laws, 60, 65, 1500, 40, 300
R. M. Smith, 75, 50, 1500, 40, 275
R. W. Ingram, 50, 37, 800, 35, 150
Thos. Deputy, 30, 3, 400, 25, 65
Shadrack Robinson, 150, 50, 3000, 45, 250
Wm. Abbott of J, 150, 50, 3000, 30, 150
Aber Willy, 75, 75, 1500, 30, 125
William Peirce, 100, 33, 2000, 60, 450
Henry J. Peirce, 100, 29, 2500, 60, 400
S.H. McColly, 60, 50, 1500, 30, 250
Robt. H. Davis, 150, 30, 5000, 150, 1000
James Dickerson, 160, 40, 3000, 75, 300
David Watson, 50, 50, 1000, 25, 150
Noah Townsend, 100, 100, 2000, 45, 250
Thos. Watson, 125, 75, 2000, 30, 175
Daniel Miland, 100, 105, 2000, 35, 250
Saml. Draper, 50, -, 5000, 25, 60
Miles Mills, 75, -, 2000, 45, 500
Jas. H. Bogan, 120, 730, 5000, 40, 350
James Deputy, 75, 75, 1500, 35, 300
B.E. Potter, 100, 200, 2500, 45, 250
Jos. Hill, 100, 180, 2000, 50, 400
Manlove Higman, 200, 200, 5000, 75, 400
Wm. Clendaniel, 60, 100, 1200, 40, 300
Henry B. Spencer, 60, 140, 1200, 45, 250

Peter Smith, 100, 60, 1000, 25, 150
Jesse Deputy, 75, 25, 1500, 35, 200
Wm. S. Williams, 50, 25, 1000, 30, 150
John Clendaniel, 150, 70, 2500, 45, 350
Wm. Vanperk, 100, 36, 1500, 35, 100
Elijah Highden, 100, 25, 1500, 35, 140
Jane Watson, 40, 20, 1000, 30, 60
Michael Shaffer, 40, 20, 1000, 35, 60
Joshua Prettyman, 100, 100, 2000, 40, 275
Wm. Porter, 40, 20, 600, 30, 150
Fred Fleuer, 50, 40, 1500, 35, 250
C. G. Greenbe__, 130, 130, 3000, 50, 400
Beniah Sharp, 120, 127, 5000, 100, 600
Levi Wilson, 100, 25, 1000, 30, 250
Zaob Deputy, 45, 55, 1000, 30, 125
Thos. Hall, 100, 10, 1000, 25, 150
Jas. H. Deputy of Z, 100, 140, 2000, 40, 225
Jacob Burton, 18, -, 200, 30, 150
Uriah Morgan, 35, 25, 1000, 25, 100
John R. Carpenter, 120, 125, 2000, 50, 250
Elias M. Daniel, 75, 25, 1000, 45, 175
R. P. Davidson, 50, -, 1000, 100, 800
Robt. Warrington, 150, 150, 3000, 100, 800
P. T. Davis, 80, 20, 2500, 75, 500
Robert Fowler, 60, 34, 2000, 40, 400
David W. Bennett, 100, 170, 3500, 60, 500
Solomon Shockly, 40, 20, 600, 25, 150
Jacob Young, 25, 5, 500, 35, 175
Anthony Shockly, 100, 80, 1500, 45, 300
Philip Civurithen, 20, 20, 500, 35, 125
Thos. E. Civurithen, 20, 20, 500, 25, 125
James Shockly, 10, -, 200, 25, 60
Nathan Young, 25, -, 200, 30, 100
Henry Davis, 12, 10, 200, 25, 50
George Davis, 15, -, 150, 25, 35
Jacob Watson, 25, -, 250, 30, 100
John W. Abbott, 50, 40, 2500, 30, 175
John W. Davidson, 30, 10, 1000, 45, 480
John Argo, 50, 40, 1200, 30, 150
Lemuel Draper, 200, 400, 6000, 150, 1800
Jacob Barton, 50, 25, 700, 25, -
Joseph Shepherd, 100, 75, 1500, 45, 400
Ben McIlvain, 75, 50, 1500, 40, 300
N. Coverdale, 100, 100, 2500, 40, 200
Francis A. Morris, 75, 25, 1500, 30, 150
Robt. Coffin, 75, 50, 1200, 40, 225
Joseph G. Morgan, 100, 16, 3000, 50, 200
Ceasar Donovan, 50, 25, 800, 35, 10
Joseph Young, 50, 50, 1000, 35, 300
Anthony Ingram, 100, 100, 2500, 100, 400
Wm. Vincent, 20, - 300, 25, 100
Henry S. Watson, 50, 50, 3000, 150, 300
Nathaniel Ingram, 150, 150, 3000, 100, 300
Wm. J. Darby, 30, -, 300, 40, 200
Sarah Draper, 40, 40, 700, 35, 100
John S. Davis, 100, 50, 1500, 75, 450
Thos. R. Ingram, 125, 100, 1500, 45, 200
Molton Young, 100, 100, 2500, 30, 125
Mary Townsend, 50, -, 400, 25, 125
Ivey Young, 50, 25, 750, 25, 75
Hester A. Davis, 25, 75, 1500, 25, 50

B. D. Burton, 125, 50, 2500, 100, 500
P. F. Causey, 350, -, 29000, 400, 2300
John W. Hudson, 16, -, 600, 35, 100
John Ratcliff, 30, -, 1500, 45, 400
Robert B. Hairston, 250, 450, 7000, 225, 1110
Benjamin Burton, 150, 300, 5000, 200, 1810
John M. Burton, -, 65, 4000, 50, 425
B.B. Jones, 100, 75, 1000, -, 28
T. S. Johnson, 10, -, 600, -, 620
Henry F. Willis, -, -, -, -, 175
John H. Messick, 75, 75, 1500, 50, 240
Joseph Morris, 100, 50, 4000, 20, 475
William S. Phillips, 200, 400, 3000, 115, 369
Elizabeth Phillips, 40, 60, 1000, -, 15
Joshua Phillips, 250, 240, 5000, -, 758
S. A. Phillips, 200, 320, 3000, -, 758
Betsey Phillips, 70, 70, 2500, -, 400
James Dickerson, 50, 65, 1000, 50, 200
James Sampson, 50, 100, 2500, 50, 125
Benj. Joseph, 80, 61, 1500, 75, 500
Joseph N. Waples (Wapler), 16, 16, 500, -, 75
Ann Wapler, 10, -, 500, -, -
Isaac Cranfield, 50, -, 1000, 5, 500
Charles Ingram, 75, 150, 2000, 20, 125
Charles Lekites, 50, 50, 500, 25, 75
William Fosque, 90, 70, 3000, 200, 570
G. H. C. Hearn, 100, 700, 8000, 100, 790
Hiram Dukes, 35, 35, 1500, 50, 529
Miles Johnson, 210, 90, 1000, 25, 60
Mitchel Scott, 55, 30, 600, 30, 278
Isham S. Pusey, 100, 200, 5000, 25, 160
Peter Ingram, 60, 40, 3000, 25, 175
Isaac Moore, 40, 250, 3000, 25, 175
Charles Ingram, 50, 100, 600, 10, 100
Dr. Jno. Martin, 50, 100, 1000, 50, 184
James Bowden, 40, 60, 1000, 25, 180
John Waples (Wapler), 95, 115, 2000, 25, 165
Ishamael sSeel, 30, 30, 500, -, 36
David Marvel, 75, 25, 1500, 30, 225
Elisha Burton, 50, 150, 1200, -, 10
Robert H. Plummer, 75, 40, 1000, 15, 55
James Carpenter, 50, 200, 1000, 20, 75
James Dukes, 40, 90, 500, 20, 130
Peter Dingle, 15, 20, 50, 20, 50
John C. Tingle, 15, 10, 40, -, 55
John Morris, 5, 8, 200, -, 20
Robert Morris, 200, 250, 5000, 100, 940
James Prettyman, 125, 150, 2000, 75, 300
John P. Parvel, 125, 75, 1000, 75, 185
Joshua W. Pepper, 70, 70, 3000, 100, 425
Robert W. Jefferson, 50, 60, 1600, 20, 120
Robert Warren, 100, 200, 2500, 50, 335
Richard Warren, 50, 150, 1500, 11, 20
Rolan P. Atkins, 60, 200, 1000, 25, 85
Penelope Warren, 30, 30, 400, -, 40
Asa W. Johnson, 60, 30, 1000, 50, 141
Joseph B. Morris, 100, 100, 3500, 100, 280
Isaac Simpler, -, -, 100, -, 80
Zipperd Morris, 65, 65, 1000, 25, 80
John Warren, 50, 60, 1000, 25, 150

Joseph Marvel, 40, 100, 2000, 75, 540
Peter Marvel, 30, 100, 150, -, 35
Robert Prettyman, 60, 80, 2000, 75, 260
Lemuel Davidson, 150, 250, 2500, 75, 300
William Jefferson, 50, 50, 200, 40, 210
Zach. Jones, 30, 40, 800, 25, 181
Jacob Prettyman, 50, 40, 1200, 30, 140
Daniel S. Rogers, 7, 8, 500, 20, 125
Philip Rogers, 150, 100, 300, 100, 220
Sarah Rogers, 60, 20, 1500, 50, 101
George W. Rogers, 50, 25, 1000, 25,100
John Rogers, 150, 100, 4000, 300, 500
John W. Rogers, 40, 120, 2500, 150, 170
Cyrus G. Ferks, 150, 150, 2500, 50, 295
George Scott, 25, 25, 1200, 25, 310
Thomas Scott, 25, 20, 1200, 25, 160
Joseph S. Carpenter, 60, 240, 2000, 10, 80
George M. Carpenter, 100, 100, 2000, 50, 65
Job Marvel, 100, 40, 1500, 500, 125
Edward Short, 40, 30, 2000, 75, 315
A. B. Marvel, 80, 128, 3000, 50, 490
Robert Jefferson, 100, 150, 2500, 50, 250
Joseph Johnson, 30, 40, 500, 10, 30
Nutter Marvel, 75, 80, 2000, 75, 275
Elizabeth Ennis, 125, 50, 500, -, 35
John P. Ennis, 125, 50, 1500, 20, 105
James E. Rogers, 60, 50, 2000, 20, 5
George Rogers, 20, 50, 700, 20, 290
Curtis Rogers, 100, 75, 1500, 25, 100
Daniel Rogers, 40, 56, 1000, 25, 100
William Jones, 15, 17, 620, 5, 50
Minos Rogers, 60, 45, 15, 95, 135
Hetty A. Short, 90, 40, 1500, 75, 241
Peter Short, 100, 100, 3000, 125, 642
James Steen, 50, 40, 2000, 50, 300
Levin Workman, 40, 5, 1000, 75, 130
Laurana Harris, 60, 20, 1500, -, 43
Clayton Conaway, 100, 40, 1500, 50, 178
J. D. Holloway, 50, 10, 500, 45, -
Benj. Johnson, 20, 25, 1000, 25, 240
Charles Phillips, 25, 15, 500, 25, 180
H. S. Sharp, 250, 150, 8500, 150, 720
Manaem Short, 100, 47, 2500, 200, 420
Thos. W. Short, 50, 20, 1000, 50, 200
B. B. Hearn, 150, 50, 2500, 125, 455
James Hitchens, 40, 30, 500, -, 8
Sampson Matthews, 60, 14, 1500, 50, 115
James Jones, 150, 150, 3000, 150, 305
Robert Prettyman, 150, 63, 2500, 100, 250
Wm. C. Wingate, -, -, -, -, 130
Jacob Harvey, -, -, -, -, 75
Levin Hopkins, 100, 200, 3000, 75, 500
John C. Phillips, 75, 53, 1500, 50, 250
Joseph Phillips, 125, 75, 3000, 100, 400
Joshua Hopkins, 4, 60, 800, 10, -
E. G. Phillips, 75, 85, 1500, 40, 241
Burton Phillips, 100, 137, 1500, 150, 300
C__ Prettyman, 75, 50, 2000, 350, 175
George Butler, 40, 75, 800, 40, 100
Robert Jirns (Jones), 35, 35, 400, 15, 150
John Marvel, 60, 40, 850, 20, 175
William Spicer, 40, 80, 1000, 20, 250
Hiram Johnson, 48, 10, 90, 25, 80

William Marvel, 40, 60, 500, 10, 10
John Shorts, 40, 15, 1000, 30, 150
Thomas Mears, 80, 65, 3000, 75, 160
Robinson Mears, 77, 75, 2500, 75, 241
Eliz. Hopkins, 30, 60, 700, -, 4
John B. Phillips, 175, 100, 3500, 50, 750
Philip Shorts, 100, 220, 3000, 100, 300
Saml. Mumford, 100, 130, 2000, 25, 200
John Mumford, 50, 53, 1500, 100, 500
Benj. S. Hudson, 50, 10, 2000, 100, 200
James Long, 40, 40, 2500, 50, 500
Capt. B. Wingate, 50, 60, 1200, 75, 300
John W. Short, 75, 75, 3000, 100, 400
Elisha Cannen, 50, 200, 2500, 150, 1100
Wm. T. Otwell, 100, 120, 2500, 100, 400
Robt. Mumford, 100, 150, 2000, 50, 150
James Green, 18, 18, 500, -, 15
David Green, 150, 150, 2500, 40, 150
Thos. Johnson, 50, 60, 1000, 50, 150
Asoc Johnson, 50, 50, 2000, 50, 100
Saml. Hopkins, 60, 65, 1200, 25, 100
J. S. Atkins, 15, 15, 700, 25, 250
Wm. B. Layton, 60, 66, 1250, 25, 600
Wm. P. Jones, 2, 104, 1200, -, 125
John Jones, 50, 100, 1000, 30, 75
Nath. Phillips, 50, 105, 3000, 75, 460
Alby Short, 50, 60, 1000, -, 50
Wm. P. Carey, 40, 12, 500, 30, 250
Wallace Thompson, 75, 100, 2000, 50, 400
Elijah Carey, 75, 60, 2000, 50, 400
Ephraim Steen, 75, 75, 2000, 50, 400
Wm. S. Melson, 25, 12, 500, 25, 200
Hetty Johnson, 20, 30, 500, 20, 65
Benj. Ellingsworth, 40, 60, 600, 25, 150
John Shirden, 100, 40, 800, 25, 12
Wm. W. Thorogard, 75, 20, 800, 15, 200
Arther S. Lekite, 50, 75, 1000, 75, 160
Peter D. Shockly, 75, 175, 2340, 75, 400
H. W. Meloy, 60, 15, 800, -, 5
William Taylor, 35, 20, 600, 10, 100
Philip Short, 100, 125, 4000, 200, 500
Manam (Manaen) Gum, 50, 7, 1800, 200, 900
Clayton Hudson, 150, 150, 3500, 100, 600
Perry M. Brasure, 45, 35, 1000, 50, 200
Burton Lockwood, 50, 50, 2000, -, 100
Amos Brasure, 60, 75, 3500, 50, 2
William Mase, 40, 160, 4000, 35, 150
William Long, 30, 70, 2000, 50, 150
William H. Hudson, 50, 50, 1000, 60, 130
Elijah Smith, 65, 65, 3000, -, 50
Solomon Tingle, 30, 30, 3000, 20, 150
Ebe Atkins, 40, 30, 500, 40, 250
Jane Walls, 200, 100, 2500, 30, 150
Elijah Davidson, 75, 375, 4000, 30, 150
Joseph Davidson, 70, 80, 3500, 30, 300
Saml. Haslett, 35, 100, 2000, 150, 212
Elijah Atkins, 20, 130, 1800, 20, 125
John C. Hazzard, 40, 30, 1500, 75, 500
George M. Long, 20, 30, 500, 20, 200

Henry Stephenson, 60, 30, 1200, 10, 75
John Hickman, 50, 500, 3000, 100, 700
William T. Clark, 15, 20, 800, 10, 25
James T. Williams, 100, 10, 1500, 100, 450
M. B. Moore, 25, 6, 500, 10, 200
Jno. Tingle, 50, 65, 4500, 200, 600
Isaac Rickards, 70, 300, 2000, 15, 30
Joshua Robinson, 50, 100, 300, 25, 150
D. J. B. Sudler, 6, -, 1500, 100, 525
N. Vickers, 30, 30, 500, 10, 35
John S. Long, 65, 80, 5000, 80, 175
L. S. Hopkins, 60, 60, 1200, 25, 300
Elisha Holloway, 40, 110, 3000, 20, 250
Sampson Cambell, 35, 40, 1000, 20, 50
Aaron B. Holloway, 200, 500, 8000, 200, 550
P. R. W. Hudson, 150, 150, 8000, 100, 175
Seth Hudson, 75, 30, 1500, 100, 250
Jane W. Lockwood, 30, 20, 1500, 20, 100
Seth Long, 100, 150, 4000, 25, 400
Lambert Campbell, 40, 40, 2000, 50, 150
Joshua Hudson, 100, 125, 900, 150, 575
Thomas Brine, 60, 40, 2000, 50, 140
Thomas Dunaway, 65, 30, 2000, 40, 200
Jonathan Baker, 30, 20, 250, 20, 100
Jos. Baker, 35, 40, 800, 25, 75
Albert Baker, 40, 20, 1500, 30, 100
George Lewis, 40, 35, 1500, 30, 250
Manean Tingle, 10, 20, 500, -, 40
Saml. Baker, 6, 40, 600, -, -
Jos. Baker, 25, 30, 1000, 40, 100
Thos. Wills, 30, 40, 1000, 25, 150
Joseph Lewis, 50, 50, 2500, 100, 475
James Wells, 40, 61, 2000, 20, 175
Wm. Tingle, 30, 50, 2000, 20, 200
Isaac Gray, 100, -, 1500, -, 175
Archable Baker, 50, 154, 3000, 125, 400
William Willey, 5, -, 600, -, 125
Thos. Shockley, 40, 60, 1000, 20, 135
Milliam Morris, 70, 70, 1000, 25, 250
Mary A. Houston, 100, 70, 2700, 100, 175
John M. Houston, 100, 128, 5000, 175, 250
Peter Frame, 50, 40, 1000, 20, 280
Gatty Rogers, 40, 20, 400, -, -
Benj. Brittingham, 40, 50, 1000, -, 150
Nat. Warrington, 50, 40, 1000, 20, 175
Handy Hasting, 60, 80, 4000, 60, 300
Brian Brittingham, 25, 60, 1500, -, 150
Elijah Murry, 75, 100, 2000, 50, 200
Saml. Rogers, 60,75, 2500, 40, 150
Elisha Vickers, 40, 45, 1000, 25, 125
James Wilkerson, 60, 75, 1500, 50, 200
George Parker, 60, 50, 2000, 50, 250
John Brasure, 100, 180, 3800, 50, 150
Mitchel Wise, 100, 114, 2000, -, 100
Littleton Dakes (Daker), 10, 1, 1000, 20, 110
Noah Lockwood, 40, 45, 1500, 50, 100
Robert White, 30, 45, 1000, 25, 210
Sarah Campbell, 40, 60, 6000, 20, 200
Sarah Campbell, 50, 70, 2200, 20, -
Elizabeth Wharton, 50, 23, 1000, 50, 125
Wm. O. Short, 60, 50, 1000, 75, 500
George Hudson, 60, 60, 2000, 25, -
Wm. F. Toomey, 110, 20, 2000, 20, 150
Stan.Short, 40, 10, 2075, 75, 300

Peter N. B. Hellum, 80, 80, 2000, 50, 215
Jane Christopher, 45, 35, 1200, 50, 120
John Woolford, 75, 50, 2000, 20, -
John Moore, 10, 7, 500, 10, 30
Hazlett Thompson, 40, 35, 1500, 25, 125
James Thompson, 23, 40, 1000, 30, 150
Purnal Perkins, 40, 60, 1000, 20, 50
Daniel Adams, 50, 65, 2000, 25, 75
John Moore, 60, 75, 2000, 40, 200
Peter Steen, 40, 20, 1000, 25, 50
J. C. West, 60, 65, 2000, 40, 300
Jacob Tingle, 40, 30, 1000, 20, 50
Robert Messick, 35, 30, 1500, 25, 150
Wm. B. Truitt, 40, 70, 1000, 30, 200
Philip Short, 60, 30, 800, 40, 180
John Taylor, 40, 35, 1500, 50, 300
J. Baily, 60, 30, 1000, 20, 49
Wm. Anderson, 50, 30, 800, 25, -
John P. Hudson, 20, 10, 500, 10, -
Thomas Coffin, 75, 50, 1000, 25, 150
Wingate Bounds, 40, 45, 1000, 30, 175
Beleth Brasure, 25, 30, 500, 20, 150
John McCabe, 40, 35, 1000, 25, 175
Joseph McCabe, 35, 30, 1000, 30, 200
David Hall, 40, 45, 1500, 50, 250
Thomas Mannier, 40, 45, 1000, 20, 50
James Carpenter, 30, 60, 800, 35, 50
Jeptha Hickman, 30, 60, 1000, 50, 200
George Davis, 50, 40, 900, 25, 60
Tolbert Matthews, 40, 60, 1000, 30, 150
Minos Johnson, -, 70, 1200, 50, 200
James Locksider, 50, 100,800, 20, 75
Ishew Burton, 35, 50, 1000, 20, 150
Peter Steen, 25, 40, 1000, 25, 100
John Fleetwood, 45, 60, 1500, 20, 150
D. Hudson, 25, 30, 700, 30, 60
John Jones, 60, 70, 700, 25, 150
Mary Watson, 25, 20, 500, 40, -
Elijah Howard, 40, 30, 1200, 50, 300
William Marvel, 40, 20, 600, 25, 50
Jacob Cramfuler, 50, 30, 800, 50, 200
Robert Lawson, 80, 40, 1200, 55, 268
Paynter Johnson, 80, 300, 3000, 10, 60
Pernel Johnson, 100, 50, 1200, 22, 80
Burton W. Hall, 100, 150, 2000, 150, 700
Eli Clark, 65, 150, 1700, 25, 192
Hanby Watson, 240, 240, 3000, 8, 218
Wm. H. Thoroughgood, 115, 185, 2400, 120, 287
Henry L. Lingo, 180, 320, 4000, 25, 312
Peter A. Rust, 140, 100, 1900, 17, 128
Charles H. Rust, 100, 300, 2400, 16, 122
William L. Morris, 275, 62, 3000, 60, 500
Garretson Harmon, 25, 65, 800, 31, 130
Unicy Johnson, 46, 15, 620, 22, 170
John D. Johnson, 67, 67, 2500, 30, 235
Thopholous Street, 67, 67, 2500, 30,2 35
Lemuel B. Lingo, 130, 230, 2800, 50, 187
Isaac Burton, 48, -, 480, 45, 21
David Miller, 47, 47, 800, 9, 8
Wing A. Pride, 200, 200, 3500, 25, 272
Robert Warrington, 80, 74, 1200, 30, 189

Levin Seekum, 150, 96, 2500, 100, 272
William W. Gorlee, 60, 40, 1200, 25, 270
Samuel C. Collins, 100, 100, 1800, 2, 58
Edward Palmer, 50, 40, 800, 10, 40
Lewis P. Reynolds, 50, 40, 900, 5, 20
Rachel B. Burton, 100, 93, 1900, 21, 258
John Burton, 75, 35, 1000, 67, 472
Daniel Lingo, 200, 35, 1000, 67, 472
John R. Burton, 200, 250, 5000, 130, 773
Mary Baylis, 420, 1160, 11000, 200, 1245
Wingate Street, 100, 15, 1200, 43, 277
Phillip Green, 50, 20, 700, 21, 178
Jacob Wilson, 125, 125, 2500, 46, 338
Siles Warrington, 125, 125, 2000, 25, 182
Peter E. Gorlee, 240, 34, 2400, 75, 793
Daniel W. Steel, 70, -, 800, 40, 361
William T. Burton, 100, 20, 1200, 28, 159
Joseph Lynch, 70, 63, 1000, 35, 233
Henry C. Hood, 140, 115, 2500, 85, 510
Ben. Rickards, 60, 60, 700, -, -
George B. Green, 60, 27, 750, 13, 15
Peter B. Lockerman, 40, 33, 650, 24, 84
John Harmon, 40, 33, 600, 16, 98
Jesse E. Joseph, 90, 125, 1600, 22, 105
James A. Atkins, 45, 45, 900, 20, 228
John W. Steel, 22, 18, 400, 44, 334
Priscilla Burton, 200, 400, 6000, 79, 936
Edward McCrary, 50, 37, 870, 28, 175
William Lingo, 10, 7, 170, 27, 246
Paynter E. Lingo, 180, 72, 1700, 40, 353
Cornelius B. Lingo, 80, 70, 1200, 55, 192
John Harmon, 70, 70, 1300, 16, 15
Isaac Harmon, 60, 87, 1300, 20, 122
William D. Waples (Wapler), 129, 100, 2500, 47, 398
Thomas P. Collins, 90, 360, 3500, 30, 305
David P. Street, 40, 204, 2000, 26, 475
John B. Thoroughgood, 150, 50, 2000, 42, 395
Jane Waples, 165, 25, 2400, 64, 620
Wesly Harmon, 50, 310, 2000, 7, 80
Isabella Walls, 60, 15, 750, 33, 122
Simon W. Thoroughgood, 187, 18, 1000, 32, 195
William Dorman, 150, 325, 4000, 45, 355
William I. Lingo, 80, 100, 1200, 36, 117
Nathaniel Cormean, 70, 30, 800, 20, 247
Nathaniel W. Burton, 160, 110, 2700, 55, 725
James P. Johnson, 30, 50, 600, 11,43
Samuel B. Norwood, 40, 50, 850, 30, 60
Mitchel Johnson, 45, 45, 500, 10, 108
John R. Johnson, 50, 70, 800, 6, 500
Purnel Johnson, 50, 90, 1000, 38, 165
Burton Johnson, 100, 200, 2400, 9, 192
Stephen S. Collins, 87, 113, 1600, 8, 115
Robert Fooks, 115, 30, 1500, 75, 301
Nehemiah Coffin, 125, 73, 1700, 40, 276
Henry C. Frame, 150, 150, 3000, 70, 440

Thomas W. Benton, 150, 200, 3500, 75, 400
Eliza Lingo, 150, 80, 1800, 30, 203
Jesse B. Stephenson, 60, 40, 800, 10, 150
John Lingo, 190, 373, 6530, 75, 400
William W. Hurdle, 150, 300, 4000, 100, 300
David M. Prettyman, 107, 107, 2000, 100, 300
Samuel Davidson, 40, 100, 1400, 10, 81
Major Miller, 40, 30, 700, 10, 70
James. B. Coffin, 12, 13, 300, 10, 100
David H. Coffin, 20, 65, 800, 15, 100
Nathan W. Davison, 45, 45, 1000, 25, 200
John I. Johnson, 60, 150, 1700, 30, 150
Nehemiah Joseph, 40, 50, 900, 10, 160
Elias J. Messie, 80, 100, 1200, 15, 50
John W. Davidson, 25, 20, 450, 35, 80
Joseph Barker, 137, 137, 2200, 100, 400
James H. Pride, 175, 175, 2400, 20, 124
William C. Joseph, 60, 20, 800, 10, 55
Burton Carey, 40, 85, 1000, 25, 137
James W. Lynch, 100, 80, 1500, 60, 700
Joseph Hendal (Hurdal), 60, 130, 1400, 20, 200
James H. Davidson, 30, 100, 1500, 25, 125
Selby Lawson, 60, 100, 1200, 25, 140
Thomas H. Joseph, 130, 100, 1580, 50, 250
Jacob F. Hurdal, 60, 65, 1000, 20, 100
James D. Crage, 100, 68, 1400, 35, 175
John W. Davidson, 30, 30, 480, 25, 75
Samuel Davidson, 80, 60, 1300, 60, 125
David Joseph, 30, 30, 600, 20, 66
William B. Rust, 75, 75, 1500, 10, 230
Isaac W. Mager (Magee), 75, 25, 800, 17, 39
James B. Simpler, 50, 50, 1000, 12, 150
Eligha Joseph, 50, 75, 1200, 18, 85
William Johnson, 70, 30, 650, 60, 155
Joseph J. Ennis, 50, 25, 800, 20, 125
Nathaniel M. Johnson, 50, 25, 800, 20, 125
Joseph C. Johnson, 40, 55, 700, 10, 150
Azle Johnson, 30, 6, 360, 20, 140
Albert J. Johnson, 30, 6, 360, 20, 140
George F. Rust, 75, 75, 2000, 75, 310
James Simpler, 42, 42, 550, 10, 80
Peter R. Simpler, 40, 60, 800, 6, 150
James Rust, 120, 100, 2000, 75, 350
Moses Magee, 40, 80, 950, 5, 100
Thomas Rust, 50, 90, 1600, 100, 600
George W. Joseph, 40, 44, 800, 140, 200
Isaac Miller, 45, 100, 1000, 6, 60
William C. Burton, 200, 200, 4000, 200, 986
James Prettyman, 35, 65, 1000, 40, 200
Wesley W. Stephenson, 75, 30, 900, 30, 130
Paynter Joseph, 25, 56, 800, 12, 70
Samuel Joseph, 25, 90, 600, 10, 50
Burton C. Prettyman, 100, 100, 2000, 50, 365
Thomas Joseph, 35, 47, 800, 8, 67
Wm. S. Warrington, 175, 187, 4000, 150, 734

John Prettyman, 100, 100, 2000, 75, 400
John B. Burton, 75, 103, 1400, 40, 142
Henry D. Joseph, 150, 290, 3200, 75, 500
William L. Magee, 100, 50, 1200, 30, 225
Truitt Pettyjohn, 100, 150, 2000, 50, 412
George Brown, 80, 15, 900, 20, 65
Noah Joseph, 30, 30, 480, 10, 50
Samuel N. Johnson, 100, 100, 1600, 20, 200
Ann Wilson, 100, 50, 1500, 25, 200
Jonathan Joseph, 40, 15, 300, 5, 6
John Edington, 70, 125, 1500, 20, 50
John Walls, 48, 20, 1000, 25, 270
Renator Walls, 45, 12, 800, 25, 200
Gideon Walls, 50, 50, 1000, 30, 200
Thomas R. Hudson, 30, 35, 500, 25, 125
William B. Ennis, 45, 55, 800, 25, 130
Stephen E. Blizzard, 50, 40, 700, 25, 150
Woolsey Carey, 59, 50, 800, 12, 45
Absolem R. Sharp, 55, 40, 780, 20, 90
James Brettenham, 40, 33, 438, 20, 20
Elizabeth J. Frame, 240, 44, 4000, 150, 1700
Jane Hunter, 150, 150, 3000, 125, 499
Siles Warrington, 75, 75, 2000, 50, 452
Alfred C.Warrington, 150, 150, 3000, 40, 852
David Showel, 35, 25, 480, 12, 100
John Sammon, 40, 40, 800, 15, 85
James T. Burton, 80, 10, 1000, 25, 197
Whittinton Johnson, 110, 50, 2000, 40, 300
Siles Smith, 125, 20, 3000, 75, 400
John M. Perry, 85, 106, 1900, 25, 140
Nathan Prettyman, 75, 105, 1400, 15, 75
Thomas Robinson, 150, 350, 4000, 50, 490
Alfred McIlvain, 200, 310, 4080, 65, 559
Hezekiah Joseph, 40, 410, 4000, 25, 75
Philip Wright, 50, 100, 1000, 30, 16
James L. Lawson, 50, 50, 800, 50, 370
John Phillops, 160, 160, 2200, 100, 400
Robert Clark, 105, 25, 1300, 50, 140
Lemuel Johnson, 45, 51, 960, 30, 140
William Hopkins, 112, 50, 1600, 50, 190
Gideon Blizzard, 112, 50, 1600, 117, 455
John B. Mustard, 35, 68, 1300, 10, 140
Henry R. Johnson, 60, 200, 2600, 60, 300
Sampson Hutson, 50, 140, 950, 5, 148
William J. Wilson, 70, 68, 1310, 45, 300
Zaceriah S. Joseph, 44, 5, 500, 30, 368
Robert D. Stephenson, 125, 345, 3500, 50, 300
David J. Hazzard, 20, 40, 480, 10, 100
Peter W. Burton, 25, -, 300, 30, 113
Eli Walls, 60, 100, 2000, 100, 450
Daniel Drain, 20, 9, 300, 30, 175
David Hopkins, 20, 30, 500, 25, 110
Muir Vepels (Vessels), 80, 70, 1800, 90, 300
Benj. Phillips, 24, 156, 2500, 40, 300
Charlotte Vessels, 100, 60, 1600, 50, 350

Lemuel P. Burton, 90, 28, 2000, 200, 500
John Webb, 75, 55, 1400, 100, 300
Francis Hazzard, 40, 40, 900, 75, 175
Charles Webb, 60, 90, 1500, 110, 179
Benj. F. Virden, 69, 73, 1400, 40, 202
Dagworthy Derickson, 350, 280, 6300, 200, 1600
Daniel B. Wilson, 70, 100, 1300, 20, 138
Hammon Lingo, 150, 175, 2580, 100, 300
John B. Hazzard, 70, 200, 2500, 30, 490
James A. Brinton, 100, 100, 1600, 50, 415
J___ S. Burton, 40, 35, 600, 25, 200
Erasmus Massey, 20, 125, 870, 12, 20
Fletcher Lacy, 18, 2, 300, 25,50
Thomas Hart, 75, 55, 1300, 40, 494
Alfred Joseph, 80, 125, 1800, 40, 215
George B. Marvel, 40, 150, 1200, 30, 170
Ajaniah Harmon, 70, 20, 720, 10, 60
John E. Hazzard, 60, 28, 1000, 30, 234
Coard Harmon, 55, -, 600, 22, 96
William Harris, 80, 30, 800, 30, 223
James Fisher, 40, 35, 600, 12, 12
William E. Hazzard, 60, 48, 1000, 50, 285
John Phillips, 100, 280, 2500, 40, 289
Memane (Memam) B. Marvel, 80, 32, 1100, 50, 150
John P. W. Marsh, 75, 95, 1700, 172, 217
George Robinson, 50, 169, 2000, 20, 200
Thomas Robinson, 50, 150, 1600, 30, 250
James Burton, 60, 140, 1500, 31, 200
William Massy, 100, 30, 1400, 25, 400
Robert Long, 140, 120, 2600, 75, 377
William Polete, 55, 45, 1000, 40, 131
William S. Robinson, 25, 35, 500, 40, 200
Amos Simpler, 25, 35, 50, 25, 138
Elizabeth Hazzard, 26, 50, 600, 20, 191
Parker Robinson, 50, 30, 800, 25, 237
Brittenham Reynolds, 160, 50, 1600, 20, 130
Nehemiah Miller, 60, 7, 670, 10, 2
John H. Burton, 75, 25, 1000, 25, 177
Jeremiah Harris, 150, -, 1200, 20, 75
Asa Walls, 60, 40, 800, 20, 126
Samuel R. Hart, 25, 5, 240, 20, 45
William B. Johnson, 54, 100, 1200, -, 5
John S. Massey, 20, 30, 500, 12, 88
Thomas Palmer, 40, 100, 1000, 14, 10
Edwin C. Burton, 350, 40, 4000, 75, 634
Clement Baylis, 100, 30, 1500, 20, 100
William Lynch, 200, 150, 2400, 42, 168
Peter R. Burton, 200, 500, 5600, 200, 2000
Theodore W. Joseph, 87, 87, 1000, 10, 7
Josiah Simpler, 30, 130, 1600, -, -
John W. Thompson, 100, 150, 1800, 50, 550
Lorenzo Dow, 30, 40, 700, 20, 255
George M. Cooper, 150, 200, 2500,75, 549
Andrew J. Marsh, 40, 160, 1800, 25, 207

James Mager (Magee), 60, -, 800, 25, 148
James C. Hudson, 100, 76, 2000, 40, 338
Cornelius Holland, 150, 150, 5000, 200, 929
Joseph H. Dodd, 300, 200, 5500, 300, 1332
John W. Walls, 80, 195, 2500, 30, 359
Saml. W. Joseph, 50, 20, 800, 75, 250
Thomas Wilson, 145, 128, 3500, 100, 400
Mills Casey, 27, -, 300, 30, 141
William Dickerson, 30, 30, 700, 40, 294
John Marsh, 20, 60, 800, 30, 382
William B. Paynter, 26, -, 300, 30, 225
Reece Painter, 59, 17, 700, 12, 144
Woolsey W. Hudson, 45, 40, 1000, 40, 241
Thomas Walker, 160, 50, 3500, 300, 1400
John N. Hood, 160, 20, 2600, 150, 594
Richard G. Paynter, 40, -, 400, 30, 151
Comfort B. Holland, 120, 160, 2800, 50, 410
Lydia Wolfe, 250, 225, 5000, 200, 1110
Elizabeth Heborn, 15, 5, 200, 20, 129
Daniel Wolfe, 110, 15, 1800, 25, 223
William Thompson, 75, 175, 2500, 40, 341
Edward McColley, 70, 170, 2400, -, -
William R. Wolfe, 80, 60, 1000, -, 1
Shepard P. Houston, 100, 60, 2000, 225, 800
Watson Otwell, 60, 50, 1200, 30, 200
Joseph Wittbank (Willbanks), 33, 40, 800, 15, -
John Sammons, 112, 20, 3000, 85, 442
John H. Parsons, 120, 15, 2500, 60, 580
Charles Albertson, 30, -, 500, 15, 100
Benj. G. Lingo, 100, 30, 2500, 100, 240
Isaac King, 14, -, 400, 30, 280
Gideon Prettyman, 45, 7, 900, 50, 420
James R. Prettyman, 100, 30, 1500, 50, 490
John Corsey, 43, -, 430, 20, 175
Elisha Wright, 70, 7, 1000, 65, 210
Warren Wright, 19, 7, 300, 80, 170
William R. White, 48, 2, 500, 40, 130
Paynter S. Turner, 150, 40, 3000, 200, 852
Erasmus D. Marsh, 15, 82, 1500, 5, 100
William Lauk, 75, 105, 1800, 25, 195
Moses Hevlow, 40, 40, 200, 20, 160
David King, 35, 125, 1200, 50, 50
Arron D. Kimmey, 70, 170, 2000 50, 276
Thomas D. Wilson, 60, 100, 1600, 20, 199
Thomas Walls, 90, 40, 1500, 30, 182
Jacob Wright, 35, 65, 1000, 20,100
Barkley Wilson, 100, 45, 1500, 200, 640
James F. Martin, 100, 150, 2200, 40, 318
William J. Hitch, 70, 120, 1700, 50, 400
John A. Marsh, 42, 18, 600, 20, 211
John M. Hopkins, 125, 70, 2000, 40, 325
Robert Hazel, 65, 150, 1600, 18, 62
James Corsey, 75, 57, 1200, 20, 164

Elkana Reynolds, 60, 60, 1100, 60, 300
Joseph Black (Block), 160, 80, 200, 45, 495
John C. Lauk, 80, 40, 1000, 25, 282
Isaac King, 80, 66, 1200, -, 7
James C. King, 245, 85, 3300, 188, 804
James Drain, 12, -, 150, 10, 157
Samuel A. Burton, 68, -, 1300, 50, 892
William S. Wolfe, 25, 25, 500, 25, 300
Zacheriah Pride, 50, 40, 1800, 25, -
William Nowood, 60, 10, 800, 20, 100
Levin Hudson, 60, 100, 1200, 20, 90
John C. Palmer, 70, -, 700, 25, 75
Samuel P. Palmer, 70, 200, 2700, 25, 91
William Crage, 80, 120, 1600, 10, 100
Philip Reece, 160, 140, 2000, -, 75
Walter Hudson, 75, 100, 2400, 50, 344
James Lauk, 75, 25, 1000, 25, 220
John Foster, 12, -, 130, 26, 44
Henry M. Jones, 100, 30, 1500, 50, 399
Joseph Holland, 75, -, 800, 25, 218
Robert R. Russel, 120, -, 2000, 200, 1272
Thomas Stockley, 70, 20, 1100, 30, 214
Robert W. McColley, 138, 100, 2500, 50, 532
George Paynter, 80, 200, 900, 10, 80
Joel Prettyman, 180, 20, 4000, 200, 1530
John Carpertson, 250, -, 2400, -, -
John White, 15, -, 200, 60, 70
Samuel J. Lodge, 200, 70, 3000, 100, 573
David H. Waples, 270, 22, 3500, 200, 500
Abreham Drain, 40, -, 600, 30, 264
Lewis Prettyman, 30, -, 500, 15, 55
Selby Hichins, 30, -, 500, 15, 55
Thomas Coleman, 7, -, 600, 50, 142
John Arnal, 50, 5, 1600, 150, 280
Edward Watson, 50, 67, 2000, 40, 525
Henry F. Hall, 96, 100, 4000, 200, 780
Jane McIlvain, 100, -, 1800, 20, 292
Henry Wolfe, 175, -, 3000, 200, 1200
John Metcalf, 25, 10, 800, 50, 626
Jacob A. Marshal, 65, 25, 1800, 100, 265
Robert West, 56, 160, 6000, 150, 429
Peter Warrington, 50, -, 800, 30,75
John P. Palmer, 45, -, 1000, 25, 60
Betsey Russel, 150, 15, 2500, 100, 650
James Maull, 18, 5, 1000, 30, 200
Sarah Rowland, 75, -, 1500, 30, 110
Maryw. Hickman, 20, -, 1000, 100, 500
Robert W. Baker, 100, -, 3000, 50, 20
Laben L. Lyons, 270, -, 8000, 170, 930
Joshua S. Burton, 54, 68, 2500, 153, 1000
Jonathan C. Collins, 20, 10, 300, -, -
Benj. H. Elliott, 64, 55, 1200, 30, 275
Jas. B. Foskey, 70, 20, 1500, 75, 500
John W. Parsons, 50, -, 1500, 100, 250
Geo. H. Hearn, 20, 80, 1500, 100, 400
Benj. Ward, 200, 200, 6000, 100, 600
Josiah Downs, 35, 25, 1000, 25, 320
Leonard Ward, 60, 120, 1800, 30, 100
Priscilla Ward, 60, -, 1300, 40, 250
Joshua S. Cannon, 60, 100, 2000, 90, 500

Jenkins Palmer, 80, 30, 1300, 30, 300
Wm. T. Elliott, 70, 10, 800, 25, 200
Levi C. Calaway, 50, 25, 1200, 40, 200
Wm. W. Elliott, 30, 20, 800, 25, 200
E. A. Warrington, 40, 120, 2000, 250, 500
Luther W. Collins, 80, 42, 800, 10, 175
Jas. H. Tyre, 45, 55, 1500, 10, 120
Benton H. Gordy, 175, 75, 4000, 250, 500
Philip W. Cannon, 100, 70, 2000, 75, 350
John Wootten, 110, 59, 1200, 20, 130
Polly Jones, 42, 10, 400, 5, 150
Isaac T. Whaley, 100, 60, 1200, 50, 300
Sylvester Wootten, 150, 50, 2000, 25, 200
Gincy Vincent, 40, 10, 700, 20, 110
Thomas Love, 40, 60, 800, 30, 150
Thos. Bacon, 130, 25, 3000, 120, 400
Wm. S. Windsor, 70, 40, 1500, 20, 100
Saml. W. Thompson, 40, 20, 900, 25, 175
Jas. H. Windsor, 54, 13, 700, 25, 150
John McGee, 160, 27, 2500, 75, 328
James Ward, 100, 35, 2500, 30, 375
Harry Calaway, 100, 50, 1500, 10, 150
Peter R. Johnson, 40, 30, 800, 25, 125
Elijah Wootten, 100, 110, 2100, 25, 225
Geo. Dixon, 70, 30, 1000, 10, 125
John H. Hosea, 100, 60, 1600, 175, 500
Asa G. Turpin, 215, 185, 3000, 25, 100
Matthew C. Wingate, 210, 83, 2500, 20, 223
Geo. Roberts, 40, 20, 1000, 25, 60
Jacob Wootten, 137, 70, 3000, 150, 400
Chas. B. Green, 95, 55, 1000, 25, 175
Thos. Carmean, 40, 18, 500, 5, 100
John B. Baker, 70, 55, 1200, 25, 200
Joshua J. Hasting, 100, 41, 2000, 25, 400
John Morris of J, 125, 125, 3000, 150, 500
Jisseah Hearn, 50, 64, 800, 10, 80
Sallie Hearn, 40, 35, 600, 25, 125
James Ward, 30, 10, 500, 26, 150
Jane (James) Purnell, 25, 14, 400, 20, 90
Jacob W. Elliott, 75, 21, 900, 30, 175
Stockley W. Elliott, 70, 68, 1100, 20, 150
John S. Dale, 40, 10, 400, 25, 150
Burton Elliott, 40, 10, 500, 30, 300
John Benson, 50, 47, 1000, 25, 150
Elias Elliott, 50, 48, 1000, 30, 150
Marshall Smith, 130, 134, 2600, 75, 275
Isaac Adams, 175, 125, 2500, 50, 200
Isaac Morris, 75, 140, 2500, 50, 200
K. P. Brittingham, 60, 11, 1500, 20, 200
Wm. L. Workman, 35, 25, 1000, 5, 150
Jacob S. Elliott, 50, 30, 1200, 20, 275
Jonathan Hearn, 110, 10, 1000, 25, 125
James Maddox, 7, 97, 1600, 75, 200
Wm. H. Melson, 50, 100, 1200, 25, 200
John W. Melson, 40, 32, 1000, 10, 100
Jos. D. Smith, 50, 51, 1000, 20, 200
Benj. S. Luster, 60, 39, 1100, 10, 75
Nathl. Wootten, 175, 180, 7000, 200, 800

Jonathan Hearn, 85, 15, 3000, 50, 150
Wm. T. Hearn of J, 60, 40, 3000, 30, 150
Isaac W. Hearn, 34, 6, 1000, 200, 350
Jas. Oliphant, 66, 34, 1500, 50, 200
Nathl. J. Elliott, 80, 70, 2000, 25, 150
John L. Elliott, 75, 78, 3000, 25, 225
Minos F. James, 80, 50, 1500, 20, 150
Danl. Hasting, 80, 37, 1500, 80, 250
Jas. Marvil, 60, 10, 1200, 25, 80
Joshua Hasting, 110, 90, 2000, 50, 325
Joseph Ellis, 175, 45, 4000, 200, 1000
Wm. T. Carmean, 30, 20, 1000, 10, 140
Wm. C. King, 175, 150, 4500, 150, 400
Wm. F. Gordy, 102, 35, 2200, 25, 125
Edwd H. James, 60, 40, 1500, 25, 225
Wm. Davis, 70, 120, 2000, 20, 180
Nutter G. Wootten, 125, 50, 4000, 150, 350
Hezekiah Hasting, 75, 50, 2000, 50, 300
Wingate Workman, 80, 70, 900, 10, 100
Augustus Davis, 60, 138, 1600, 20, 100
Levin Hasting, 100, 50, 1500, 25, 110
Thos. Ward, 75, 78, 2000, 40, 150
Wm. Carmean of J, 100, 50, 1500, 25, 180
Henry James, 60, 20, 1000, 50, 275
Geo. W. Jones, 75, 70, 1500, 40, 240
Jas. Morris, 95, 110, 2000, 25, 275
Wm. Semar of L, 125, 100, 2000, 50, 412
John Elliott, 100, 240, 2500, 20, 200
Leonard Hasting, 110, 125, 2500, 25, 400
Nehemiah Morris, 70, 130, 2500, 20, 175
Jacob Hearn, 60, 54, 1000, 75, 400
Joseph Hearn, 100, 81, 2000, 75, 100
Benj. Hearn, 100, 56, 1500, 40, 350
Elijah Williams, 80, 220, 3000, 164, 400
Hetty Lynch, 100, 103, 3000, 50, 250
Wm. Oliphant, 160, 80, 4800, 40, 180
Alfred Adams, 150, 115, 2200, 100, 225
John Cordry, 140, 55, 2000, 50, 250
Wm. Gordy, 200, 150, 3500, 50, 400
James H. Lynch, 75, 25, 1500, 30, 200
Perry Cordry, 75, 45, 1800, 30, 90
Elijah M. Oliphant, 150, 49, 3000, 60, 340
Burton W. Calaway, 90, 70, 3000, 30, 250
Hannah Morris, 175, 225, 3200, 75, 350
Jas. R. Sirmon, 80, 120, 1200, 30, 175
Aaron Gordy, 70, 130, 1200, 25, 225
Chandler Game, 80, 53, 1000, 50, 275
Wm. W. Elliott, 100, 130, 2000, 25, 100
Winder Hasting of M, 109, 60, 1600, 75, 400
Archelus Hasting, 182, 190, 4000, 100, 400
Jon. J. B. Fooks, 130, 64, 2500, 60, 300
Washington B. Calaway, 80, 40, 1200, 100, 350
Job Sermon (Sirmon), 91, 12, 1100, 50, 400
Wm. Hearn of Wm., 70, 80 1500, 50, 325

Jonathan Hearn of S, 60, 30, 900, 20, 80
Jos. S. Calaway, 50, 10, 600, 10, 140
Levin Calaway, 100, 50, 1500, 60, 300
Gordy Culver, 60, 10, 600, 50, 300
Isaac Hearn, 100, 102, 2500, 100, 450
Benj. B. Hasting, 50, 10, 600, 20, 150
Danl. Hasting, 30, 16, 350, 20, 100
Elihu Hasting, 75, 27, 1200, 50, 200
Winder Hasting, 125, 125, 500, 75, 275
Hardy Culver, 80, 70, 2000, 25, 300
Danl. Culver, 80, 87, 2000, 75, 4000
Chas. Culver, 175, 200, 5500, 30, 325
Wm. Hearn, 160, 40, 2000, 25, 275
Thos. C. Lecat, 60, 90, 1500, 70, 300
Kendall B. Hearn, 100, 116, 4000, 75, 600
Saml. Bacon, 130, 50, 2800, 100, 400
Chas. Palmer, 150, 130, 3000, 70, 300
James Palmer, 208, 113, 6400, 70, 350
George O'Neal, 100, 55, 2000, 50, 350
Harry Sharp, 170, 80, 5000, 75, 400
Levin J. Hill, 70, 38, 1800, 150, 300
Jas. Ellis of L, 70, 130, 1600, 10, 120
Isaiah Beach, 50, 30, 700, 15, 250
Minos Lecat, 50, 17, 1000, 20, 1400
John W. Mills, 113, 113, 3400, 25, 300
Wm. T. Cooper, 200, 100, 4500, 50, 300
Sallie Calaway, 74, 77, 1500, 20, 175
Winder Hearn, 60, 95, 1200, 100, 400
Peter White, 75, 79, 2000, 20, 275
Jonathan Waller, 150, 50, 2400, 60, 550
Moses Hasting, 100, 180, 2000, 40, 375
Andrew J. Hearn, 35, 17, 400, 30, 75
Danl. Boyce, 80, 70, 2500, 90, 750
Isaac Carmean, 40, 20, 500, 20, 40
John N. E. Calaway, 40, 40, 700, 10, 200
John Game, 95, 43, 2200, 75, 400
Cotmas Miller, 60, 62, 1300, 20, 70
Wm. N. Hastings, 75, 75, 1500, 60, 400
Elizabeth Rickards, 130, 33, 1700, 75, 450
Thos. S. Rickards, 80, 32, 1000, 25, 200
Saml. Kismey of S, 75, 25, 1500, 300, 500
Elihu Calaway, 36, 41, 700, 20, 175
Jas Ellis of E, 30, 28, 800, 25, 200
Beacham Hasting, 48, 43, 2000, 20, 150
Elijah C. Kinney, 100, 100, 3000, 150, 275
Wm. S. Kinney, 50, 14, 650, 25, 150
Handy (Hardy) Beach, 66, 34, 1000, 40, 225
Phillis Ralph, 100, 80, 1800, 20, 175
Saml. Kinney, 150, 50, 2400, 100, 500
Wm. Ellis of S, 80, 50, 1500, 25, 125
Angetetta Ellis, 54, 21, 800, 20, 240
Wm. J. Chase, 150, 50, 4000, 20, 325
John B. Collins, 60, 38, 1800, 50, 300
Anthony Collins, 60, 140, 3000, 25, 150
Daughty Collins, 125, 35, 3000, 75, 300
Elihu Hasting, 150, 50, 4000, 50, 300
Eli Hasting, 150, 50, 4000, 100, 300
Michael Hearn, 125, 128, 5000, 75, 350
James Hasting, 125, 128, 5000, 40, 150

Martin Collins, 125, 128, 5000, 40, 150
Benj. Hitch, 140, 100, 3600, 75, 525
John Hasting of M, 150, 50, 2000, 10, 150
Elias Culver, 37, 70, 1000, 20, 150
Andrew J. Horsey, 150, 125, 4000, 100, 750
Cornelius Ellis, 60, 115, 1700, 125, 375
Sarah Ellis, 100, 275, 3750, 100, 275
George Ellis, 100, 127, 2800, 90, 450
Rachel Waller, 110, 116, 2700, 20, 140
Burton Culver, 60, 140, 2000, 25, 175
James Ellis of W, 70, 105, 1600, 75, 250
Wm. Lowe, 100, 180, 2800, 70, 350
Martin M. Ellis, 100, 70, 2000, 200, 450
Nutter Hasting, 50, 150, 2500, 40, 175
Michael Hasting, 60, 40, 1500, 20, 140
Wm. J. Ralph, 110, 33, 1500, 50, 300
Wm. Workman, 125, 75, 3000, 30, 320
Noah Carmean, 160, 70, 4600, 25, 275
Geo. W. Kinney, 100, 50, 1500, 20, 200
Spencer M. Cordry, 200, 100, 3000,50, 200
Saml. Ralph, 50, 100, 1500, 25, 225
Sarah Ralph, 50, 21, 1000, 40, 125
Wilmore Culver, 110, 72, 5000, 20, 200
James C. Lowe, 90, 33, 800, 25, 740
Isaac N. Henry, 80, 83, 1600, 30, 175
Wm. A R. Phillips, 200, 150, 3500, 150, 300
Wm. P. Cooper, 115, 56, 1700, 50, 350
Elizabeth Ellis, 55, 27, 900, 50, 175
Margaret Cooper, 165, 96, 2700, 75, 400
Saml. J. Phillips, 60, 100, 2000, 50, 325
Wm. T. Howard, 70, 50, 1700, 25, 120
Sovereign D. Bradley, 35, 15, 800, 25, 150
Charles Wright, 50, 50, 1500, 15, 60
Noah C. Cooper, 90, 68, 3000, 40, 350
Badger Phillips, 60, 60, 1500, 50, 200
Saml. Phillips, 200, 150, 5000, 150, 800
James Pritchet, 80, 90, 1600, 30, 300
Burton R. Hearn, 60, 60, 1200, 25, 175
James Elzey, 150, 50, 2000, 12, 150
Elisha Pennel, 175, 75, 3000, 50, 450
Jane Cooper, 200, 100, 8000, 50, 180
Nathl. Horsey, 200, 50, 6000, 100, 800
Nathl. Horsey Jr., 150, 50, 4000, 100, 600
Wm. L. James, 160, 10, 5000, 50, 250
Jonathan A. Hearn, 150, 80, 9000, 200, 535
Elzey Hill, 60, 40, 2000, 40, 450
Wm. Kinakin, 120, 60, 1800, 30, 200
Elijah Hitch, 150, 75, 4000, 150, 400
Wm. S. Phillips, 100, 120, 5000, 100, 350
Wm. Culver, 100, 40, 1400, 50, 400
Wm. Lloyd, 170, 140, 4000, 30, 250
Chas. M. Walston, 125, 125, 3000, 50, 300
Whitefield Moore, 50, 50, 1000, 30, 150
Saml. Spencer, 70, 30, 1000, 40, 150
Isaac G. Phillips, 200, 100, 3000, 100, 300
Wm. H. Anderson, 90, 200, 5000, 75, 400

Josiah Collins, 150, 50, 2000, 30, 100
Isaac Henderson, 70, 100, 2000, 20, 150
Henry J. Culver, 150, 150, 4000, 50, 400
Jonathan Bailey, 160, 82, 4000, 100, 1000
Clayton Owens, 80, 85, 2000, 25, 150
Barnabas Beach, 80, 110, 3000, 20, 1100
Jesse A. D Bradley, 80, 85, 2100, 150, 700
James Howard, 50, 40, 800, 25, 200
Solomon Elliott, 70, 60, 2000, 25, 225
Richard Knowles, 80, 50, 1500, 40, 375
Phebe Bradley, 80, 27, 1500, 30, 300
Isaac Giles, 250, 150, 8000, 200, 2000
Griffith Goater (Gootee), 70, 30, 1200, 20, 100
Chas. N. Moore, 120, 60, 2800, 75, 500
Ebenezer Waller, 120, 96, 4000, 100, 1000
Benj. Waller, 65, 32, 1000, 30, 200
Henry Douglas, 80, 95, 2000, 25, 150
Darius M. Phillips, 110, 85, 3000, 40, 325
Robert Twilley, 140, 80, 3300, 50, 500
Aaron Owens, 200, 190, 6000, 300, 1000
Levin Twilley, 150, 175, 4500, 50, 650
Joseph P. Twilley, 75, 50, 2000, 30, 350
James Twilley, 75, 70, 2500, 50, 350
Robert C. Twilley, 75, 95, 2500, 30, 225
Thomas Morris, 50, 50, 1500, 30, 100
Stephen G. Kinnikin, 70, 17, 1000, 50, 300
Jacob Marine (Morine), 180, 120, 6000, 50, 300
Alexr. C. Morine, 100, 60, 3000, 100, 450
Burton Dunn, 130, 154, 4200, 200, 500
Ada Henry, 60, 34, 800, 30, 150
Caldwell Moore, 70, 105, 1500, 30, 175
Julia A. Walston, 90, 65, 2500, 20, 300
Lambert A. Walston, 60, 45, 1500, 10, 100
Saml. Kinney of J, 49, 19, 500, 20, 80
James Collins, 100, 100, 2000, 100, 200
Robert Hitchens, 100, 50, 1800, 20, 100
Joseph Ellis of T, 80, 162, 3500, 100, 500
Wm. T. English, 80, 20, 1100, 30, 220
John C. Bradley, 50, 50, 700, 25, 250
Jonathan A. Nichols, 74, 7, 700, 35, 150
George Vincent, 96, 8, 800, 20, 150
Joseph Phillips, 100, 350, 9000, 200, 1100
Wm. B. Richards, 90, 35, 2000, 50, 400
Joshua Hill, 90, 35, 1800, 20, 120
Thos. Phillips of J, 150, 68, 4000, 40, 150
Elizabeth C. Collins, 150, 50, 3000, 120, 300
Levi L. Collins, 35, 55, 1950, 20, 110
Jeremiah Adams, 80, 68, 3000, 100, 500
Stephen W. Ellis, 55, 17, 700, 30, 250
Eben Hasting, 30, 50, 400, 15, 200
Polly Burbush, 45, 40, 800, 20, 100

Mary Adams, 200, 430, 7500, 175, 1200
Susan A. Collins, 75, 79, 3000, 25, 200
Wm. Condry, 70, 60, 1600, 21, 100
Jane E. Ellis, 150, 125, 5500, 150, 300
Denny Shropshire, 55, 10, 700, 30, 50
Jas. C. Carmean, 50, 60, 2000, 40, 175
John Workman, 70, 71, 850, 30, 150

Pages in last part of the county are out of sequence—pgs 21-30 & 1-20

Wm. P. Lay, 100, 40, 1000, 20, 144
Thos. Wilkins, 125, 75, 2000, 35, 275
Bob Wolford, 30, 120, 1000, 8, 46
David L. Wilson, 70, 80, 1200, 15, 85
Jos. Hurley, 75, 75, 1500, 60, 250
U. T. James, 200, 100, 6000, 125, 900
John P. Conaway, 60, 100, 200, 80, 400
Jas. Conaway, 100, 60, 2000, 100, 400
Wm. Fleetwood, 75, 100, 2000, 80, 380
Thos. Jones, 100, 60, 1200, 12, 100
Cyrus Fleetwood, 150, 150, 2000, 50, 300
David Ellensworth, 80, 40, 1000, 14, 37
Nathanl. Conaway, 300, 300, 6000, 30, 300
Arragus Lamden, 300, 229, 5000, 60, 225
Cyrus Jefferson, 60, 30, 1000, 37, 130
Widow B. Jefferson, 50, 100, 1500, -, 40
Lewis Jefferson, 60, 40, 1200, 20, 70
Miles Tindal, 140, 90, 2800, 60, 350
Stephen Breeden, 140, 40, 1400, 20, 100
John M. Short, 150, 140, 3000, 75, 210
John C. S. Short, 65, 85, 2000, 45, 285
Gilley M. S. Short, 65, 85, 2000, 45, 300
Elisha Dickerson, 85, 28, 11000, 20, 90
Wm. Swain, 80, 25, 1000, 25, 175
Walter Swain, 70, 30, 1000, 30, 200
John Day, 100, 40, 2000, 100, 325
P. W. Short, 87, 88, 2500, 60, 345
Abe Swain, 100, 30, 1500, 25, 200
Gilley C. Short, 75, 70, 1600, 55, 200
Jim Coverdale, 100, 60, 1400, 20, 100
Sallie Jones, 100, 25, 1500, 5, 40
Jas. Messick, 70, 80, 1500, 14, 50
Stephen M. Morgan, 135, 72, 3000, 50, 500
Thos. L. Rawlin, 125, 100, 4000, 60, 350
Wm. Allen, 148, 100, 4000, 62, 140
Byard Jonson, 100, 95, 2000, 40, 155
Jno. J. Short, 140, 160, 3000, 80, 350
Daniel Short, 100, 60, 2500, 100, 300
B. Tindal Farm, 100, 60, 1600, -, -
Loe Jackson, 160, 150, 2000, -, -
Jobe Sharp, 100, 100, 1800, 20, 160
Jas. W. Spicer, 150, 196, 2900, 45, 360
Chas. A. Rawlins, 150, 100, 1800, 100, 800
David B. Owens, 130, 20, 1200, 70, 275
George Reynolds, 100, 60, 1600, 25, 175
Whitly Merrideth, 50, 50, 1000, 25, 100
Thos. H. Brown, 250, 200, 8000, 100, 500

Baletest Cornwell, 175, 130, 2500, 75, 275
Noah Isaacs, 125, 80, 1800, 100, 360
Wm. W. Sanals, 50, 75, 800, 8, 60
Joseph Issacs, 100, 200, 2500, 60, 300
Minos Isaacs, 110, 100, 2000, 75, 300
Peter Lynch, 50, 50, 800, 8, 120
Wm. Holston, 50, 150, 1500, 25, 130
Joseph Wilson, 70, 125, 1800, 50, 300
Jersey Banning, 200, 175, 4000, 60, 400
Wilson Farm, 50, 50, 700, -, -
Cornelius Lofland, 100, 50, 1000, 45, 140
Widow Dunfries, 50, 45, 800, 5, 40
David Reynols, 60, 60, 1000, 30, 100
John Greenley, 75, 20, 1000, 70, 275
Joseph Russel, 80, 80, 1000, 12, 300
Josiah Prettyman, 100, 100, 2000, 60, 320
Joshua W. Davis, 100, 100, 2000, 70, 500
Louisa Russel, 100, 109, 1800, 40, 180
James Samuels, 100, 138, 1000, 20, 50
Purnel Walls, 150, 200, 2500, 25, 100
Nutter Ratcliffe, 120, 64, 1600, 60, 300
Nube Short, 100, 100, 1200, 20, 75
Chas. Macklin, 100, 70, 2000, 63, 350
Curtis Houston, 125, 120, 1800, 60, 400
Thos. Polk, 25, 100, 700, 10, 25
James B. Lynch, 80, 50, 1000, 25, 70
Robert Tucker, 50, 40, 500, 25, 40
Benton Sharp, 150, 125, 1800, 50, 350
Jos. Sharp, 75, 100, 1500, 60, 350
Byard Joseph, 100, 200, 1800, 14, 100
John Carlisle, 75, 120, 1000, 15, 50
Robert Willey, 75, 75, 1500, 16, 175
Wm. W. Sharp, 137, 137, 4000, 75, 600
Joseph L. Sharp, 50, 40, 1200, 30, 175
Wm. Conaway, 150, 130, 1800, 60, 325
Burton Walls, 100, 100, 1200, 50, 200
Sander Kellum, 100, 100, 1000, 8, 50
Wm. Johnson, 35, 60, 500, 12, 100
Even Morgan, 200, 200, 1800, 20, 140
Benjamin Morgan, 60, 40, 500, 6, 120
Isaiah Webb, 175, 225, 4000, 100, 600
John Heramous, 65, 65, 800, 20, 230
Charlton Smith, 75, 128, 1500, 75, 500
James M. Walls, 100, 200, 3000, 75, 450
Boos Wharton, 50, 48, 700, 6, 40
Sebastian Passwater, 50, 50, 800, 25, 175
Eli Sharp, 100, 100, 1500, 10, 156
Byard Sharp, 200, 100, 2000, 100, 500
Zeb Nutter, 20, -, 200, 12, 100
Edward Samuels, 60, 50, 1000, 15, 225
Nehemiah Dickerson, 75, 70, 1400, 24, 300
David R. Smith, 150, 150, 3000, 100, 450
James W. Welch, 60, 40, 800, 24, 350
Elijah Smith, 150, 150, 2500, 25, 300
Mitchel Turner, 100, 100, 1600, 25, 110
Millie Fisher, 150, 70, 3000, 100, 500
Wm. T. Jones, 150, 200, 1600, 7, 40

Dan Turner, 200, 167, 2200, 25, 115
Levin B. Brown, 350, 200, 4000, 80, 600
Tob. Coverdale, 125, 75, 2000, 24, 140
James H. Layton, 200, 196, 3500, 75, 450
James Polk, 175, 200, 2000, 12, 140
Jonathan Owens, 180, 60, 2000, 50, 340
Saml. G. Milley, 270, 270, 6000, 45, 400
Joshua Sharp, 300, 140, 5000, 125, 600
Ann Milley, 100, 75, 1500, 25, 150
William Lynch, 150, 120, 3000, 80, 400
Isaac Owens, 50, 70, 1000, 50, 325
Alex Owens, 70, 50, 1200, 25, 260
Leonard Cooper, 300, 150, 6000, 40, 250
Edward Owens, 70, 90, 800, 8, 140
John O. Dawson, 100, 97, 15200, 6, 50
Waitman Milley, 80, 45, 2000, 25, 340
James Scott, 50, 30, 1000, 50, 140
Job Lecatt, 70, 70, 1500, 24, 175
James O'Day, 60, 30, 1000, 20, 20
Smith Rhoton, 200, 200, 4000, 50, 300
John Hays, 60, 30, 1000, 20, 160
John Spanish, 100, 50, 1500, 25, 350
Wm. H. Lines, 100, 140, 2000, 18, 130
Robert B. Owens, 140, 65, 3000, 30, 140
Nehemiah Messick, 40, 90, 2000, 25, 150
Chas. Rickards, 350, 150, 5000, 50, 325
Margaret Fowler, 120, 120, 2000, 35, 330
Wm. B. Ryan, 100, 65, 2000, 50, 375
Sallie Tatman, 200, 170, 4000, 40, 180
David Lines, 150, 150, 3000, 25, 385
James Webb, 100, 95, 1500, 65, 250
Thos. C. Slaton, 150, 150, 4000, 75, 550
John Lynch, 70, -, 800, 50, 275
James Johnson, 80, 70, 1600, 40, 160
David Johnson,, 40, 50, 1000, 20, 100
John Brown, 25, 8, 350, 25, 60
Wesley Carlisle, 75, 30, 1100, 6, 500
Wm. E. Davison, 50, 40, 1000, 30, 340
Joshua Griffith, 175, 175, 3000, 50, 475
James Hatfield, 100, 75, 1600, 24, 100
John H. Satterfield, 100, 80, 2000, 24, 165
Elisabeth Murphy, 50, 40, 1000, 20, 200
Jesse Lindal (Tindal), 40, 50, 1000, 18, 150
John Lecatt, 50, 40, 1000, 17, 60
John Pardee, 25, 33, 1500, 40, 350
Daniel G. Stevens, 20, 10, 800, 20, 60
John Johnson, 150, 100, 1800, 15, 40
John W. Scott, 100, 90, 1500, 18, 135
Zach Griffith, 100, 81, 1800, 75, 400
Curtis Morris, 125, 75, 2000, 50, 280
Joseph Nichells, 75, 50, 1800, 50, 400
John W. Collins, 150, 150, 3000, 25, 180
Albert Curry, 200, 200, 6000, 120, 700
John W. Carlisle, 100, 59, 2000, 75, 360
Joel Carlisle, 100, 100, 4000, 75, 500
Edward Jones, 200, 150, 6000, 80, 750
James Ratcliffe, 20, 60, 2000, 40, 160
Richard Lines, 60, 40, 1000, 8, 65

Mary Griffith, 200, 100, 3000, 75, 400
Wm. G. Carlisle, 80, 125, 4000, 60, 450
Solomon Davis, 400, 100, 6000, 40, 300
James Lawless, 150, 150, 3600, 55, 250
John H. Lyons, 100, 75, 1800, 60, 250
David R. Smith, 125, 75, 2000, 15, 100
David Lord, 225, 170, 5000, 140, 50
Betsy Vincent, 100, 100, 2000, 25, 175
Thos. W. Layton, 200, 125, 4000, 100, 480
Jeremiah Long, 200, 96, 2000, 90, 375
Nicholas W. Adams, 100, 110, 1000, 10, 200
Wm. Taylor, 100, 150, 1200, 75, 285
Chas. Morgan, 100, 50, 2000, 70, 500
James Twilley (Milley), 120, 20, 1500, 50, 250
Loxley Twilley, 150, 150, 4000, 100, 400
John Twilley, 150, 150, 5000, 100, 400
John Conaway, 150, 50, 6000, 150, 600
Zach Clifton, 100, 150, 3000, 15, 45
Thos. A. More, 250, 150, 8000, 100, 40
Revil Bosman, 100, 125, 3000, 85, 375
John Dutton, 200, 200, 5000, 160, 500
Wesley Jackson, 200, 100, 3000, 125, 400
Ann M. Milley, 40, 40, 1000, 40, 300
John Coats, 150, 100, 3000, 50, 550
John G. Collins, 150, 150, 3000, 60, 400
Elisabeth Carmean, 125, 175, 5000, 100, 500
Wm. E. Cannon, 187, 187, 6000, 60, 500
Theodore Price, 150, 150, 7000, 300, 800
Wm. N. Cannon, 200, 115, 8000, 50, 300
Thos. B. Spicer, 100, 50, 1000, 30, 60
Wm. Rawlins, 90, 40, 1000, 25, 175
John M. Rawlins, 100, 110, 1500, 45, 375
Thos. B. Starkey, 10, -, 400, 25, 200
Charles Wright, 350, 185, 20000, 700, 2000
Jacob Williams, 100, 50, 3000, 100, 500
Wm. Huffington Conner, 3350, 350, 15000, 600, 1100
Wm. H. Rose, 400, 450, 50000, 150, 2800
Isaac Willin, 13, -, 600, 75, 150
Wm. P. Brown, 40, -, 800, 50, 180
Michael Colbourn, 8, -, 300, 20, 400
Ben. Stockley, 60, -, 2000, 60, 250
Henry Little, 100, 85, 3000, 100, 240
Frank Brown, 200, 138, 4000, 50, 2000
James Darber, 50, 4, 3000, 50, 340
Hugh Martin, 240, 100, 20000, 1500, 2000
John L. Colbourn, 42, 2, 2200, 60, 270
Levin Cannon, 100, 60, 3000, 50, 250
J. W. Merlin, 40, 60, 1000, 50, 75
Wm. E. Rogers, 150, 40, 500, 150, 700
John H. Brown, 200, 100, 8000, 250, 600
Hugh Brown, 125, 75, 3000, 300, 800
Sovereign Brown, 118, 100, 2500, 40, 150
Zillah Houston, 10, -, 500, 20, 175

Thos. Millin, 100, 50, 2000, 50, 250
Wm. W. Dulaney, 200, 100, 10000, 200, 1800
Thos. Short, 150, 150, 6000, 50, 500
Theopholes Messick, 75, 50, 2000, 30, 125
Wm. Lingo, 100, -, 1500, 20, 100
Chas. Thomas, 150, 150, 8000, 200, 500
Isaac Obier, 160, 80, 2500, 150, 600
Wm. A. Allen, 80, 60, 1800, 60, 212
Wm. Allen, 100, 75, 2200, 80, 300
Kitty Cannon, 193, 100, 4000, 70, 450
John Simpson, 75, 15, 1300, 40, 200
Charles Willins, 100, 50, 2000, 60, 250
De Witt Hasting, 100, 50, 2000, 30, 90
Dennis Cannon, 75, 50, 1000, 30, 80
Levin Williams, 100, 150, 2500, 60, 350
Wm. R. Phillips, 100, 150, 2000, 20, 250
Wesley Thomason, 100, 295, 5000, 50, 175
Thos. Brown, 70, 30, 1000, 40, 230
Joseph B. Allen, 48, 10, 700, 15, 280
Wm. B. Adams, 75, 75, 1500, 32, 340
Jas. S. Wainwright, 70, 30, 1000, 35, 230
Edward Towers, 100, 50, 1500, 15, 130
Charles Hooper, 30, 2, 300, 10, 25
Joseph Allen, 70, 40, 1500, 75, 300
Wm. Betts, 25, -, 400, 15, 55
John H. Allen, 69, 30, 1000, 25, 250
Jesse Allen, 80, 10, 1000, 25, 220
Isaac Cannon, 150, 115, 5000, 60, 450
Levi Cannon, 166, 100, 5000, 35, 500
Wm. T. Cannon, 120, 30, 3000, 75, 350
Jesse Brown, 150, 50, 3500, 175, 600
Wm. Neal, 180, 100, 3000, 20, 500
Wm. W. Spicer, 130, 135, 6000, 165, 690
Joseph Neal, 250, 264, 8000, 223, 985
Major W. Allen, 250, 250, 7000, 275, 8560
Joshua Obier, 275, 125, 6000, 170, 700
Truston Cannon, 130, 50, 2500, 50, 450
Newton Williams, 200, 125, 400, 120, 538
John L. Block__, 125, 100, 2200, 75, 500
David L. Harm, 50, 4, 800, 25, 200
Saml. Eskridge, 150, 50, 3000, 40, 800
John E. Spicer, 220, 180, 4500, 130, 600
Perry Darbee, 125, 100, 2500, 90, 550
Levin Lecatt, 110, 50, 2000, 50, 500
Burton W. Hurley, 150, 58, 2500,70, 450
Silas C. Wainwright, 50, 70, 3000, 100, 300
John M. Wainwright, 75, 25, 1500, 75, 300
Henry Wallace, 70, 25, 1500, 75, 300
John G. Allen, 60, 31, 1000, 25, 215
Zach Fleetwood, 20, 10, 350, 10, 50
Wm. Fleetwood, 200, 75, 3000, 75, 300
Elisabeth Benson, 50, 8, 600, 15, 100
David Shiler (Shiles), 100, 50, 2000, 30, 250
Henry Powell, 60, 40, 1500, 22, 65
Mary Allen, 70, 40, 1200, 20, 68
Wm. Tull, 50, 10, 500, 25, 175
Robert Tull, 30, 12, 500, 20, 100
John A. Tull, 30, 10, 500, 30, 90
John Holt, 60, 125, 2000, 15, 100

James Moore, 30, -, 360, 14, 140
Geo. Lloyd, 60, 140, 3400, 25, 220
Martin D. Merlean (Morlean), 40, 15, 550, 30, 200
Wm. B. Huston, 100, 100, 2500, 45, 200
Wm. W. Wright, 35, 80, 2000, 200, 650
Den Truitt, 75, 325, 4800, 15, 40
Wm. Ellis, 150, 165, 6000, 200, 550
E. J. Ellis, 60, 68, 2500, 50, 75
Thos. J. Phillips, 5, 145, 3000, 30, 250
Wm. O'Day, 150, 4, 2500, 50, 300
James Gordy, 50, 241, 3500, 40, 225
Gwen Flowers, 50, 100, 2000, 20, 50
L. D. Morris, 100, 200, 6000, 20, 500
Ezekiel Reed, 60, 2, 1900, 90, 280
Gillis S. Ellis, 140, 143, 2700, 200, 600
Ananias Vincent, 100, 100, 2800, 18, 200
David Hessey, 120, 150, 4000, 50, 300
Elijah Colbourn, 100, 100, 2500, 10, 300
Wm. Harris, 150, 250, 4000, 35, 250
Garrison Marine, 6, -, 400, -, 40
Leonard Hatfield, 50, 40, 1500, 50, 200
John R. Sadler, 1400, 100, 25000, 400, 600
John Victer, 100, 40, 1000, 100, 265
Robert Brown, 200, 200, 4000, 150, 500
Charles Marine, 150, 150, 2500, 30, 300
Robert P. Swain, 100, 63, 2500, 50, 150
Joshua Brown, 90, 30, 1200, 60, 252
Martin Mayman, 150, 180, 400, 75, 400
Raymon Coates, 150, 75, 3000, 115, 575
Algier Russum, 70, -, 700, 30, 95
Ross Smith, 100, 100, 2500, 75, 500
Burton Short, 140, 100, 3000, 60, 400
Silas Hollis, 200, 200, 4000, 80, 350
Henry Beauchamp, 100, 70, 2000, 30, 275
Marin Frampton, 100, 25, 1500, 30, 400
Ben S. Melson, 200, 230, 5000, 110, 600
Saml. Melson(Melser), 100, 52, 2000, 75, 175
Mark Horsey, 100, 45, 500, 200, 500
Jerry Cole, 170, 80, 3000, 50, 400
Thos. Jacobs, 250, 150, 8000, 175, 960
Thos. W. Dawson, 100, 20, 1800, 120, 300
Alice Kinder, 175, 50, 4000, 140, 468
S. T. Noble, 100, 50, 3000, 100, 400
Daniel Andrews, 60, 70, 2000, 75, 320
Peter Bradley, 30, 15, 4000, 20, 60
Jerry Cannon, 60, 20, 700, 300, 230
Jonathan Noble, 130, 57, 3000, 100, 650
Stephen Corbin, 250, 70, 3000, 100, 500
Wm. Corbin, 75, 75, 1000, 12, 100
John Lemmas, 100, 140, 2000, 25, 100
Jesse W. Williams, 120, 10, 1500, 28, 300
Wesley Smith, 90, 90, 2500, 100, 400
Laureneson Smith, 80, 50, 1500, 40, 275
Thos. Brown, 120, 100, 3500, 30, 250
Curtis Jacobs, 200, 100, 3500, 50, 235
Robert H. Marvel, 60, 20, 600, 20, 200
Richard Greenwell, 150, 200, 4400, 250, 550

Waitman Jones, 250, 300, 8000, 400, 420
Littleton Lankford, 200, 149, 4000, 100, 800
Levin Allen, 75, 15, 1200, 60, 200
Henry Messick, 100, 30, 1500, 100, 550
Tyrus Phillips, 200, 100, 6000, 150, 400
Edward Coulbourn, 62, 40, 2200, 75, 175
Saml. Sherwood, 100, 20, 1500, 18, 75
David Williams, 120, 40, 2000, 95, 450
Henry Roach, 9, -, 500, 6, 100
John A. Twiford, 200, 100, 5000, 100, 335
Dan Adams, 100, 50, 2000, 20, 50
John S. James, 220, 100, 7000, 275, 1100
Hugh Brown, 180, 100, 2000, 80, 200
Matthew Davis, 130, 60, 4000, 150, 200
Joseph Noble, 100, 50, 2000, 20, 250
Peter R. Cannon, 100, 60, 3000, 25, 125
Jerome Layton, 200, 95, 4000, 25, 300
Holstein Masten, 150, 75, 5000, 130, 525
Jesse W. Layton, 172, 120, 5000, 125, 600
James B. Layton, 260, 160, 6000, 125, 1000
Warren Kinder, 300, 200, 8000, 160, 700
Tilghman Kinder, 100, 70, 2400, 125, 500
James Reed, 80, 20, 1400, 8, 125
Stansbery Wingate, 100, 20, 2200, 85, 375
Wm. Parker, 100, 60, 2000, 90, 340
Solomon O'Day, 150, 200, 3500, 85, 300
Nathaniel Clifton, 100, 125, 2000, 40, 160
R. G. Dillsworth, 225, 200, 12000, 500, 890
Henry Lord, 200, 78, 4000, 75, 350
Archie Satterfield, 80, 50, 2000, 55, 325
Jerry Redden, 38, 18, 900, 20, 60
John Sedgwick, 125, 375, 6000, 35, 175
Hugh W. Taylor, 38, 18, 1000, 15, 275
Loxley R. Jacobs, 160, 140, 5000, 100, 500
Luther C. Workman, 50, -, 1000, 25, 800
Geo. Larmer, 10, -, 600, 50, 125
Wm. Cannon, 100, -, 5000, 200, 1200
Garrett S. Layton, 10, 15, 500, 10, 80
Manlove Adams, 3, -, 300, 25, 240
Joshua Willey, 250, 30, 2000, 65, 275
John C. Cannon, 200, 100, 5000, 75, 200
Daniel Todd, 150, 150, 2400, 40, 260
Thos. Todd, 150, 150, 2400, 15, 200
Jas. Gosling, 150, 50, 1600, 40, 300
Wm. Ross, 125, 300, 3000, 200, 600
Roger Adams, 300, 300, 10000, 1000, 150
John H. Hignut, 100, 60, 1000, 20, 200
Archie Evans, 100, 50, 800, 20, 65
James W. Smith, 100, 40, 800, 14, 50
Job Willoughby, 120, 120, 2000, 125, 400
Daniel C. Adams, 100, 70, 1000, 35, 400
Theodore Wilson, 75, 60, 900, 25, 100
Henry Todd, 70, 40, 900, 27, 225
Henry Black, 24, 4, 200, 3, 400

Jasan Vyland, 100, 116, 1000, 30, 400
Elijah Hignut, 80, 24, 1000, 27, 200
Tilghman Reed, 100, 50, 1000, 8, 100
Sinah Twiford, 34, 6, 200, 15, 75
Wm. B. Adams, 150, 50, 2000, 50, 350
Gove Adkisson, 100, 30, 2000, 30, 500
H. Atkinson, 45, 45, 1000, 10, 35
Preston Chaffin, 140, 70, 2000, 30, 500
Jas. H. Simpson, 45, 45, 1000, 10, 35
Wm. Ward, 100, 50, 700, 15, 130
Thos. J. Cannon, 150, 150, 2000, 73, 360
David Walls, 100, 65, 1200, 12, 75
Wm. Collinson, 200, 100, 2000, 30, 375
Wm. N. Coats, 120, 75, 3000, 100, 675
John W. Higman, 150, 140, 2000, 60, 250
Peter Cannon, 400, 400, 8000, 100, 500
Jacob Kinder, 200, 150, 5000, 200, 700
Elias Hooper, 140, 140, 3000, 30, 130
Ezekiel Eaton, 60, 200, 2000, 28, 160
John Kinder, 210, 210, 7000, 200, 625
John Shepherd, 10, -, 200, 20, 150
Lewis N. Wright, 290, 190, 5700, 150, 600
Isaac Wright, 130, 65, 2000, 70, 280
Mark H. Coats, 120, 126, 3000, 90, 300
Hat. R. Jacobs, 200, 51, 2500, 100, 700
Joseph Rickards, 175, 90, 6000, 125, 500
Youst Cassman (Cannon), 150, 130, 2500, 75, -
George M. Davis, 30, -, 2000, 50, 250
Willard Hatfield, 60, 12, 1500, 40, 230
Lemuel Smith, 70, -, 1400, 27, 180
Cannon Willey, 150, 50, 4000, 100, 650
Ager Andrew, 75, 25, 1500, 70, 600
Thos. Greenley, 120, 40, 3000, 55, 300
Clement Jones, 200, 120, 4000, 80, 600
Wm. Nichols, 50, 50, 1000, 8, 20
Burton Layton, 30, 30, 1000, 20, 60
Daniel Brown, 20, 37, 1000, 25, 200
John Layton, 6, -, 350, 22, 75
Jos. Swain, 200, 100, 5000, 75, 725
Wm. Swain, 140, 47, 3000, 100, 400
Wm. Todd, 200, 50, 4000, 100, 500
George Warren, 25, -, 300, 8, 65
Henry Smith, 175, 100, 3000, 70, 300
John Duling, 80, 130, 2000, 50, 250
Wm. J. Stuart, 143, 1453, 3050, 150, 200
Edward P. McColley, 200, 100, 3000, 60, 225
Stansbury Reed, 200, 100, 3000, 75, 350
Thos. Lord, 70, 70, 1400, -, 50
Wm. Rose, 80, 40, 1200, 12, 75
Jebez Fisher, 180, 30, 3000, 120, 640
John Cade, 140, 70, 1800, 90, 300
Ambrose Kinnaman, 146, 60, 3000, 100, 500
John Jones, 150, 98, 2500, 50, 200
Wm. Jones Hains, 145, 100, 3000, 40, 25
Thos. Carney, 60, 20, 800, 10, 50
Levin Todd, 125, 78, 3000, 100, 550
Lewis Slayton, 66, 100, 900, 10, 75
John C. Hamilton, 100, 65, 2000, 25, 300
Elias Morris, 100, 45, 1800, 20, 200
Thos. Grey, 90, 70, 1800, 50, 340
Saml. Knox, 175, 40, 2000, 125, 300

Isaac Willey, 100, 110, 4000, 50, 480
Sallie Collison, 100, 30, 1800, 25, 260
John R. Prichard, 300, 200, 8000, 120, 750
Luke Griffeth, 40, 60, 1000, 6, 75
Thos. Wadkins, 30, 20, 400, -, 45
Wingate Passwater, 40, -, 400, 4, 38
Jerrey (Jenney) Norris, 50, 20, 1000, 8, 175
Stouton O'Day, 200, 75, 5000, 65, 400
Purnel Fleetwood, 90, 130, 3800, 75, 450
David Smith, 75, 75, 1800, 45, 300
Wm. Cole, 50, 20, 1000, 60, 260
Mary Willey, 100, 30, 1400, 35, 175
Alex Jones, 125, 80, 4000, 125, 350
Dilworth farm, 150, 150, 3000, -, 70
John Ellensworth, 200, 80, 2000, 100, 400
John R. Rickards, 150, 10, 4000, 50, 550
Thos. Jacobs, 150, 100, 2700, 25, 100
John Fleming, 120, 80, 2500, 60, 400
John Stevens, 150, 100, 1600, 40, 375
Simon Pennewell, 500, 400, 18000, 500, 2000
Wm. Bradley, 100, -, 2000, 12, 54
John M. Collinson, 100, 100, 2000, 150, 450
Robert Lord, 250, 50, 1000, 100, 800
Robert Smith, 6, -, 500, 15, 140
Pierson Spence, 8, -, 200, 8, 60
Thos. Lord, 140, 30, 5000, 80, 400
Curtis Scott, 70, 20, 800, 25, 140
Thos. Bullock, 100, 40, 2000, 50, 650
Nathan J. Barwick, 100, 100, 2500, 45, 300
Saml. B. Raughley, 260, 188, 5000, 80, 360
George Pratt, 200, 240, 4000, 65, 300
George W. Collinson, 100, 63, 3500, 100, 500
Henry Pratt, 50, 30, 1000, 75, 300
Edward Morris, 110, 156, 3000, 60, 400
Smith Jacobs, 120, 125, 2800, 90, 175
Hezekiah Morris, 100, 70, 1500, 40, 280
John Hayes, 150, 170, 2500, 75, 450
George W. Jones, 100, 60, 1500, 6, 130
Nehemiah _. Stayton (Slayton), 125, 175, 4000, 50, 450
Danl. Clifton, 100, 140, 3200, 200, 420
Wm. Wright, 200, 70, 4000, 110, 450
Grarretson Adams, 80, 80, 2000, 50, 400
Wm. Williams, 125, 50, 3400, 100, 550
Wm. Collinson, 30, 10, 550, 18, 60
Wm. O. Redden, 200, 200, 7000, 200, 400
Whit Smullen, 150, 40, 2800, 20, 280
Ann Milson, 130, 100, 3000, 18, 140
Alexandria Jacobs, 140, 140, 400, 75, 600
John B. Walker, 100, 175, 1200, 20, 300
Isaac M. Fisher, 150, 80, 3000, 100, 800
Wesly Perkins, 100, 150, 3000, 35, 200
Lot Rawlins, 200, 200, 3000, 100, 350
Geo. Stuart, 40, 18, 700, 20, 110
Geo. W. Collins, 200, 75, 3800, 75, 350
John Spicer, 100, 100, 2000, 20, 100
Thos. A. Jones, 175, 200, 5000, 100, 300

Ezekiel Jones, 115, 100, 4000, 45,325
Wm. Simpson, 100, 50, 1200, 25, 120
Thos. Coffin, 25, 25, 500, 12, 60
James Outten, 20, -, 1000, 15, 150
Philip Neal, 80, 200, 2000, 12, 50
Wm. C. Wheatley, 125, 117, 4000, 120, 400
James Boyce, 80, 10, 1200, 80, 400
Thos. Calhoun, 130, 50, 2500, 150, 500
Wm. Calhoun, 100, 40, 1000, 20, 275
Thos. Messick, 100, 80, 1800, 33, 200
Wm. Calloway, 85, 5,1800, 20, 800
Andrew J. Dolby, 100, 50, 1800, 75, 380
Frederick Gibbons, 65, 65, 1000, 40, 150
Sam Messick, 50, -, 1000, 5, 80
Geo. B. Tindal, 125, 75, 200, 8, 150
Sarah Conaway, 100, 100, 2000, 16, 140
Minor P. Conoway, 200, 100, 4000, 85, 600
B. Brown, 51, 40, 3000, 75, 400
Elias Maxfield, 800, 100, 2000, 15, 90
Nat Wharton, 100, 20, 1400, 45, 180
Seth L. Outten, 160, -, 1500, 40, 165
Miles Messick, 200, 160, 8000, 170, 1100
Nat Messick, 100, 30, 1300, 65, 360
John Tindal, 140, 50, 1500, 40, 175
David Calhoun, 150, 250, 6000, 15, 480
Samuel Morgan, 100, 60, 3000, 95, 425
Thos. A. Long, 100, 80, 1800, 65, 220
Nancy Conaway, 300, 300, 5000, 80, 600
Wm. F. Daveson, 75, 217, 4000, 100, 340
Noble Conaway, 136, 200, 3000, 100, 800
Levin Conaway, 120, 100, 2000, 60, 300
Allasuras Tindal, 100, 80, 1500, 30, 2100
Benton H. Tindal, 75, 150, 2500, 95, 600
Thos. H. Fooks, 160, 200, 5000, 150, 800
Josiah P. Marvel, 180, 60, 4000, 200, 800
Mitchell Smith, 80, 60, 1600, 25, 75
Lou Mitchell, 40, 40, 900, 8, 40
Robert P. Barr, 125, 125, 4000, 100, 500
Davis W. Barr, 100, 100, 4000, 100, 1000
Wm. C. Smith, 90, 100, 1800, 75, 375
Joseph Williams, 100, 50, 1200, 20, 184
Job C. Dolby, 80, 40, 1700, 55, 450
Curtis S. Fleetwood, 100, 40, 2500, 65, 400
Dickson Conaway, 60, 36, 800, 12, 120
Geideon Mitchell (Pritchard), 57, 12, 700, 17, 160
Manlove D. Hill, 30, 40, 1400, 9, 120
Manlove Hill, 70, 100, 3000, 75, 425
Stephen Warrington, 100, 100, 2500, 20, 1800
Peter W. McColley, 70, 30, 800, 25, 175
Stanley Messick, 100, 100, 2000, 20, 181
James Huffington, 200, 200, 6000, 125, 765
Robert Smith, 75, 170, 3000, 60, 130
John Swain, 75, 50, 1500, 20, 270
John W. McColley, 100, 125, 3000, 75, 275
Benton Dickerson, 25, 5, 500, 25, 60

Robert Donnervan, 100, 50, 1500, 24, 85
Lemuel Smith, 100, 100, 2000, 10, -,
John Swain, 12, -, 300, 20, 160
Milman farm, 20, 40, 600, -, -
Thos. W. Willin, 110, 159, 2500, 150, 435
Phillip West, 180, 200, 3700, -, 85
Gilly C. Smith, 100, 60, 1200, 60, 160
John Messick, 50, 6, 700, 24, 170

Delaware 1860
Index

Other Books by Linda L. Green:

Alabama 1850 Agricultural and Manufacturing Census: Volume 1
Alabama 1850 Agricultural and Manufacturing Census: Volume 2
Alabama 1860 Agricultural and Manufacturing Census: Volume 1
Alabama 1860 Agricultural and Manufacturing Census: Volume 2
Delaware 1850-1860 Agricultural Census: Volume 1
Delaware 1870-1880 Agricultural Census: Volume 2
Delaware Mortality Schedules 1850-1880: Delaware Insanity Schedule 1880 only
Dunklin County, Missouri Marriage Records, Volume 1: 1903-1916
Dunklin County, Missouri Marriage Records, Volume 2: 1916-1927
Florida 1860 Agricultural Census
Georgia 1860 Agricultural Census: Volume 1
Georgia 1860 Agricultural Census: Volume 2
Kentucky 1850 Agricultural Census
Kentucky 1860 Agricultural Census: Volume 1
Kentucky 1860 Agricultural Census: Volume 2
Kentucky 1860 Agricultural Census: Volume 3
Kentucky 1860 Agricultural Census: Volume 4
Louisiana 1860 Agricultural Census: Volume 1
Louisiana 1860 Agricultural Census: Volume 2
Maryland 1860 Agricultural Census: Volume 1
Maryland 1860 Agricultural Census: Volume 2
Mississippi 1860 Agricultural Census: Volume 1
Mississippi 1860 Agricultural Census: Volume 2
Missouri 1890 Special Enumeration Union Veterans Census
Montgomery County, Tennessee 1850 Agricultural Census
New Madrid County, Missouri Marriage Records: 1899-1924
Pemiscot County, Missouri, Marriage Records, January 26, 1898 to September 20, 1912, Volume 1
Pemiscot County, Missouri, Marriage Records, November 1, 1911 to December 6, 1922, Volume 2
South Carolina 1860 Agricultural Census: Volume 1
South Carolina 1860 Agricultural Census: Volume 2
South Carolina 1860 Agricultural Census: Volume 3
Tennessee 1850 Agricultural Census: Volume 1
Tennessee 1850 Agricultural Census: Volume 2
Tennessee 1860 Agricultural Census: Volume 1
Tennessee 1860 Agricultural Census: Volume 2
Texas 1850 Agricultural Census: Volume 1
Texas 1850 Agricultural Census: Volume 2
Town Crier: The Jeffreys Clan
1890 Union Veterans Census: Missouri Counties
Virginia 1850 Agricultural Census: Volume 1
Virginia 1850 Agricultural Census: Volume 2
Virginia 1860 Agricultural Census: Volume 1
Virginia 1860 Agricultural Census: Volume 2

LaVergne, TN USA
07 October 2009
160220LV00001B/29/P

9 780788 438264